T0239876

HYBRID ANISOTROPIC MATERIALS FOR STRUCTURAL AVIATION PARTS

HYBRID ANISOTROPIC MATERIALS FOR STRUCTURAL AVIATION PARTS

YOSIF GOLFMAN

CRC Press
Taylor & Francis Group
Boca Raton London New York

CRC Press is an imprint of the
Taylor & Francis Group, an **informa** business

CRC Press
Taylor & Francis Group
6000 Broken Sound Parkway NW, Suite 300
Boca Raton, FL 33487-2742

First issued in paperback 2017

© 2011 by Taylor and Francis Group, LLC
CRC Press is an imprint of Taylor & Francis Group, an Informa business

No claim to original U.S. Government works

ISBN-13: 978-1-4398-3680-4 (hbk)
ISBN-13: 978-1-138-11234-6 (pbk)

This book contains information obtained from authentic and highly regarded sources. Reasonable efforts have been made to publish reliable data and information, but the author and publisher cannot assume responsibility for the validity of all materials or the consequences of their use. The authors and publishers have attempted to trace the copyright holders of all material reproduced in this publication and apologize to copyright holders if permission to publish in this form has not been obtained. If any copyright material has not been acknowledged please write and let us know so we may rectify in any future reprint.

Except as permitted under U.S. Copyright Law, no part of this book may be reprinted, reproduced, transmitted, or utilized in any form by any electronic, mechanical, or other means, now known or hereafter invented, including photocopying, microfilming, and recording, or in any information storage or retrieval system, without written permission from the publishers.

For permission to photocopy or use material electronically from this work, please access www.copyright. com (http://www.copyright.com/) or contact the Copyright Clearance Center, Inc. (CCC), 222 Rosewood Drive, Danvers, MA 01923, 978-750-8400. CCC is a not-for-profit organization that provides licenses and registration for a variety of users. For organizations that have been granted a photocopy license by the CCC, a separate system of payment has been arranged.

Trademark Notice: Product or corporate names may be trademarks or registered trademarks, and are used only for identification and explanation without intent to infringe.

Library of Congress Cataloging-in-Publication Data

Golfman, Yosif.
    Hybrid anisotropic materials for structural aviation parts / by Yosif Golfman.
      p. cm.
    Includes bibliographical references and index.
    ISBN 978-1-4398-3680-4
    1. Airframes--Materials. 2. Composite materials. 3. Anisotropy. I. Title.

TL671.6.G65 2011
629.134'3--dc22   2010010719

Visit the Taylor & Francis Web site at
http://www.taylorandfrancis.com

and the CRC Press Web site at
http://www.crcpress.com

Contents

Preface

The development of nanoscale hybrid composite materials with anisotropic properties and high strength-to-weight or stiffness-to-weight ratios with high-wear resistance, thermal conductivity, and dynamic stability has been a problem in the 20th and 21st centuries. High oil prices make it necessary to reduce the weight of gas and turbine engines and aviation structures. *Hybrid Anisotropic Materials for Structural Aviation Parts* include articles published in the U.S. (e.g., *Journal of Reinforced Plastics and Composites, SAMPE Journal, Journal of Advanced Materials,* and *JEC Composites Magazine*) from 1991 to 2009. This book analyzes how mechanisms work and fail, and how current manufacturing techniques can improve the strength of aviation parts. Analysis of the dynamic stability of hybrid structures in aviation and space vehicle parts helps to reduce vibration and icing in aircrafts by developing alternative electronic and ultrasonic systems. Designers use strong anisotropy composites in lamination molding, braiders, and pultrusion processes, and lay up a maximum number of fibers in the direction of the basic load, so that the load direction coincides with the planar direction of the fiber. The optimal design will be successful if designers know the properties of strong anisotropic materials, the theory of laminates, and basic technologies.

Titanium aluminum alloys and titanium–aluminum–graphite laminate prepregs are promising structural materials for high-temperature applications, such as gas turbine blades in aircraft or satellite space vehicles, because of their low densities, high melting points, excellent strength and modulus properties, and creep and oxidation resistance.

The advantages of using carbon nanotube (CNT)-reinforced aluminum–titanium alloys are reduced weight and minimized oil expenses, better resistance to corrosion and erosion, and improved dynamic balance. CNT-reinforced aluminum significantly improves the mechanical properties of aluminum powder metallurgy. For example, the hardness of composite aluminum is several times greater than that of unalloyed aluminum; tensile strengths comparable to those of steel can be achieved; and the impact strength and thermal conductivity of the lightweight metal can be improved significantly. The density of CNT-reinforced aluminum is only around one-third that of steel. Therefore, the material can be used in any number of applications in which the goal is to reduce weight and energy consumption. With its combination of high strength and low weight, Baytubes®-reinforced aluminum is a welcome alternative to steel and expensive specialty metals such as titanium and carbon-fiber reinforced plastics. CNT-reinforced epoxy and liquid polymers are promising materials that are 1.5 times lighter than aluminum, have erosion and corrosion resistance, and are promising materials for military and civil applications. Graphite–epoxy composite is a super hybrid epoxy resin matrix with a high epoxy content, which is reinforced with graphite or carbon particles.

Carbon–carbon composites must be capable of maintaining high geometric stability and low thermal expansion when they are used in engine turbine blades, rocket nozzles, and seal rings in a liquid propeller rocket engine. Carbon–carbon composites were developed using high-pressure impregnation/carbonization techniques.

This book consists of seven chapters. Chapter 1 is devoted to the use of carbon–silicon nanotubes and ceramic technology in satellites and space vehicles.

Chapter 2 details the development of the impregnation process for prepregs, braided composites, hybrid polymers carbon fibers, and continuous molding and pultrusion combined with low cost manufacturing .

Chapter 3 focuses on strength criteria, which are used to develop a validated design and life prediction methodology for polymeric matrix composites. Also it analyzes the dynamic aspects and stability of jetliners and lattice aviation structures.

Chapter 4 describes interlaminar shear stress analysis and possible failure as a result of low shear strength matrix composites. The use of a carbon fiber–epoxy sandwich is examined for leading and trailing panels on aircraft. New technologies such as braider winding, pultrusion, and vapor-phase deposition open new horizons and transform to practical realization projects like the carbon–carbon fuselage in Boeing jetliners or the carbon fiber used in the interstage structures of satellites.

Chapter 5 considers fatigue strength and vibration analysis. This chapter also focuses on strength analysis of turbine engine blades including the effect of thermoelasticity on a composite turbine disk.

Chapter 6 evaluates nondestructive methods that control technological parameters and reduce technological defects such as microcracks that reduce fatigue strength and durability.

Lastly, Chapter 7 discusses coating processes applied to the protection of aviation parts. It also analyzes coatings for helicopter rotor blades, nonthermal antiicing and deicing systems, and thermoplastics-reinforced carbon fibers used in large ground based radomes.

This book is very appropriate and recommended for graduate and undergraduate students, industrial professionals, and researchers working in national laboratories and industrial companies.

Acknowledgments

The editor of *SAMPE*, Dr. Mel Schwartz, has helped me with editing many of my articles published in *SAMPE* and the *Journal of Advanced Materials* during the last 10 years. Thank you very much for this help and for recommendations of my book.

Dr. Jerome Fanucci, president/CEO of KaZaK Composites, Inc., a leader of pultrusion technology, encouraged me, and gave a recommendation to publish this book.

Dr. Scott W. Beckwith, technical director of *SAMPE*, and all achievements in composites, gave me a lot of advice and also encouraged me to write this book; thank you very much for this.

I would like to thank Dr. A. Brent Strong, editor in chief of the *Journal of Advanced Materials*, who sent my articles to reviewers and helped me update articles.

I would like to thank the following staff of CRC Press: Jonathan W. Plant, senior editor; Richard Tressider, project editor; Amber Donley, project coordinator; Arlene Kopeloff, editorial assistant; and Scott H. Hayes, prepress supervisor, for his technical advice on preparation of the book manuscript.

I worked with the president of Neo-Advent Technologies, Nelson Landrau, and vice president Dr. Alex Nevoroshkin for DoD proposals that gave me many ideas for structuring this book.

Acknowledgments

1

Nanocomposite Automation Process

1.1 Ceramic Technology in Space Programs

1.1.1 Introduction

Ceramic technology was enhanced through NASA's space programs and used in heat-protective tiles on rockets and most recently the turbine engine components for space vehicles.

The Joint Strike Fighter F-35 and other military platforms are targeting ceramic matrix composites (CMCs) for exhaust and engine applications with an ultimate goal of weight reduction. However, concerns exist over acquisition cost, reliability, durability, and life expectancy. CMCs are typically fabricated with two-dimensional (2-D) woven ceramic grade (CG) Nicalon fabric reinforcement, which is coated with a boron nitride (BN) interface coating. 2-D CMC components have been found to be life-limited in high thermal gradient environments due to inherently low matrix dominated interlaminar shear strength, but cost less than 3-D fiber architectures. 3-D fiber architectures offer the promise of increased durability by enhancing the interlaminar and through-thickness mechanical properties specially developed for 3TEX Inc.'s 3-D orthogonal weaving machines. The purpose of this research is to develop effective low-cost BN interface coatings for 2-D reinforced CMC components and to investigate the fractographic model prediction deformations and fatigue strength.

SiC/SiC CMCs are targeted for use as advanced aerospace turbine engine components that will be exposed to temperatures of 2400 to 2700°F [1,2]. The current understanding and models of CMC behavior are based on extensive work in laboratory environments and limited efforts under representative environments such as steam and burner rigs. Accurate prediction of durability and usable life of CMCs requires an in-depth understanding of the environmental effects on the long-term deformation and failure in aerospace turbine engine combustion environments [3–5].

Matrix cracking combined with oxidation-induced damage has been shown to be responsible for the reduction in life at elevated temperatures. Hence, these advanced SiC/SiC composites will be protected with environmental barrier coatings (EBCs) to significantly enhance the durability during service. We seek validated physics-based models that can predict the life of SiC/SiC CMCs under expected service environmental and thermomechanical loading conditions. The proposed effort should include fundamental characterization methodologies such as high vacuum testing and detailed fractographic studies to understand and model the effect of environment on damage accumulation. Validation of the model's ability to predict the deformation, damage characteristics, growth of damage zones, and total life should be addressed. Developing and validating physics-based long-term deformation and life prediction methods for advanced SiC/SiC CMCs under aerospace gas turbine engine environmental conditions are also very important.

An understanding of the elevated temperature tensile creep, fatigue, rupture, and retained properties of CMCs envisioned for use in gas turbine engine applications is essential for component design and life prediction. To quantify the effect of stress, time, temperature, and oxidation for a state-of-the-art composite system, a wide variety of tensile creep, dwell fatigue, and cyclic fatigue experiments were performed in air at 1204°C for the SiC/SiC CMC system consisting of Sylramic-iBN (in situ boron nitride) SiC fibers, BN fiber interface coating (border between Sic and BN), and slurry-cast melt-infiltrated (MI) SiC-based matrix. Tests were either taken to failure or interrupted. Interrupted tests were then mechanically tested at room temperature to determine the residual properties. The retained properties of most of the composites subjected to tensile creep or fatigue were usually within 20% of the as-produced strength and 10% of the as-produced elastic modulus. It was observed that during creep, residual stresses in the composite are altered to some extent that results in an increased compressive stress in the matrix upon cooling and a subsequent increased stress required to form matrix cracks. Microscopy of polished sections and the fracture surfaces of specimens, which failed during stressed oxidation or after the room temperature–retained property test, was performed on some of the specimens in order to quantify the nature and extent of damage accumulation that occurred during the test. It was discovered that the distribution of stress-dependent matrix cracking at 1204°C was similar to the as-produced composites at room temperature; however, matrix crack growth occurred over time and typically did not appear to propagate through the thickness except at the final failure crack. Failure of the composites was either due to oxidation-induced unabridged crack growth, which dominated the higher stress regime (179 MPa) or controlled by degradation of the fibers, probably caused by intrinsic creep-induced flaw growth of the fibers or internal attack of the fibers via Si diffusion through the chemical vapor infiltration (CVI) SiC and/or microcracks at the lower stress regime (165 MPa). Effects of loading rate and temperature on tensile behavior have been studied in air using two types of orthogonal 3-D woven Si–Ti–C–O fiber-reinforced Si–Ti–C–O matrix composites, processed by polymer infiltration and pyrolysis (PIP) and CVI. Since the interface and porosity of the two composites are controlled in as similar a manner as possible, the effect of matrix processing method is understood. The strength of the PIP composite is greater than that of the CVI composite at room temperature, but they are almost the same at high temperatures. It was found that the PIP composite is more sensitive to loading rate than the CVI composite due to more glassy phases in the PIP composite.

The monotonic tension, fatigue, and creep behavior of SiC-fiber–reinforced SiC matrix composites (SiC/SiC) has been reviewed. Although the short-term properties of SiC/SiC at high temperatures are very desirable, fatigue and creep resistance at high temperatures in argon was much lower than at room temperature. Enhanced SiC/SiC exhibits excellent fatigue and creep properties in air, but the mechanisms are not well understood. The present Hi-Nicalon/SiC has requirement properties to enhanced SiC/SiC, but at higher cost. Improvement of Hi-Nicalon/SiC therefore seems necessary for the development of a high-performance SiC/SiC material.

1.1.2 Process and Development

Three types of reinforced fiber—Sylramic, CG Nicalon, and Nextel N720 impregnated by silicon carbide matrix—are represented in Table 1.1.

Thermomechanical properties of Hi-Nicalon and Sylramic fibers are shown in Table 1.2.

TABLE 1.1

Typical Properties of Ceramic Composites

Material Name	S300 (Nonoxide)	S200 (Nonoxide)	AS/N720-1 (Oxide)
Fiber	Sylramic	CG Nicalon	Nextel N720
Fiber coating	Boron containing	Boron containing	None
Matrix	SiC +Si_3N_4	SiNC	Aluminosilicate
Filler	Various	Various	—
Typical ply thickness, mils	7.5	12.5	9.1
Fiber volume fraction	0.42	0.42	0.45
Lay-up	5HS Fabric, warp-aligned symmetric	8HS Fabric, warp-aligned symmetric	8HS Fabric, 12 ply, 0/90 symmetric
Bulk density, g/cm³	2.6	2.3	2.6
Open porosity, %	<5	<5	2.5
Maximum use temperature continuous, °C	1315	1000	1000
Maximum use temperature, short term, °C	1650	1250	1100

TABLE 1.2

Thermomechanical Properties of HI-Nicalon and Sylramic Fibers

Property	Sylramic	Hi-Nicalon
Density, g/cm^3	3.0	2.74
Diameter, μm	10	14
Tensile strength, GPa	3.4	2.8
Elastic modulus, GPa	386	269
Thermal expansion coefficient, 10^{-5}/K	5.4 (20–1320°C)	3.5 (25–500°C)
Thermal conductivity, W/m K	40–45	7.77 (25°C)
Specific heat, J/g K	075	0.67 (25°C); 1.17 (500°C)

Analysis shows that Hi-Nicalon fiber has five times less thermal conductivity coefficient than Sylramic fiber.

The simplest boron hydride is borane (BH_3), which interacts with glass silica ($SiCl_4$) to form a silicon–boron bond [6].

Boron fibers are five times as strong and twice as stiff as steel. A chemical vapor deposition (CVD) process in which boron vapors are deposited onto a fine tungsten or carbon filament makes them [7–8].

Boron provides strength, stiffness, is lightweight and possesses excellent compressive properties as well as buckling resistance.

Special Materials Inc. (previously known as Textron) uses the CVD process to create the boron layers. The process uses fine tungsten wire for the substrate and boron trichloride gas as the boron source [8].

The boron manufacturing process is precisely controlled and constantly monitored to assure consistent production of boron filaments with diameters of 4.0 and 5.6 mil (100 and 140 μm). The mechanical properties of boron are represented in Table 1.3.

Combining the boron fiber with graphite prepreg, a high-performance material, Hy-Bor, has been produced with exceptional properties (see Table 1.4).

TABLE 1.3

Mechanical Properties of Boron

Mechanical Properties	Values
Tensile strength	520 ksi (3600 MPa)
Tensile modulus	58 msi (400 GPa)
Compression strength	1000 ksi (6900 MPa)
Ksi	2.5 ppm/°F (4.5–3200)
Density	0.093 lb/in^3 (2.57 g/cm^2)
Temperature performance	350°F

TABLE 1.4

Material Properties of Typical Hy-Bor Laminate (4.0 mil–100 μm)

Mechanical	Values
Tensile strength	275 ksi (1896 MPa)
Tensile modulus	35 msi (241 GPa)
Flexural strength	350 ksi (2413 MPa)
Flexural modulus	31 msi (214 GPa)
Compression strength	400 ksi (2756 MPa)
Compression modulus	35 msi (241 GPa)
Interlaminar shear strength	15 ksi (103 MPa)
Strain	0.86%
Short beam shear	17 ksi (117 MPa)

Other types of materials, such as low thermal conductivity ceramics, can offer advantages for protective coatings on the space shuttle. Thermal barrier coatings (TBCs) have thin ceramic layers, generally applied by plasma spraying or by physical vapor deposition, and are used to insulate air-cooled metallic components from hot gases in gas turbine and other heat engines [9]. The ceramic layer consists of 95.4 at.% zirconia ZrO_2 + 4.6 at.% yttria Y_2O_3. However, these coatings have porous and microcracked structures. Recently, scandia was identified as a stabilizer that could be used in addition to yttria [10]. A composition of 3 mol % scandia and 2.5 mol % yttria may confer the desired phase stability at 1400°C. Superior Technical Ceramics Corp. has manufactured extensive CNC machining components such as MSZ-100 or Y-TZP (zirconia, yttria, titanium) that keep temperatures at 1832 to 2730°F (1000–1500°C).

The Institute for Operation Research and the Management Sciences of NASA Space Strategic Affairs published a report about the content and consistency for the high-temperature tiles. This report said that the damage on the left wing of the shuttle *Columbia* indicated the loss of ceramic tiles, which resulted in the subsequent tragedy.

The aluminum skin of the shuttle needs to be protected from oxidation. It is proposed that layers of adhesive, silica fiber for heat resistance, and finally glass coating could be deposited on the skin.

Spray gun carbon silicon coating technology is shown in Figure 1.1. The thermoprotective layers (pos. 1) consist of silica and carbon nanotube coating. Between Nicalon fibers layers (pos. 2 and pos. 4) accommodate flexible carbon silicon matrix (pos. 3).

However, carbon silicon coating can protect from UV radiation.

A thermoprotective system of aluminum oxidation is shown in Figure 1.2.

A thermoprotective vapor deposition system is shown in Figure 1.3.

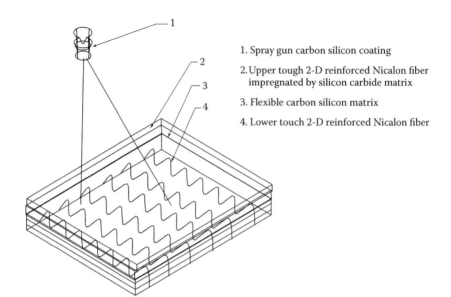

1. Spray gun carbon silicon coating

2. Upper tough 2-D reinforced Nicalon fiber impregnated by silicon carbide matrix

3. Flexible carbon silicon matrix

4. Lower touch 2-D reinforced Nicalon fiber

FIGURE 1.1
Spray gun for carbon silicon coating technology.

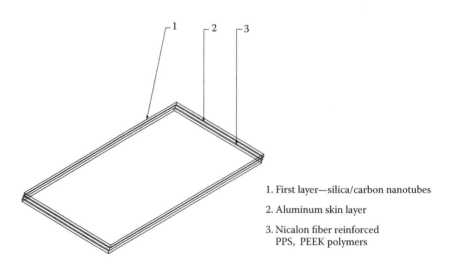

1. First layer—silica/carbon nanotubes

2. Aluminum skin layer

3. Nicalon fiber reinforced PPS, PEEK polymers

FIGURE 1.2
Thermoprotective system.

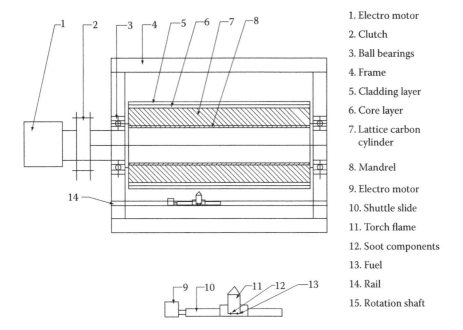

1. Electro motor
2. Clutch
3. Ball bearings
4. Frame
5. Cladding layer
6. Core layer
7. Lattice carbon cylinder
8. Mandrel
9. Electro motor
10. Shuttle slide
11. Torch flame
12. Soot components
13. Fuel
14. Rail
15. Rotation shaft

FIGURE 1.3
Thermoprotective vapor deposition system.

Microscopic observations of a number of cracks using fiber-optic sensors (see Figure 1.4) gives as summations linear deformation and damage accumulation during load environment conditions.

We can observe cracks delaminating on a computer using fiber-optic wires embedded in an SiC/SiC package.

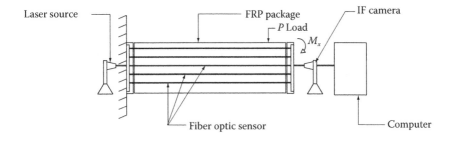

FIGURE 1.4
Stand for delaminating control.

1.1.2.1 Physical Characteristics of Prepreg

The typical prepreg tapes manufactured by Neo-Advent Technologies, LLC (NAT) have a 25-mm width and a thickness of 0.15 mm. We estimated that our final laminate would include 10 to 12 silicon–carbon- and 6 to 8 boron-based layers impregnated in silicon–carbon nanotubes matrix, with a total thickness of 2.0 to 3.0 mm. The lamination sequences of the prepreg layout would include a cross-ply configuration, $[0,90]_{2s}$ and two quasi-isotropic configurations, $[0/+45/-45/90]_s/[0/+45/90/-45]_s$, in alternating order. The pyrolysis process would be carried and synthesized including absorption and pyrolysis, and optimized process within a pressure range of 20 to 50 psi and a curing temperature of 800–1000°C.

1.1.2.2 Testing Mechanical and Thermal Properties of the Prepreg Laminates

We will test the key mechanical properties of the laminated plates (25 × 2.4 mm) of the different lengths, including tensile and compression strength (modulus) at 0, ±45°, and 90° configurations; shear strength/modulus, and interlaminar shear strength. Thermal properties of interest, thermal expansion coefficient (CTE) and thermal conductivity, will be examined shortly. The combined test panel will be based on the following ASTM standards: ASTM D638 (ref. D3039/D3039M) "Test Method for Tensile Properties of Polymer Matrix Composite Materials"; ASTM D696 (ref. D3410) "Test Method for Compression Properties of Polymer Matrix Composite Materials"; ASTM D732 "Shear Strength of Plastics by Punch Tool"; ASTM D903 "Peel or Stripping Strength of Adhesive Bonds"; and ASTM D696 "Coefficient of Linear Thermal Expansion of Plastics." We will also examine the microscopy of the laminate surface after thermal cycling for presence of microcracking.

1.2 Fractographic Model Prediction Deformations and Fatigue Strength

Large aviation and marine components fabricated from composites (i.e., fiber glass or graphite–epoxy materials) have a significantly lower strength than on samples. Some authors call these "scaling effects" [9], and this strength construction is shown below:

$$\sigma = \frac{\sigma_0}{N_i^{1/2}} \tag{1.1}$$

where σ_0 is a sample strength, and N_i is the number of layers.

Strength reduction can be explained as a result of the defects in the laminate.

Cracks develop from a number of structural factors in the laminate, including shrinkage and warpage from thermal stress occurring during the molding process.

Bailey et al. [10] have shown, using a simple equilibrium model, that the thermal residual stress transverse to the fibers in a constrained 90° ply can be expressed as

$$\sigma_{th} \quad \frac{\Delta T t^c E_2 E^c (\alpha^c - \alpha_2)}{E_2 t_2 + E^c t^c} \tag{1.2}$$

where ΔT is the change in temperature and t^c, α^c, and E^c are the thickness, thermal coefficient of expansion, and stiffness, respectively, of the constraining plies, and t_2, α_2, and E_2 are the thickness, thermal coefficient of expansion, and stiffness, respectively, of the 90° piles. This stress is introduced upon cooldown from the curing temperature due to the mismatch in the coefficient of thermal expansion of the adjacent piles in a laminate.

From the prediction strength of every layer we can approximate the average strength of all construction and answer the question of how long this construction will be serviceable [11]. We consider that every layer of construction has strong orthotropic properties, and the construction has a homogeneous structure and is equally impregnated by epoxy or other isotropic resin.

Another approach that the fractographic model predicts is the deformations and fatigue strength. SiC/SiC CMCs are based on the virtual deformation approach.

Turbine engine blade manufacturing from CMCs are shown in Figure 1.5.

The virtual deformation approach of the CMC components of gas turbine blades depends on the centrifugal forces, bending and torsion moments, and rigidity of material blades (see Equation 1.3).

This system follows the theory of Kerhgofa–Klebsha and has been transformed on five independent relationships:

$$\varepsilon_z = \frac{F_z}{E_1 S_1}; \quad \varepsilon_y = \frac{F_y}{E_2 S_2}; \quad \gamma_z = \frac{M_z^R}{E_1 I_z}; \quad \gamma_R = \frac{M_y^R}{E_1 I_R}; \quad \gamma_\theta = \frac{M_x^R}{G_{xz} T_x} \tag{1.3}$$

where ε_z, ε_y, γ_z, γ_R, γ_θ are deformations in x, y, z directions; E_1, E_2 are the modulus of elasticity; S_1, S_2 are the sections area of gas turbine blades; and F_z, F_y are the centrifugal forces that coincide with ply Nicalon fiber in 0/90 directions.

M_z^R, M_y^R, M_x^R are bending moments; I_z, I_R are the moments of inertia for axes x, R; G_{xz} is a shear modulus; and T_x is a geometrical stiffness for torsion (T_x is a moment of inertia for axes x).

These parameters can be determined as:

$$I_z = \int y^2 \, dS; \quad I_R = \int z^2 \, dS \tag{1.4}$$

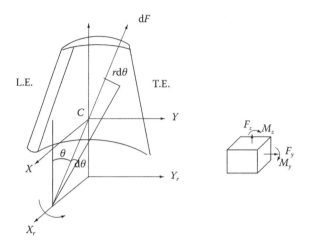

FIGURE 1.5
Turbine blade manufacturing from SiC/SiC.

The geometrical stiffness for torsion can be determined as:

$$T_x = \int \left(R^2 \theta dS \right)$$

Fatigue stress prediction for a CMC model in x (longitudinal) tape direction coinciding with centrifugal force is:

$$\sigma_{-1x} = \int_1^n E_x \varepsilon_x^2 e^n \partial n + \int_1^n E_x \alpha_x e^n \partial n \qquad (1.5)$$

where σ_{-1x} is the SiC/SiC model fatigue stress, E_x is the modulus of elasticity for Hi-Nicalon fiber (see Table 1.2), ε_x^2 is a nonlinear strain applied in the x direction, n is a number of stress cycles per minute, e^n is an exponential function, T is a temperature gradient, and α_x is a coefficient of thermal expansion (see Table 1.2).

We replace the integrals of Equation (1.5) with summations:

$$\sigma_{-1x} = \sum_1^n E_x \varepsilon_x^2 e^n + \sum_1^n E_x \alpha_x T e^n \qquad (1.6)$$

The natural logarithm of an exponential function is represented in Equation (1.7):

$$\log \sigma_{-1x} = n * \log e \left(\sum E_x \varepsilon_x^2 + \sum E_x \alpha_x T \right) \qquad (1.7)$$

$\log e = 0.434$, and the number of stress cycle varies from 1 to 1000.

The maximum use temperature is 1000°C (see Table 1.2). The coefficient of thermal expansion $\alpha_x = 3.5 \times 10^{-5}/K$ and the modulus of elasticity $E_x = 269$ GPa.

The level of correlation stress ($\log \sigma_{-1}$) when the load varied from 100 to 1000 lb, the deformation calculated by Equation (1.3), and temperature increases in the process of loading from 20°C to 500°C are shown in Figure 1.6.

The number of symmetrical cycles also varied from 1 to 1000. This process is connected with damage accumulation in the matrix and interface zone between the matrix and fiber. The natural logarithm of the exponential function in the transverse direction is represented in Equation (1.8):

$$\log \sigma_{-1y} = n * \log e \left(\sum_{n}^{} E_y \varepsilon_y^2 + \sum_{n}^{} E_y \alpha_x T \right) \qquad (1.8)$$

where
σ_{-1y} = Fatigue stress SiC/SiC in 90 grad direction
E_y = Modulus of elasticity of Hi-Nicalon fiber
ε_y^2 = Nonlinear strain applied in y direction
n = Number of stress cycles per minute
e^n = Exponential function
T = Temperature gradient
α_y = Coefficient of thermal expansion

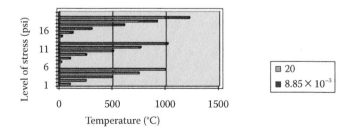

FIGURE 1.6
Level of stress correlated with temperature.

We assume that every stress cycle is provided per minute, so $n = 1/f$, where f is frequency oscillation. We can find a correlation between temperature gradient and frequency oscillation for different stress levels from Equation (1.9).

$$T = \frac{f * \log \sigma_{-1y} - \log e\left(\sum E_y \varepsilon_y^2\right)}{\log e\left(\sum E_y \alpha_y\right)} \tag{1.9}$$

In biaxial stress conditions relations, $\log \sigma_{-1x}/\log \sigma_{-1y}$ represents a coefficient of cracks accumulation (see Figure 1.7).

The variation method to predict fatigue strength can be described as [11]:

$$\sigma_{-1} = \sigma_s \Phi(\sigma)$$

where σ_{-1} (fatigue strength) and σ_s (compression strength) are in the x, y, z directions. The function $\Phi(\sigma)$ can be shown as the Weibull distribution function:

$$\Phi(\sigma) = 1 - P(t) \tag{1.10}$$

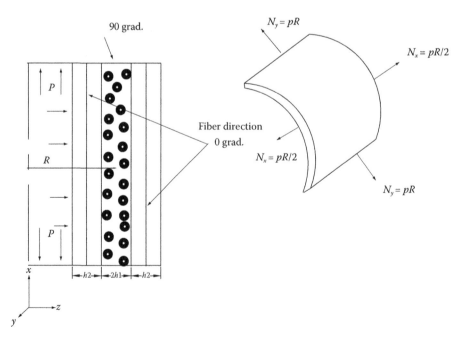

FIGURE 1.7
Cylindrical laminate under biaxial load.

where $P(t)$ is the probability of collapse in the local part of the construction from compression strength.

We assume that the general strength equals unity and $P(t)$ has been subordinated to the normal distribution law. General strength equals unity and $P(t)$ has been subordinated to the normal distribution law.

$$P(t) = \frac{1}{(2\pi)^{1/2}} e^{-t^2/2} \qquad (1.11)$$

where parameter t is

$$t = \frac{\sigma_{bi} - \sigma_{bm}}{S_j} \qquad (1.12)$$

where
σ_{bi} = Current strength in the x, y, z directions
σ_{bm} = Middle strength in the x, y, z directions
S_j = Sample standard deviation for each environment via

$$S_j^2 = \frac{1}{n_j - 1} \sum_{l=1}^{n_j} \left(\sigma_{bi} - \sigma_{bm} \right)^2 \qquad (1.13)$$

where n_j is the number of testing samples.
The sample mean σ_{bm} is calculated as:

$$\sigma_{bm} = \frac{1}{n_j} \sum_{n=1}^{n_j} \sigma_{bi} \qquad (1.14)$$

For a single test condition (such as 0^0 compression strength), data were collected for each environment being tested. The number of observations in each environmental condition was n_j, where j represents the total number of environments being pooled. If the assumption of normality was significantly violated, the other statistical model should be investigated to fit the data. In general, the Weibull distribution provides the most conservative basic value.

In the work of Talreja [12], the Weibull distribution function is given by:

$$\Phi(X, A, B, C) = 1 - \exp\left\{ -\left(\frac{X - A}{B} \right)^C \right\} \qquad (1.15)$$

where the parameters are:

$$X > 0;\ B > 0;\ C > 0$$

X, A, B, C are each equal to a discrete symbol.
For strength distribution, we designate:

$$X = \sigma_{bi};\ A = \sigma_{bm};\ B = S_j;\ C = N$$

where N is the base of testing.
Therefore, Equation (1.14) will be shown as:

$$\Phi(\sigma_{bi}, \sigma_{bm}, S_j, N) = 1 - \exp\left\{-\left(\frac{\sigma_{bi} - \sigma_{bm}}{S_j}\right)^N\right\} \tag{1.16}$$

If we consider that

$$1 - P(t) = \exp\left\{-\left(\frac{\sigma_{bi} - \sigma_{bm}}{S_j}\right)^N\right\} \tag{1.17}$$

We get the logarithmic Equation (1.18):

$$\ln[1 - P(t)] = N\ln(\sigma_{bm} - \sigma_{bi}) - N\ln S_j \tag{1.18}$$

Equation (1.18) shows a straight line in logarithmic coordinates. Base of testing N can determine the inclination of this straight line.

The Weibull distribution function Φ provides damage accumulation defects when we test compression strength σ_{bm}, a middle strength in the x, y, z directions. Therefore, the period of testing N will be determined as:

$$N = \frac{\ln\left[1 - P(t)\right]}{\ln\left(\sigma_{bm} - \sigma_{bi}\right) - \ln S_j} \tag{1.19}$$

Life prediction methodology is based on the model that accumulates damage (cracks) during the thermomechanical loading conditions and high vacuum test process.

The damage mechanics approach was proposed by Kachanov [13] and Rabotnov [14]. The damage parameter D_k is given by:

$$D_k = 1 - \frac{E}{E_0} \tag{1.20}$$

where E_0 is the elastic modulus of the undamaged material and E is the elastic modulus of the damage material at time t (second modulus of the hysteresis loops). We compare this equation with the Weibull distribution function (Equation 1.10).

The probability of collapse

$$P(t) = \int_t^0 \frac{E}{E_0} = \frac{1}{(2\pi)^{1/2}} e^{-t^2/2} \tag{1.21}$$

where parameter t is a parameter of stress distribution (see Equation 1.12).

Following the work of Barbero [15] and Mallick et al. [16], the crack damage accumulation coefficient D_k is determined as:

$$D_k = \frac{\varepsilon_x^0}{\varepsilon_x^d} = \frac{pR\left(1 - 2\mu_{xy}\right)}{2E_x h \varepsilon_x^d} \tag{1.22}$$

where p is the change load, R is the radius of the cylinder of the microcracked laminates, μ_{xy} is the Poisson ratio, E_x is the effective modulus of the [$\pm\theta$/90] laminate along the x axis, $h = 2(h_1 + h_2)$ is a total thickness of cylinder laminate, and ε_x^d is the axial strain proportional to crack delaminating:

$$\varepsilon_x^d = \frac{\Delta_1}{l} \tag{1.23}$$

where Δ_1 is the crack length and l is the total length. Equation (1.20) has been spread for biaxial stress conditions. The schematic mechanism of a CMC ceramic matrix reinforced with Hi-Nicalon fibers is shown in Figure 1.8 [4].

First, we have a matrix microcracking in a silicon–carbide matrix (Figure 1.8a, b, and c), and second, if these cracks spread, the fiber will be damaged only in the last observation of loading (Figure 1.8d, f, and g).

1.2.1 Conclusions

1. Boron nitride is the appropriate interface between Hi-Nicalon fiber layers reinforced with silicon carbide matrix.
2. Spray boron can coat Hi-Nicalon fiber and increase life prediction.
3. Life prediction methodology is based on the fractographic model, which predicts deformations, cracks and fatigue strength, including accumulation damage (cracks) during the thermomechanical loading conditions and high vacuum test process.

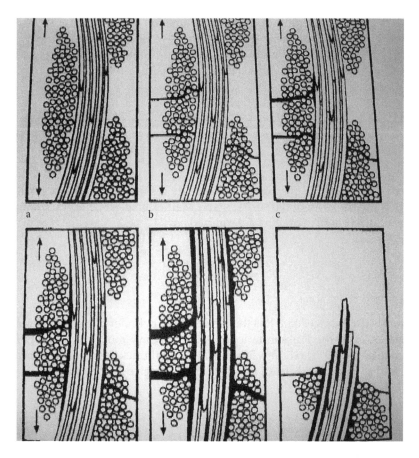

FIGURE 1.8
Schematic mechanism of a CMC ceramic matrix reinforced with Hi-Nicalon fibers. (From Chermant, J. L. et al., *Journal of the European Ceramic Society*, 22, 2443–2460, 2002. With permission.)

1.3 Fiber Draw Automation Control

1.3.1 Introduction

The noncontact system for deformation of fiber draw parameters was recently developed at Spectran, Inc. Its primary advantages are automatic product inspection without distortion of a free measurement and elimination of human measurement error. Scanners can accurately detect physical parameters in rod draw boules in a single or dual axis or in circle diameter.

An optical fiber is a thin glass fiber whose diameter is about 150 μm (approximately 0.006 in). Optic fibers are used to replace copper conductors in telecommunications, since the data transmission capacity of optic fibers is much higher than that of copper conductors.

Most ordinary glasses are based on silica, with other materials added to modify their properties. Research carried out by Hecht [17] shows that high-loss optical properties of plastic fibers have limited their applications to short-distance communications and to flexible bundles for image transmission and illumination. Liquid core fibers and mid-infrared fibers are finding a new life transmitting visible light short distance for illumination. Fabrication of standard optical glasses inevitably leaves traces of impurities such as copper and iron, which absorb some visible light. Communication fibers are coated by a system called flame hydrolysis and then subjected to vapor-phase axial deposition (VAD), outside vapor-phase axial deposition (OVD), hybrid vapor deposition (HVD), or modified chemical vapor deposition (MCVD). These systems are currently used worldwide in fiber preform manufacturing. These processes yield porous soot boules as an intermediate product that are subsequently dehydrated and sintered to obtain preform from which fiber is drawn. Control technological parameters such as draw diameter, temperature, and density are considered by Golfman [18], and present a nanotechnology automatic process.

1.3.2 Core/Cladding Covering Preform

After the glass sintering process, preform is covered by deposits of core glass silica $SiCl_4$ and germanium ($GeCL_4$). Upper cladding layers have deposits of $SiCL_4$ and oxygen (O_2). Figure 1.9 shows outside vapor deposition layers of fiberglass soot on the rotation mandrel.

The rotation mandrel can be installed in horizontal or vertical directions.

The glass preform (boule) rotates with flexible speed that correlates during the core/cladding process with a power flame.

The distance between torch soot components and preform surfaces are regulated by a slide moving in the x, z direction due to servomotor. The volume of components delivery during deposition can also be changed, because there can be possible losses in the system delivery. Gas temperature correlated with the temperature on surface preforms.

1.3.3 Control Gas Preform Diameter

The glass rod geometry analyzer (GRGA) was designed after the core/cladding coating process for preliminary control of the glass preform diameter.

The glass preform was fabricated as a glass bar 25.4 to 50.8 mm in diameter and 1.5 m length with top and bottom ends. The top end was connected by a clamp drive with a stepping motor developed by New England Affiliated Technologies. This was installed in a positioning table with 2° of perform (x, y). The bottom end was supported by a bearing fixed in a frame (Figure 1.10).

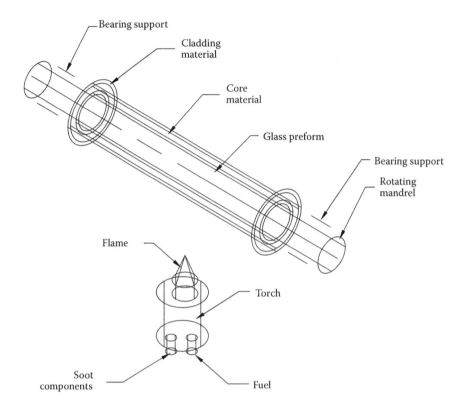

FIGURE 1.9
Outside vapor deposition layers on glass soot on a mandrel.

A laser scanning micrometer moves up and down in z directions of about 1.75 m driven by a Thomson liner system. The carriage is driven by a geomotor with an encoder installed on the top side of the frame.

1.3.4 Control Rod Draw Technological Parameters

Optic fiber is prepared in a so-called draw tower, which is a large-sized tower with a height of about 8 to 10 m. At the upper end of the tower, a glass preform is pulled through a furnace molten and drawn downward into a thin fiber. The hot fiber cools as it progresses downward. Figure 1.11 shows three blocks (entry, central, and exit), which are used for coating and cooling.

A laser scanning micrometer was connected with a Thomson linear motion system to automatically control all the fiber length. The laser scanning micrometer was also equipped with a photo detector that automatically controls fiber diameter. After the coating step, the fiber passed over a tension meter, which was equipped with a load cell and measured the force tension.

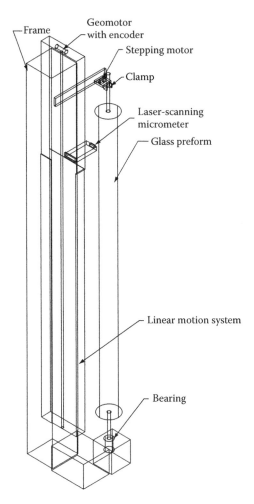

FIGURE 1.10
A glass rod geometry analyzer.

Infrared thermometers measured temperature in the furnace. Infrared thermometers have the ability to measure temperature without physical contact. The ability to accomplish this is based on the fact that energy and the intensity of this radiation is a function of its temperature.

Temperature on the glass fiber surface can be determined as:

$$T = \int_{L_0}^{L_1} T_f x \, dL = \frac{T_f}{2(L_1 - L_0)} \qquad (1.24)$$

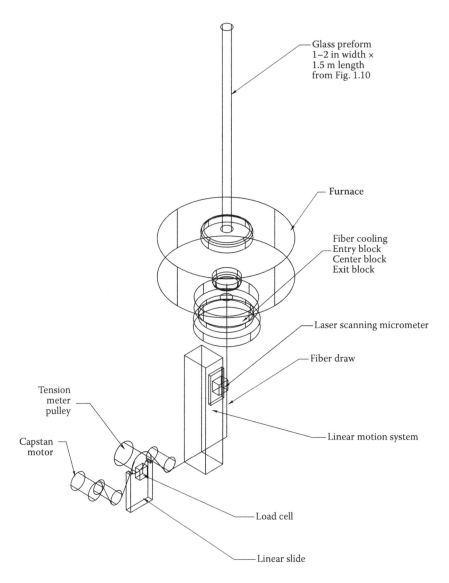

FIGURE 1.11
Rod draw automation control.

Density of the glass fiber is found using the algorithm:

$$g = \frac{F\pi d^2}{2C^2(1-\mu^2)\alpha T} \tag{1.25}$$

where
    F = Measuring tension force by load cell
    d = Glass fiber diameter measured by laser micrometer
    μ = Fiber Poisson ratio
    C = Velocity of ultrasonic light propagation
    α = Coefficient of fiber expansion
    T = Temperature on the glass fiber surface

Correlation between modulus of elasticity E_z and velocity of propagation C along the fiber was established in the work of Golfman [19].

$$E_z = C^2 g(1 - \mu^2) \tag{1.26}$$

In matrix form the correlation between stress and strain is shown as [20]:

$$\sigma_{ij} = Q_{ij}(\varepsilon_{ij} - \alpha_{ij}T) \tag{1.27}$$

where
    Q_{ij} = Stiffness constants
    α_{ij} = Coefficient of temperature expansion
    ε_{ij} = Strain glass fiber
    T = Temperature of glass fiber

We assume that the fiber has tension only in direction z and stiffness will be determined as:

$$Q_{ij} = \frac{E_z}{1 - \mu_{12}\mu_{21}} \tag{1.28}$$

where μ_{12}, μ_{21} is the Poisson ratio of material. The first symbol designates the direction of force and the second symbol designates the direction of the transverse deformations. In our case, $\mu_{12} = \mu_{21}$. On the other side, stress σ correlates with tension force:

$$\sigma = F\frac{\pi d^2}{2} \tag{1.29}$$

Replace σ in Equation (1.27) by Equation (1.29) using Equations (1.26) and (1.28), we get Equation (1.25). We determine the fiber force expansion using a tension device with the load cell (Figure 1.12).

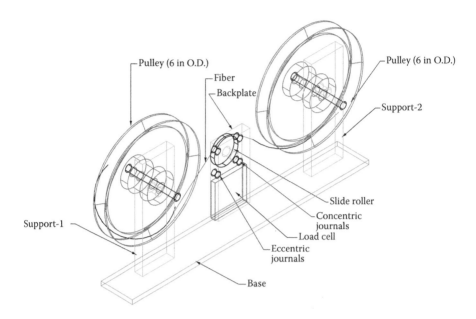

FIGURE 1.12
Tension device.

The equation for forced vibration without damping is:

$$m\frac{\partial^2 \varpi}{\partial z^2} + Q_{11}z = F_0 \sin \Omega t \qquad (1.30)$$

where Ω is the forcing frequency and t is the time of wave propagation.

We assume that a periodic force of magnitude $F = F_0\sin\Omega t$. In the case of free vibration, when $F_0\sin\Omega t = 0$, Equation (1.30) has as solution:

$$z = C_1\sin \varpi t + C_2\cos \varpi t \qquad (1.31)$$

where circular frequency $\varpi = (Q_{11}/m)^{1/2}$. Here, Q_{11} fiber stiffness is determined using Equation (1.28), where m is the mass of fiber, and C_1 and C_2 are arbitrary constants. We assume

$$C_1 = A\cos\phi; \quad C_2 = A\sin\phi$$

where A is amplitude of fiber vibration and ϕ is a phase angle of fiber. We input this in Equation (1.31). So, distance z will be:

$$Z = A\cos \varpi\sin \varpi t + A\sin \varpi\cos \varpi t \qquad (1.32)$$

or $z = A \sin(\varpi t + \varpi)$. We replace circular frequency $\varpi = 2\pi f$, where f is the motion frequency.

$$f = 1/2\pi(Q_{11}/m)^{1/2} \tag{1.33}$$

By differentiating Equation (1.32), we can determine velocity and acceleration of the fiber:

$$V = A(\varpi t + \phi)\cos(\varpi t - \phi) \tag{1.34}$$

$$a = A(\varpi t + \phi)^2\cos(\varpi t - \phi) \tag{1.35}$$

The appropriate equation of motion in this case becomes

$$m\frac{\partial^2 f^2}{\partial z^2} + c\frac{\partial f}{\partial z} + Q_{11}z = F_0 \sin \Omega t \tag{1.36}$$

where m is the mass of fiber, c is the critical damping coefficient, and $c = 2$ mw.
  The particular solution that applies to the steady-state vibration of the system should be a harmonic function of time such as:

$$z_p = A\sin(\Omega t - \phi) \tag{1.37}$$

where A and ϕ are constant.
  Substituting z_p in Equation (1.36), we get:

$$-m\Omega^2 A\sin(\Omega t - \phi) + c\Omega A\cos(\Omega t - \phi) + Q_{11}A\sin(\Omega t - \phi) = F_0\sin \Omega t \tag{1.38}$$

Substituting two boundary conditions $(\Omega t - \phi) = 0$ or $(\Omega t - \phi) = \pi/2$ results in:

$$(Q_{11} - m\Omega^2) = F_0 \sin \Omega t$$
$$c\Omega A = F_0 \sin \Omega t \tag{1.39}$$

The phase angle ϕ reflects a different phase between the applied force and the resulting vibration and is determined as:

$$\tan \phi = \frac{c\Omega}{Q_{11} - m\Omega^2} \tag{1.40}$$

The sine and cosine functions have been eliminated from Equation (1.36) by summing the squares of Equation (1.40):

$$A^2[(c\Omega)^2 + (Q_{11} - m\Omega^2)] = F^2 \tag{1.41}$$

From Equation (1.39), the forcing frequency Ω will be determined as $\tan\phi = 1$.

$$\Omega = [(Q_{11}A - F)/mA]^{1/2} \tag{1.42}$$

where force vibration F is determined by using the load cell (Figure 1.12).

Fiber stiffness Q_{11} (Equation 1.28) and the modulus of elasticity E_z were found using an ultrasonic detector. The ultrasonic detector was installed in the same panel as the laser scanning micrometer. A laser beam will not penetrate through the fiber, but only goes around the fiber.

The transverse displacement of the fiber will be determined as:

$$\frac{\partial^2 S}{\partial t^2} = F/m \frac{\partial^2 S}{\partial^2 z^2} \tag{1.43}$$

where S is the transverse displacement of the fiber from the z axis, t is the time, m is the mass per unit length of the fiber, and F is the tension force of fiber.

The general solution of this equation for an arbitrary driving force and arbitrary initial conditions can be written as a sum of a harmonic having the frequencies:

$$\Omega_n = (n/2L) + (F/m)^{1/2} \tag{1.44}$$

where Ω_n is the frequency of the harmonic and L is the length of the fiber.

In particular, the first harmonic frequency of the fiber motion is given by:

$$\Omega_1 1/2L(F/m)^{1/2} \tag{1.45}$$

A study [21] found that the transverse motion of an optical wave guide fiber during drawing can be broken up into a series of harmonics at least to a first approximation.

It is known that bare optic fiber–uncoated optic fiber cannot bend but will break apart. However, after the silica coats the fiber, the fiber is very resistant to bending. In conventional draw towers, the draw rate is generally about 2 to 4 m/s. The draw rate of the fiber and thickness of the fiber are adjusted by means of a tension meter. A tension device shown in Figure 1.12 consists of two pulleys 6 in. in diameter and a slide roller. The sideway is guided by concentric and eccentric journals. Fibers transmit force to the roller and load cell. The frame consists of three elements: support 1, support 2, and the base. It is not possible to measure the tension stress of the fiber before the fiber is coated.

1.3.5 Fiber Draw and Bending in Flexible Directions

At present, fiber draw and bending in different directions represents a problem that has not been solved.

The fiber is not flexible and loses its optical properties and damages the covering core glass. All fibers are flexible and can bend if subjected to a furnace temperature of 2192°F (1200°C). Figure 1.13 shows a proposed schematic reflecting a rod draw from a preform through an electronic oven in different directions.

Fiber was pulled from the dancer roller through guide rollers by the drive dancer roller, which was connected with the coupler and servomotor. All rollers fabricated from weaving carbon–carbon have an air cooling system, which affords the opportunity to rotate without contact fiber. All optical

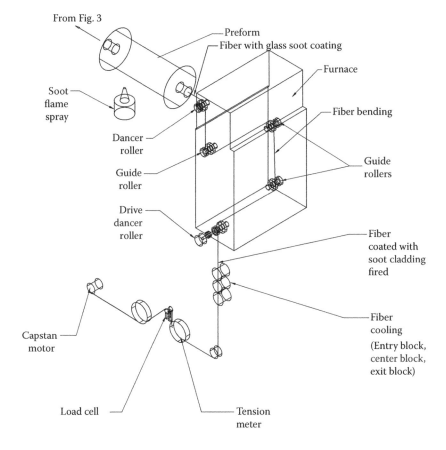

FIGURE 1.13
Rod draw bending in flexible directions.

properties appeared after the fiber coating core/cladding soot and slow cooling when the fiber was pulled through three blocks (entry, center, and exit). After cooling, the fiber force was determined by a typical system using a tension meter and load cell. Finally, a capstan motor was used for weaving the glass fiber on the pulley.

1.3.6 Conclusions

Quality control of MFG fibers was solved by using nondestructive methods to determine the diameter, density, temperature, and modulus of elasticity of fiber pulled through a tower. Control frequency freedom and force vibration give us an opportunity to increase the speed of fiber productivity. Further increase can prove that the productivity of fibers will be possible if we solve the problems of rod draw and bending in flexible directions.

1.4 Spray Deposition of Aerogels as a Thermal Insulation for the Space Shuttle Fuel Tanks

1.4.1 Introduction

The NASA investigation regarding the crash of the space shuttle *Columbia* discussed the loss of insulation of the fuel tank. Pieces of foam detached from the fuel tank area 81 s into the flight, smashing into tiles on the underside of the left wing.

A leading theory of the accident is that the foam insulation may have damaged the heat-protecting tiles during liftoff and was enough to trigger a breach that caused the spaceship to break up, with the resulting loss of the shuttle and its crew. The insulation epoxy foam attached to the tank, by hand depended on the quality of adhesive and premolded technological conditions. Instead of using epoxy resin on the top of the external tank, we propose changing the technology to spray deposition. The purpose of this research is to improve the quality of insulation form and limit the risk of pieces breaking off.

Aerogels are exceptional thermal insulators, characterized by a thermal conductivity of less than 0.02 W/mK. They are used to bond to high-strength metals such as 7075 aluminum and 350 special steel. A typical silica aerogel has a total thermal conductivity of ≈ 0.017 W/mK (R10/in) [22].

The passage of thermal energy through an insulating material occurs through three mechanisms: solid conductivity, gaseous conductivity, and radiation (infrared) transmission.

Minimizing the solid component of thermal conductivity means increasing the overall porosity of the material, which requires a higher vacuum to achieve the maximum performance. The matrix has a low rigidity and brittle-ness character. Minimizing the gases (nitrogen, oxygen) inside aerogel means increasing the solid portion and thermal conductivity. Minimizing the radio-active component of thermal conductivity of silica aerogels means adding black carbon. Carbon is an effective absorber of infrared radiation and actu-ally increases the mechanical strength of the aerogel. Investigation by the Berkeley lab [23] showed that silica aerogel with 9% (wt/wt) carbon black lowers the thermal conductivity from 0.017 to 0.035 W/mK. The minimum value for the carbon composite of 0.0042 W/mK corresponds to R30/in.

Both thermal spray technology spread of atmospheric plasma spray [24] and atmospheric or vacuum spray deposition [25] processes are very expen-sive and require sophisticated equipment.

Consequently, the Navy is seeking a new cost-effective spray process to make aerogel for space shuttle missiles and projectiles flight for functional viability in high-*g* launch and hypersonic flight environments.

We will focus on the following specific objectives:

1. Predict the aerogel strength and silicon resin adhesive properties according to the following parameters: a melt temperature of 1450 K, thermal shock of 1000 K/s, ability to withstand 40 kG accelerations, and thermal conductivity less than 0.02 W/mK.

2. Formulate aerogel compositions.

3. Determine the mechanical and thermal properties of new aerogels.

4. Evaluate the adhesion properties of silicon resins to aerogel and high-strength metals such as 7075 aluminum and 350 mar aging steel.

5. Predict the strength of aerogel and silicon resin adhesive properties according to the following parameters: a melt temperature of 1450 K, thermal shock of 1000 K/s, ability to withstand 40 kG accelerations, and thermal conductivity less than 0.02 W/mK.

6. Tensile, shear strength, and adhesion properties of the devel-oped formulation will be satisfied to the following parameters: melt temperature of 1450 K, thermal shock of 1000 K/s, ability to withstand 40 kG accelerations, and thermal conductivity less than 0.02 W/mK.

1.4.2 Theoretical Prediction of Mechanical and Thermal Aerogel Properties

To provide a quantitative analysis of heat transfer across the complex layered system consisting of aerogel coating and an adhesive layer, it is proposed to do a computational analysis based on finite element calculations. The entire material system will be modeled, as a composite two-component coating, covering the proper combination of parameters of layers and thickness will be performed. The goal will be to identify such a combination of system parameters that ensures that the requirements of thermal conductivity (less than 0.02 W/mk), thermal shock (1000 K/s), and so forth, are met. It is proposed to utilize one of the commercially available finite elements codes.

A correct model should allow some random voids in the scattering medium. This would be rather tedious to calculate. Therefore, we prefer to create a periodic lattice of voids. We approximate the unit cell of this lattice by a sphere. This "isotropic inhomogeneous scattering model" can account rather well for the measuring coefficient of thermal conductivity.

1.4.2.1 Thermal Barrier Coatings

TBCs have thin ceramic/carbon layers, generally applied by plasma spraying or by physical vapor deposition [24,25].

The formation of aerogels in general involves two major steps; the formation of a wet gel and the drying of the wet gel to form an aerogel. Originally, wet gels were made by the aqueous condensation of sodium silicate or a similar material. The vast majority of prepared silica aerogels utilize silicon alkoxide precursors. The most common of these are tetramenthyl orthosilicate (TMOS, $Si(OCH_3)_4$), and tetraethyl orthosilicate (TEOS, $Si(OCH_2CH_3)_4$) [26].

However, many other alkoxides, containing various organic functional groups can be used to impart different properties to the gel. The balanced chemical equation for this formulation of a silica gel from TEOS is:

$$Si(OCH_2CH_3)_{4(liq)} + 2H_2O_{(liq.)} = SiO_{2(solid)} + 4HOCH_2CH_{3(liq.)} \qquad (1.46)$$

The preceding reaction is typically performed in ethanol, with the final density of the aerogel dependent on the concentration of silicon alkoxide monomers in the solution. The stoichiometry of the reaction requires 2 mol water per mole of TEOS. In practice, this amount of water leads to incomplete reaction and weak, cloudy aerogels. Therefore, most aerogel recipes use a higher water ratio than is required by the balanced equation (anywhere from 4 to 30 equivalents). Aerogels prepared with an acid catalyst often show more shrinkage during supercritical drying and may be less transparent than base catalyzed aerogels. The macrostructural effects of various catalysts are harder to describe accurately, as substructure of the primary particles of aerogels can be difficult to image with electron microscopy. All show small (2–5 nm diameter) particles that are generally spherical or egg-shaped. With acid catalyst, however, these particles may appear "less solid" (looking something like a ball of string) than those in base catalyzed gels.

As condensation reactions progress, the sol will set into a rigid gel. At this point, the gel is usually removed from the mold. However, the gel must be kept covered by alcohol to prevent evaporation of the liquid contained in the pores of the gel. Evaporation causes severe damage to the gel and will lead to poor-quality aerogels.

1.4.3 Concept of Thermal Conductivity

The coefficient of thermal conductivity of the silica/carbon aerogels is calculated as:

$$\alpha = \rho\beta C \tag{1.47}$$

where
ρ = Density of silica/carbon aerogels (kg/m^3)
β = Coefficient of thermal diffusivity of silica/carbon aerogels (m^2/s)
C = Specific heat capacity; J/kg K. 1 J/s = 1 W, so α = w/mK

The thermal diffusivity directly depends on the effect of porosity [27,28].

$$\beta' = \beta\frac{1-P}{1+P} \tag{1.48}$$

where β' is the coefficient of thermal diffusivity with porosity aerogels, β is the coefficient of thermal diffusivity without porosity, and P is the porosity of aerogels.

Thermal shock requires a cycle from increasing the temperature to 1000 K and reducing to absolute zero (273.15 K). C = K(1000 grad) – 273.15 grad = 726.85 grad.

The melting temperature of aerogels is 1450 K, C = K(1450 grad) – 273.15 grad = 1176.85 grad.

1.4.3.1 Design a Spraying Process with the Low-Thermal Conductivity Ceramic/Carbon Spray Aerogels

Design of low-thermal conductivity ceramic/carbon spray aerogels is shown in Figure 1.14, and the thermal insulator is shown in Figure 1.15.

TBCs have thin ceramic/carbon layers, generally applied by plasma spraying or by physical vapor deposition, and are used to insulate air-cooled metallic components from hot gases in gas turbine and other heat engines [29]. The ceramic layer consists of 95.4 at.% zirconia ZrO$_2$ and 4.6 at.% yttria Y$_2$O$_3$. However, these coatings have porous and microcracked structures. Recently, scandia was identified as a stabilizer that could be used in addition to yttria [30]. A composition of 3 mol % scandia and 2.5 mol % yttria may confer the desired phase stability at 1400°C. Our cost-effective process is very simple: After spraying adhesive layers of liquid poly(dimethylsiloxane) polymer, we spray silica/carbon liquid. Temporary pivots are located on the nest of pivot ring. We use wax for easy separation of the pivots from the aerogel. After spraying aerogel layers and solidifying them, the pivots will

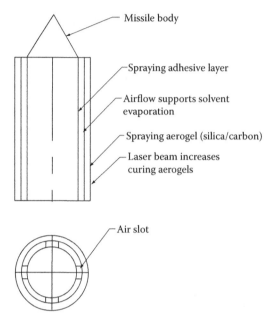

FIGURE 1.14
Design of a low thermal conductivity spray system.

be removed. We take dry air flow to the air channels to support the solvents' evaporation.

Aerogel polymerization will occur faster if we use a laser (or infrared) beam (see Figure 1.16).

Spray silica/carbon aerogels on a poly(dimethylsiloxane) layer.

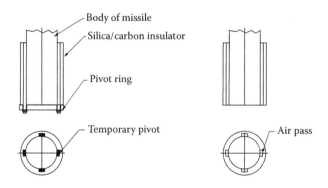

FIGURE 1.15
Thermal insulator.

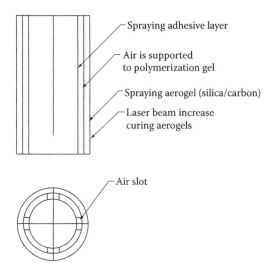

FIGURE 1.16
Spray adhesive technology.

$$SiH_4 + SiC = 2SiCH_4 \qquad (1.49)$$

The poly(dimethylsiloxane) SiH_4 links chemically with carbon to make a strong spatial structure.

Silicone fluids are usually straight chains of poly(dimethylsiloxane) (PDMS), which are terminated with a trimethylsilyl group (or groups). PDMS fluids come in all viscosities, from waterlike liquids to intractable fluids. All of these are essentially water-insoluble. PDMS fluids may be further modified with the addition of organofunctional groups at any point in the polymer chain.

Silicone gels are lightly cross-linked PDMS fluids, where the cross-link is introduced either through a trifunctional silane such as CH_3SiCl_3, giving a "T-branched" silicone structure, or through a chemical reaction between a Si-vinyl group on one polymer chain with a hydrogen bonded to silicon in another. This chemical "tying" of siloxane chains produces a 3-D network that can be swollen with PDMS fluids to give a sticky, cohesive mass without form.

1.4.3.2 Single-Step Base Catalyzed Silica Aerogel

This step will produce an aerogel with a density of approximately 0.08 g/cm³. The gel time should be 60 to 120 min, depending on temperature.

1. Silica solution containing 50 mL of TEOS, 40 mL of ethanol.
2. Catalyst solution containing 35 mL of ethanol, 70 mL of water, 0.275 mL of 30% aqueous ammonia, and 1.21 mL of 0.5 M ammonium fluoride and 9% (wt/wt) carbon black.
3. Slowly add the catalyst solution to the silica solution while stirring.
4. Pour the mixture into an appropriate mold until gelation.
5. Process as described above.

1.4.4 Experimental Investigation Results

The primary objective of the base program is to establish the feasibility of ceramics/carbon spray aerogel by completing the following tasks of selecting the adhesion layers and demonstrating the ability of the formulation to be applied by sparing on the adhesive layers.

For adhesives with aluminum or steel we use silicon fluid poly(dimethylsiloxane) polymer.

1. Establish the ceramic/carbon spray aerogel requirements.
2. Develop an aerogel formulation.
3. Test and evaluate mechanical and thermal properties of ceramic carbon spray aerogels.

For testing aerogel properties, we use nondestructive methods [31–34].

We investigated the combination of thermal protection deposition of substrates and basic aluminum or titanium attachment parts and conclude:

1. The carbon–boron deposition on the steel/aluminum semimonocoque cylindrical surfaces has more adhesion than a spray-on foam of phenol thermal insulation. The result of the diffusion process increases the fluctuation of durability on the surface 1.18 to 3.5 times.
2. Silicon carbide is initially formed by the injection of gaseous silicon and carbon on the carbon monofilament. The boron deposition (borane BH_3) combines with the glass silica $SiCl_4$ to form a silicon–boron bond, which has higher surface strength and temperature resistance than silicon carbide.
3. The new attachment titanium plates with the silicon–boron insulation can protect the ET tank from liquefaction of the air-exposed metallic attachments.

We develop a model of durable, fiber-reinforced refractory composites for protection systems (TPSs) [35–37]. TPSs on leading edges, control surfaces, and over large areas of the skin represent a crucial need for next-generation reentry and military space vehicles. The reusable TPSs currently employed by NASA on the space shuttle require rehabilitation after each mission, offer no multifunctionality, and are very susceptible to impact damage.

The loss of the space shuttle *Columbia* emphasizes the need for more impact-resistant leading edge designs. The U.S. Air Force is currently developing durable leading edge concepts that will meet the quick turnaround requirements needed for military reusable launch systems. This design employs carbon–carbon, carbon–silicon–carbide, and/or silicon carbide–silicon–carbide aeroshells. The majority of the research to date has been on the integration of different materials to meet the aerothermal and mechanical requirements. We are developing impact-resistant, durable material solutions and design concepts for leading edge TPSs.

1.4.5 Conclusions

1. We developed a low-cost aerogel process. For spraying an adhesive layer we selected poly(dimethylsiloxane) polymer. We used the temporary pivots located on the nest of pivot ring. We created a dry air pressure channel to support solvent evaporation.

2. We sprayed single-step base catalyzed silica aerogels with carbon component. We sprayed a thermal barrier coating. Two technologies proposed for the outside TBCs of the shuttle structure consist of ceramic layers zirconia (ZrO_2) + scandia + yttria (Y_2O_3), and boron nitride (BN) + strontium chromate ($SrCrO_4$).

3. The flux heat of internal energy transfer was investigated in different directions. The low coefficient of thermal conductivity can guarantee the permanent thermal stability of the lattice structure, and therefore will be able to protect future shuttles from failure.

4. When the shuttle enters the dense layers of the atmosphere, the temperature gradient on the wing leading edge increases and in the study we used two carbon–carbon aeroshells. The aeroshells open and protect the leading wing edges from thermo heat and radiation.

1.5 Self-Sealing Fuel Tank Technology Development

1.5.1 Introduction

Current self-sealing fuel tank technology is able to withstand a small arms attack.

Commercially available self-sealing fuel tank coatings have been explored and incorporated into EVP external fuel tanks. This effort seeks an integrated solution that goes beyond current fuel tank coatings. Explosion-proof fuel tanks have been manufactured with a flexible polyurethane (either polyester or polyether) foam with fully open pores, composed of a skeletal network of tiny lightweight interconnecting strands that act as a three-dimensional fire screen [38].

Fuel systems for military and commercial aircraft use Kevlar fuel tanks to provide an additional 140 gallons of oil [39]. Built-up layers of synthetic and rubber sheeting on plaster foam have been used to make self-sealing fuel tanks for airplanes; a brushed solution on specified plaster foam provides a smooth surface. Brushed cement between layers of rubber provides an adhesive surface. Cement fittings on fuel tanks, rubber reinforcement strips over the base of fittings, and reinforcement strips of gum rubber seams help seal fuel tanks [40,41].

1.5.1.1 Principal Concept of a Self-Sealing Fuel Tank Design

The principal concept for a self-sealing fuel tank design is based on the creation of a buffer zone between a flexible polymer shell and a rigid hull of the fuel tank. A load impact explosion can hide a flexible polymer shell that can bend and change its temporary form and then return to its previous condition. A compression spring is installed in a buffer zone in the x, y, z directions.

In Figure 1.17, the compression force is designated as F_x, F_y, F_z, where the compression springs fuse to the hull of the fuel tank by contact joining by a CO_2 laser in the x, y, z directions.

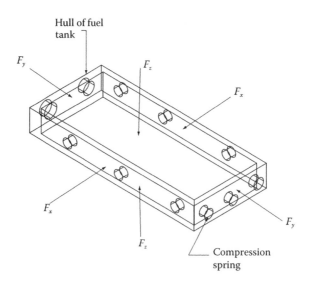

FIGURE 1.17
Principal concept for fuel tank design.

The hull is manufactured from Kevlar. The fuel tank assembly is shown in Figure 1.18.

Multilayers for the shell protection system are divided into tough and plastic layers. Every layer has its own frequencies and own stiffness. The full energy from the impact loads is distributed between the tough and plastic layers. Glass polyester is a typical plastic layer and ceramic alumina epoxy layers are typical tough-type layers. Four thermoset resins were used as the matrix: orthophthalic polyester, isophthalic polyester, vinyl ester, and reinforced epoxy [42].

Coinjection resin transfer molding (CIRTM) and diffusion-enhanced adhesion were two processes that were created and developed to address the cost and performance barriers that hindered the introduction of composite materials for combat ground vehicle application [43]. When applied in tandem, these two composite processing technologies enabled the manufacture of lightweight composite/ceramic integral armor, offering significant cost reduction and performance enhancement over existing defense industry practices. CIRTM was developed for the single-step manufacturing of integral armor through simultaneous injection of multiple resins into a multilayer preform.

The process achieved excellent bonding between the layers, which is an important aspect of the CIRTM process. Also developed was an understanding of the resin flow and cure kinetics to aid in the process optimization. Furthermore, the study enabled the production of new composite structures, including stitched structures with improved ballistic response.

In Figure 1.19 we see a sample of multilayer protection systems. From the unwound rollers (pos. 1, 2, 3) that simultaneously can draw polyethylene

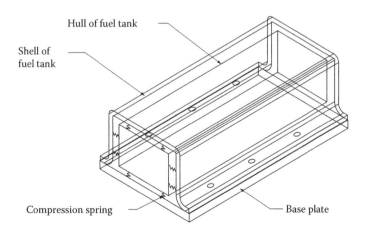

FIGURE 1.18
Fuel tank assembly.

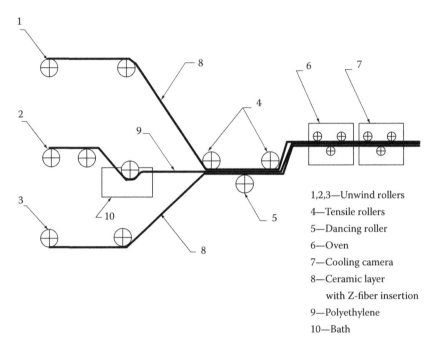

1,2,3—Unwind rollers
4—Tensile rollers
5—Dancing roller
6—Oven
7—Cooling camera
8—Ceramic layer
    with Z-fiber insertion
9—Polyethylene
10—Bath

FIGURE 1.19
Automation process for manufacturing the fuel tank shell.

(pos. 9) and the ceramic prepreg (textile + ceramic on organic polycarbosilane resin) (pos. 8). Three layers are pulled by tensile rollers (pos. 4) and a dancer roller (pos. 5). Curing this multilayer system occurs when the layers are pulled through the oven (pos. 6). For the cooling process, a camera is used in (pos. 7).

A stress analysis of a self-sealing fuel tank was carried out by Golfman [49].

1.5.2 Experimental Results

The hull of the fuel tank was manufactured from Kevlar and had a specific density of 1.45 g/cm³, the mass of the tank = 90.8 kg, the stiffnesses were $Q_{11} = 0.81 \times 10^5$ MPa, $Q_{22} = 0.646 \times 10^5$ MPa, and $Q_{33} = 0.81 \times 10^5$ MPa. The impact force frequencies varied from 200 to 1000 Hz.

The compression dynamic forces were determined in Table 1.5 [44,45,46].

Correlation between the amplitude of vibration and the compression impact forces is shown in Figure 1.20.

The coefficient of thermal expansion has a different significance for different directions that prove the anisotropic character of the selected materials. We used CTE, linear 1000°C, $\alpha_1 = 2.22$ µin/in °F, $\alpha_2 = 1.67$ µin/in °F, and $\alpha_3 = 2.22$ µin/in °F.

TABLE 1.5

Amplitude of Vibration (cm)			Damping Coefficients			Compression Impact Forces (kg)		
A_x	A_y	A_z	δ_x	δ_y	δ_z	$F_x\,10^{-6}$	$F_y\,10^{-6}$	$F_z\,10^{-6}$
0.466	0.440	0.466	0.322	0.16	0.322	0.980	0.735	0.980
0.932	0.881	0.932	–1.7	–2.4	–1.7	1.98	1.47	1.98
1.397	1.32	1.397	–5.1	–6.68	–5.1	2.95	2.21	2.95
1.864	2.016	1.864	–9.8	–12.6	–9.8	3.93	3.35	3.93
2.476	2.202	2.476	–15.9	–20.3	–15.9	4.90	3.68	4.90

The coefficient of thermal conductivity in x, y directions is $k = 55$ W/m K, whereas in the z direction $k = 33$ W/m K.

The shell of the fuel tank protection for this study was selected from GE advanced ceramic materials whose maximum service temperature range is 2100–2910°F.

The physical, mechanical, electrical, and thermal properties of GE advanced ceramic materials are shown in Tables 1.6, 1.7, 1.8, and 1.9.

1.5.3 Conclusions

1. A concept was developed for a self-sealing fuel tank design based on a fire-resistant thermal ceramic shell that consists of tough and plastic layers.

2. The hull of the fuel tank was manufactured from Kevlar, whose specific weight and anticorrosion properties increase the life of service of the tank.

3. The buffer zone between the shell and the fuel tank are supported by the installation of compression springs.

4. The compression forces created by the explosive impact load can show how failure of the fuel tank can be predicted by calculated

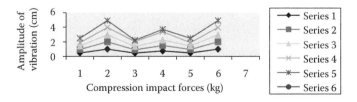

FIGURE 1.20
Correlation between amplitude of vibration and compression impact forces.

TABLE 1.6

Properties of HBR Hot-Pressed Boron Nitride

Property	Metric
Physical Properties	
Density	2 g/cm^3
Binder melting point	1150°C
Water absorption	1%
Open porosity	11%
Mechanical Properties	
Hardness, Knoop	26
Modulus of elasticity	48.2 GPa
Modulus of elasticity	62 GPa
Flexural strength	41.3 MPa
Flexural strength	51.7 MPa
Compression yield strength	62 MPa
Compression yield strength	68.9 MPa
Electrical Properties	
Electrical resistivity	Min. 1e + 0.15 Ω cm
Dielectric constant	4.1
Dielectric strength	53 kV/mm
Dissipation factor	Max. 0.0002
Thermal Properties	
CTE, linear 1000°C	3 μm/m °C
CTE, linear 1000°C	4 μm/m °C
Heat capacity	0.808 J/g °C
Thermal conductivity	33 W/m K
Thermal conductivity	55 W/m K
Maximum service temperature, air	850°C
Maximum service temperature, inert gas	1150–1600°C

Source: Momentive Performance Materials, Inc., Hot-pressed boron nitride shapes (QTZ-81507), www.momentive.com, 2007.

forces from motion equations, which include damping coefficients and linear extensions received by variation in impact loads.

5. The gradient of temperature can be predicted using calculated compression forces, material stiffness, and coefficients of thermal expansion.

6. The material of composite elements can be selected using calculated forces and thermal properties.

1.6 Deposition of the Thermal Insulation Fuel Tank of the Space Shuttle

1.6.1 Introduction

An external tank (ET) contains liquid hydrogen fuel and liquid oxygen oxidizer and supplies them under pressure to the three space shuttle main engines (SSMEs) in the shuttle orbital during liftoff and ascent. When the SSMEs are shut down, the ET is jettisoned, enters the Earth's atmosphere, breaks up, and impacts in a remote ocean area. It is not recovered [50]. The shuttle orbiter with external fuel tank and two solid rocket boosters (SRBs) are shown in Figure 1.21.

The largest and heaviest (when loaded) element of the space shuttle, the ET has three major components: the forward liquid oxygen tank, an unpressurized intertank that contains most of the electrical components, and the aft liquid hydrogen tank (see Figure 1.22).

The liquid oxygen tank (pos. 1) is an aluminum monocoque structural component composed of a fusion-welded assembly of preformed, chemically milled gores, panels, machined fittings, and ring chords. It operates in a pressure range of 20–22 psig. The tank contains antislosh and antivortex provisions to minimize liquid residuals and damp fluid motion. The tank feeds into a 17-in-diameter feed line that conveys the liquid oxygen through the intertank, then outside the ET to the aft right-hand ET/orbiter disconnect

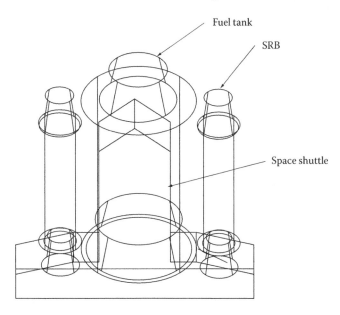

FIGURE 1.21
Space shuttle assembly.

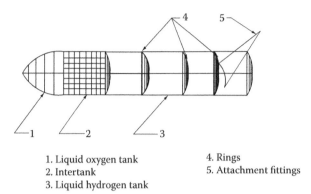

1. Liquid oxygen tank   4. Rings
2. Intertank   5. Attachment fittings
3. Liquid hydrogen tank

FIGURE 1.22
Fuel tank.

umbilical. The liquid oxygen tank's double-wedge nose cone reduces drag and heating and contains the vehicle's ascent air data system (for nine tanks only) and serves as a lightning rod. The intertank is a steel/aluminum semimonocoque cylindrical structure with flanges on each end joining the liquid oxygen and liquid hydrogen tanks (see pos. 2). The intertank houses ET instrumentation components and provides an umbilical plate that interfaces with the ground facility arm for the purge gas supply, hazardous gas detection, and hydrogen gas boil-off during ground operations. The intertank is vented during flight.

The liquid hydrogen tank is an aluminum semimonocoque structure of fusion-welded barrel sections, five major ring frames, and forward and aft ellipsoidal domes. Its operating pressure range is 32–34 psi. The tank contains an antivortex baffle and siphon outlet to transmit the liquid hydrogen from the tank through a 17-in line to the left aft umbilical. At the forward end of the liquid hydrogen tank is the ET/orbiter forward attachment pod strut, and its aft end is the two ET/orbiter aft attachment ball fittings as well as the aft SRB-ET stabilizing strut attachments. The new aluminum lithium hydrogen tank has been designed and tested [51].

This new tank has high strength and lower density properties than currently used material. The walls of the hydrogen tank will be manufactured in an orthogonal wafflelike pattern and the new ET will be the same size as the current one but 7500 lb lighter.

The ET is 153.8 ft long and has a diameter of 27.6 ft [51]. The weight reduction was accomplished by eliminating portions of stringers using fewer stiffener rings and by modifying major frames in the hydrogen tank. Also, significant portions of the tank are milled differently to reduce thickness, and the weight of the ET's aft SRB attachments were reduced by using a stronger, yet lighter and less expensive titanium alloy. After propellant loading, data from ground tests and the first few space shuttle missions were assessed and the antigeyser line was removed for STS-5 and subsequent missions. The total length and diameter of the ET remain unchanged. The ET is attached to the orbiter at one forward attachment point and two aft points. In the aft attachment area, there are also umbilicals that carry fluids, gases, electrical signals, and electrical power between the tank and the orbiter. Electrical signals and controls between the orbiter and the two SRBs also are routed through those umbilicals.

1.6.2 Design Features

A new design (Figure 1.23) has been proposed to move the orbiter forward far from the insulation tank and engines. An additional proposal has increased the cooling water system during the period of burning gases. In a typical STS launch, the two SRBs burn in parallel with the three liquid oxygen/liquid hydrogen engines for 128 s. Then, the solid boosters separate from the rest

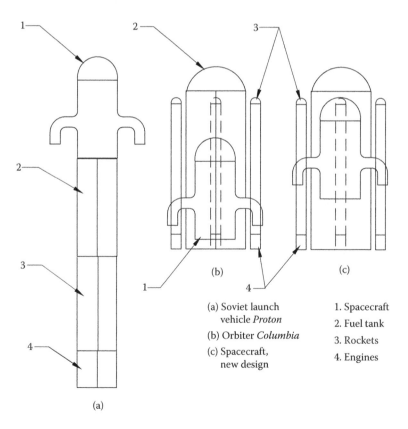

(a)

(b)   (c)

(a) Soviet launch   1. Spacecraft
   vehicle *Proton*   2. Fuel tank
(b) Orbiter *Columbia*   3. Rockets
(c) Spacecraft,   4. Engines
   new design

FIGURE 1.23
Spacecraft assembly.

of the vehicle and drop to the ocean and are recovered and refurbished for use again during a later launch. The three SSMEs continue burning for 480 s after separation of the SRBs.

1.6.3 Requirements for Thermal Protection Systems

1. Reversibility of the thermal protection system is the basic element of the liquefaction of the air-exposed metallic attachments. The heat flow is reduced by the liquid hydrogen.
2. The thermal protection systems work when the range of temperature is $-10°F$ to $+95°F$.
3. The combined thermal protection deposition covers the substrate and basic aluminum or titanium attachment parts.

1.6.4 ET Thermal Protection System and CVD Process

The ET thermal protection system consists of a spray-on foam insulation and a premolded ablative material. The system also includes the use of phenolic thermal insulators to preclude air liquefaction. Thermal insulators are required for the liquid hydrogen tank attachments to preclude the liquefaction of air-exposed metallic attachments and to reduce the heat flow into the liquid hydrogen.

A CNN report [52] describes the shrink wrap for the external fuel tank; the insulation for the tank is on the inside rather than the outside to prevent chunks of foam from breaking off and striking the shuttle, as happened with *Columbia.*

In one study [53], a boron deposition called borane (BH_3), which interacts with glass silica ($SiCl_4$), forms a silicon–boron bond.

The new attachment fittings are attached to the fuel tank. The attachment fittings are manufactured from titanium alloys and the boron was deposited directly onto the titanium surface.

We recommended the thermoprotective vapor carbon–boron deposition model as an alternative to the spray-on form of insulation shown in Figure 1.16. Cladding layer (pos. 5) and core layer (pos. 6) deposits to lattice carbon cylinder (pos. 7). Torch flame (pos. 11) has a programmable speed and moves through the shuttle slide (pos. 10) by electromotor (pos. 9).

Our recommendation was verified by Suplinskas and Hauze [54]. The deposition took place by passing the carbon monofilament through a furnace into which gaseous silicon (entry block) and carbon (center block) are injected. At a deposition temperature of about 2370°F, a deposit of beta crystals of silicon carbide grains is formed. Then boron gas (exit block), is injected onto the surface of the silicon carbide filament. A laser scanning micrometer was connected with a Thomas linear motion system to automatically control all of the filament length. The laser scanning micrometer has a photo detector that automatically controls the filament diameter.

After the coating step, the filament passed over a tension meter, which was equipped with a load cell. This measured the force tension. Temperature in the furnace was measured by infrared thermometers. Infrared thermometers have the ability to measure temperature without physical contact. The ability to accomplish this is based on the fact that the energy of an object emits radiant energy, and the intensity of this radiation is a function of temperature.

In conventional draw towers, the draw rate is generally about 2 to 4 m/s. The draw rate of the filament and thickness of the filament were adjusted by means of the tension meter. Currently, filament draw and bending in different directions is shown in present works.

The braider, shown in Figure 1.24, took up the roll with the carbon–boron filament.

Wound carbon–boron insulation on the cylinder, which represents the model of the fuel tank (cylindrical elements) are shown in Figure 1.24. All rollers were installed on the braider axis. The carbon–boron filament wound from the roll to the surface of the cylinder was fixed on the mandrel. The cylinder was able to rotate in x, y directions and move in the z direction by a shaft and electric motor. Thus our coating of carbon–boron filament weaving in x, y, and z directions is shown as a 3-D solid protection system.

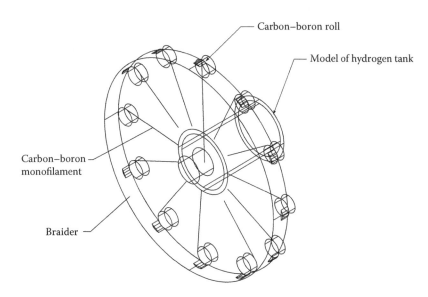

FIGURE 1.24
Wound carbon–boron insulation on the cylinder (model of fuel tank).

1.6.5 Durability of Adhesion Deposition

The deposition process covered the intertank, liquid oxygen, and liquid hydrogen tanks that transit gas to the liquid and solid surfaces. We consider deposition to be a chemical process when diffusion gases penetrate into active electrons on the metal surface. This process is intensified when pressure and temperature increase.

The intertank, liquid oxygen, and liquid hydrogen tanks are made of the steel/aluminum semimonocoque cylindrical structures while the carbon–boron insulation works under pressure and temperature.

Linear approximation between pressure, density, and temperature of the oxygen and hydrogen gases are [55]:

$$\frac{P}{\rho} = RT$$

(1.50)

where
  P = Pressure outside the fuel tank
  ρ = Gas density
  R = Universal gas constant; R = 287 J/kg K

The wave equation is:

$$\nabla^2 p - \frac{1}{c^2}\frac{\partial^2 p}{\partial t^2} = 0$$

(1.51)

where ∇^2 is the Laplacian sum of the second derivatives with respect to the three Cartesian coordinates, x, y, z.

$$\nabla^2 = \frac{\partial^2 p}{\partial x^2} + \frac{\partial^2 p}{\partial y^2} + \frac{\partial^2 p}{\partial z^2}$$

(1.52)

If we select the spread pressure in the x direction, Equation (1.51) will be assigned as:

$$\frac{\partial^2}{\partial x^2} = \frac{1}{c^2}\frac{\partial^2 p}{\partial t^2}$$

(1.53)

We consider $\partial^2 p/\partial x^2$ as a wave function in the x direction

$$\frac{\partial^2 p}{\partial x^2} = \sin(\omega t - \phi)$$

(1.54)

where ω is the circular frequency and ϕ is the delay constant [56].

$$\frac{\partial^2 p}{\partial t^2} = \frac{1}{\tau} \int_{t_0}^{t_0 + \Delta t} p^2 \partial t \qquad (1.55)$$

So we input Equation (1.53) and (1.54) into Equation (1.52), and we get:

$$\sin(\omega t - \phi) = \frac{1}{\tau} \int_{t_0}^{t_0 + \Delta t} p^2 \partial t \qquad (1.56)$$

where τ is the time of wave propagation.

After integration of Equation (1.56), we get:

$$\sin(\omega t - \phi) = \frac{p^2}{\tau} \Delta t \qquad (1.57)$$

where Δt is the decrement of temperature fluctuation.

The influence adhesion force to metal surface can be described by Equation (1.58) [57].

This is the Zhurkov equation:

$$\tau_p = \tau_0 \exp \frac{U_0 - \gamma p}{RT} \qquad (1.58)$$

where

τ_p = Durability of adhesion strength (s)

τ_0 = Constant value of carbon–boron deposition close to the period oscillation atomic molecules, equal to 10^{-11} to 10^{-13} s

U_0 = Value of energy associated with adhesion links of diffusion process (J/mol)

γ = Coefficient of ratio of special heat

p = Pressure outside the fuel tank (kg/m²)

R = Universal gas constant; $R = 287$ J/kg K

T = temperature (K)

1.6.6 Experimental Investigation

The value of energy associates with adhesion links to correlate with outside pressure and temperature. All the atoms of a tank's surface are the free radicals that interact with carbon–boron deposited layers.

TABLE 1.7

Physical Properties of Gases

Gas	Temperature, T (°C)	Density, ρ (kg/m³)	Ratio of Specific Heats, γ	Speed (m/s)
Air	0	1.293	1.402	331.6
Air	20	1.21	1.402	343
Oxygen, CO_2	0	1.43	1.4	317.2
Low frequency, CO_2	0	1.98	1.304	258
High frequency, CO_2	0	1.98	1.4	268.6
Hydrogen	0	0.09	1.41	1269.5
Stem	100	0.6	1.324	404.8

For calculation, we selected the following value: $U_0 = 100{,}000$ J/mol. In Krivopal's study [58], $U_0 = 74{,}000$ J/mol for polyamid-12 covering metal; $\gamma = 1.4$ (see Table 1.7); $\tau_0 = 10^{-12}$ s.

Following Equation (1.58), the fluctuation of durability increases 1.18 to 3.5 times. The kinetic theory also gives the ideal gas equation in the form:

$$P = NkT \tag{1.59}$$

where N is the number of molecules per unit volume and k is Boltzman's constant ($k = 1.381 \times 10^{-23}$ J/K). Thus, R in Equation (1.58), $R = R_0/M$ and $R_0 = k/M_a = 8314$ J/kg K is the universal gas constant. M_a is the average mass per molecule ($M_a = 1.661 \times 10^{-27}$ kg). The corresponding value of R is $8314/29 = 287$ J/kg K. $T = 0°C = 273.16$ K. If $(1/4RT)^{1/2} = 331$ m/s, in accordance with the accepted experimental value.

$$C = 331 + 0.6T_c \tag{1.60}$$

The physical properties of gases are shown in Table 1.7 [59].

Since Equation (1.50) is the ideal gas resulting from Boyle's law and from definition of absolute temperature T, one has:

$$C^2 = \gamma RT \tag{1.61}$$

where γ is a specific heat ratio (see Table 1.7).

The value of velocity C according to the Laplace adiabatic assumption for an ideal gas is:

$$C^2 = \frac{\gamma P}{\rho}$$

$$(1.62)$$

We compare Equations (1.61) and (1.62), and return to Equation (1.50).

Thus, finally Equation (1.50), if we replaced velocity wave propagation from Equation (1.61) will be assigned as:

$$P = \frac{(331 + 0.6T_c)^2}{\rho}$$

$$(1.63)$$

The liquid oxygen tank operated in the range of 20 to 22 psig and the pressure inside the tank increased linearly with heating.

1.6.7 Conclusions

1. New designs for spacecraft protection are described and proposed.

2. The carbon–boron deposition on the steel/aluminum semimonocoque cylindrical surfaces has more adhesion than a spray-on foam of phenolic thermal insulation. A result of the diffusion process is that the fluctuation of durability increases 1.18 to 3.5 times.

3. Silicon carbide is initially formed by the injection of gaseous silicon and carbon on the carbon monofilament. The boron deposition (borane BH_3) combines with the glass silica $SiCl_4$ to form a silicon–boron bond, which has higher surface strength and temperature resistance than silicon carbide.

4. The braided carbon–boron monofilament weaved onto the cylindrical models was done in x, y, z directions and completed the 3-D solid protection system.

5. The new attachment titanium plates with the silicon–boron insulation can protect the ET tank from liquefaction of the air-exposed metallic attachments.

1.7 Vapor-Phase Deposition of the Thermoprotective Layers for the Space Shuttle

1.7.1 Introduction

The deposition of thin Al_2O_3 coating by using controlled atmosphere plasma spray systems is very beneficial [60].

However, thin Al_2O_3 coating with a thickness of 10 μm has a porosity of about 2% and a low-roughness deposit surface.

The microstructure of advanced metallic components produced by laser engineered net shaping has been improved using optical microscopy [61]. A fine-grain microstructure has been obtained for deposited 316 stainless steel and Ti-6-4 alloy.

The oxidation behavior of alloy PM200 consists of base Fe, Cr 20%, Al 5.5%,Ti 0.5%, and Y_2O_3 0.5%, and was investigated in the temperature range of 880–1400°C [62].

The presence of deposited yttria Y_2O_3 has a beneficial effect on the oxidation resistance of alloys.

Communication fibers that are coated by a system called flame hydrolysis and other coating processes can be divided into VAD, OVD, HVD, or MCVD [63].

1.7.2 Mechanism of Deposition in the Lattice Structures

General requirements for thermal barrier coatings are:

1. To create thermodynamically stable highly detectable lattice structures with tailored ranges of defect-cluster sizes. To exploit the effectiveness of such structures, they must be capable of attenuating and scattering photons, thus reducing thermal conductivity.

2. To produce highly distorted lattice structures with essentially immobile defect clusters and/or nanoscale ordered phases, which effectively reduce the concentrations of mobile defects and movement of atoms, thus increase sintering to enhance creep resistance.

3. To exploit the formulation of complex nanoscale clusters of defects to increase the measure of such desired mechanical properties as fracture toughness.

Dynamic aspects of the lattice structures behavior in the manufacturing of carbon fiber–epoxy composites for interstate structures in launch vehicles have been developed [64]. The mechanism of deposition of boron, carbon, and silicon was described by Thomas [65].

The boron–hydrogen bond (hydroboration) to either the carbon–carbon double bond of the alkene or the carbon–carbon triple bond of an alkyne is shown in Equations (1.64) and (1.65).

$$\begin{array}{ccccc} | & | & & | & | & | & | \\ C = C + H - B & \rightarrow & H - C - C - B \\ | & | & & | & | & | & | \end{array} \tag{1.64}$$

$$-C \equiv C + H - B \rightarrow \overset{\displaystyle |}{\underset{\displaystyle H}{C}} = \overset{\displaystyle |}{\underset{\displaystyle B-}{C}} \tag{1.65}$$

The simplest boron hydride is borane, BH_3, which dimerizes to diborane B_2H_6 [65]. The carbon has four valences and the boron has three valences.

On heating of the boron–hydrogen bond the boron atom moves to the position where steric interactions are minimized.

$$+H\text{-}BR_2 \qquad\qquad BR_2 \tag{1.66}$$

$-H\text{-}BR_2$

The silica represents the four-valence silicon, which is connected using the same principles.

1.7.3 Bond Strength

The relative strength of the bonds that form silicon, boron, and carbon with some other elements are shown below. The factors in parentheses indicate the approximate increase or decrease in strength between the silicon, carbon, and boron.

$$\begin{aligned}
&Si - O \gg C - O\,(\times 2.4 - 1.6) \quad Si - C < C - C\,(0.95)\\
&Si - F \gg C - F\,(\times 1.8) \qquad\quad Si - H < C - H\,(0.95)\\
&Si - Cl > Cl\,(\times 1.4)\\
&Si - Br > C - Br\,(\times 1.5)\\
&Si - I > C - I\,(\times 1.5)
\end{aligned} \tag{1.67}$$

It can be seen above that the silicon–boron forms stronger bonds than carbon–boron, so the silicon–boron bond strength is 1.5 times that of carbon–boron. The silicon–oxygen strength is much stronger than silicon–hydrogen.

Some typical bond dissociation energies (KJ mol^{-1}) for bonds of silicon and the corresponding bonds to carbon are given below.

$$\begin{aligned}
&Si - O\ 530 \qquad C - O\ 340\\
&Si - F\ 810 \qquad\ C - F\ 450\\
&Si - C\ 320 \qquad\ C - C\ 335
\end{aligned}$$

1.7.4 Bond Length

The bonds between silicon and other atoms are generally significantly longer than those between carbon and the corresponding atoms. The relative increases in bond length between selected atoms attached to silicon and the corresponding bond to carbon are shown below.

$$Si-C > C-C\,(\times 1.25)$$

$$Si-H > C-H\,(\times 1.35) \tag{1.68}$$

$$Si-O > C-O\,(\times 1.15)$$

A typical Si–C bond length is 1.89 Å, whereas a typical C–C bond length is 1.54 Å.

1.7.5 Thermal Conductivity Aspect

The general conduction equations of the first law of thermodynamics are [66]:

$$\frac{\partial^2 T}{\partial x^2} + \frac{\partial^2 T}{\partial y^2} + \frac{\partial^2 T}{\partial z^2} + \frac{q''}{\alpha} = \frac{1}{\beta} \times \frac{\partial T}{\partial t} \tag{1.69}$$

$$\beta = \frac{\alpha}{\rho c} \tag{1.70}$$

where
   T = Temperature conduction
   t = time conduction
   β = Coefficient of thermal diffusivity for gases
   q'' = Flux heat of internal energy transfer
   α = Coefficient of thermal conductivity
   ρ = Density of the material
   c = Specific heat per unit mass

If we assume that the temperature spread is only in the x, y directions, the conduction equation for an orthotropic lattice structure is seen as:

$$\frac{\partial^2 T}{\partial x^2} + \frac{\partial^2 T}{\partial y^2} + \frac{q''}{\alpha} = \frac{1}{\beta} \times \frac{\partial T}{\partial t} \tag{1.71}$$

From Equation (1.71), we can find a temperature gradient ∂T:

$$\partial T = \beta \partial t \left[\frac{\partial^2 T}{\partial x^2} + \frac{\partial^2 T}{\partial y^2} + \frac{q''}{\alpha} \right] \tag{1.72}$$

The flux of the steady state for the lattice structures design differential is seen in Equation (1.71) as:

$$\frac{\partial T}{\partial t} = \beta \left[\frac{\partial^2 T}{\partial h^2} + \frac{1}{h}\frac{\partial T}{\partial h} + \frac{q''}{\alpha} \right] \tag{1.73}$$

where
$T(r,t)$ = Temperature (K)
h = Current coordinate in $(0 < h < H)$
$t > 0$ = Time (s)
H = Thickness of coating layer (in)
β = Coefficient of thermal diffusivity (in^2/s)

The boundary conditions are

$$T(h,0) = 0; \quad T(h,T) = bt;$$

where b, the velocity of temperature growth (K/s), is the unknown quantity;

$$\frac{\partial T(0,t)}{\partial h} = 0; \quad T(0,t) < 4 \tag{1.74}$$

Equation (1.73) can be solved if the internal energy q'' is neglected [67].

$$T(r,t) = \frac{bR^2}{\beta}\left\{ \frac{\alpha t}{H^2} - \frac{1}{4}\left(\frac{h^2}{H^2}\right) + \sum_{n=1}^{0} \frac{A_n}{\mu_n^2}\left(\mu \frac{h}{H}\right) e^{-\mu_n^2 \alpha t / H^2} \right\} \tag{1.75}$$

where A_n $(=2/\mu_n I_1(\mu_n))$ is a permanent constant and μ_n is the root of the Bessel function order zero.

The temperature gradient can be determined as:

$$\frac{\partial}{\partial h}\{T(h,t)\} = \frac{bh}{2\beta} - \frac{2br}{\beta} 3 \frac{1}{\mu_n^2} e^{-\mu_n^2 n \alpha t / H^2} \tag{1.76}$$

The maximum layer of the temperature gradient can be determined as:

$$\in T = T(H,t) - (T_0 + bt) = \frac{bH^2}{\beta}\left\{ \frac{1}{4} 3 \frac{A_n}{\mu_n^2} e^{-\mu_n^2 \alpha t / H^2} \right\} \tag{1.77}$$

If the lattice structure has a long length and temperature is only a function of thickness H, for a steady-state regime we need to solve Equation (1.73) as:

$$\frac{\partial^2 T}{\partial h^2} + \frac{1}{h}\frac{\partial T}{\partial h} + \frac{q''}{\alpha} = 0 \tag{1.78}$$

The linear differential Equation (1.78) is solved as:

$$\frac{\partial T}{\partial h} = e^{I\frac{\partial h}{h}}\left(C - I\frac{q''}{\alpha}e^{I\frac{\partial h}{h}}\right); \quad \partial h = \frac{1}{h}\left(C - \frac{q''}{2\alpha}h^2\right) \tag{1.79}$$

If the gradient temperature on the lattice structure is known, we assume that $C = 0$. So,

$$T = \frac{q''}{2\alpha}h^2 + C' \tag{1.80}$$

C' was selected in order that the boundary conditions would be satisfied:

$$\frac{\partial T}{\partial h} = \frac{\alpha}{\alpha}(T - T_c) \quad \text{for } h = H; \tag{1.81}$$

Therefore

$$C' = -\frac{q''}{2\alpha}H + \frac{q''}{2k}(H^2 - h^2) \tag{1.82}$$

and temperature distribution can be shown as:

$$T(h) = T_c - \frac{q''}{2\alpha}H + \frac{q''}{2\alpha}(H^2 - h^2) \tag{1.83}$$

Here, $\dfrac{q''}{2\alpha}H$ is a flux heat of internal energy transfer, and $\dfrac{q''}{2\alpha}(H^2 - h^2)$ is a flux heat of thermal conductivity.

The distribution temperature in three directions (x, y, z) equals zero in the steady-state condition and results in the following:

$$\frac{\partial^2 T}{\partial x^2} + \frac{\partial^2 T}{\partial y^2} + \frac{\partial^2 T}{\partial z^2} = 0 \tag{1.84}$$

Therefore, we can find an electrical analogy that is afforded by the fact that the electrical potential E also obeys the Laplace equation:

$$\frac{\partial^2 E}{\partial x^2} + \frac{\partial^2 E}{\partial y^2} + \frac{\partial^2 E}{\partial z^2} = 0 \qquad (1.85)$$

Consequently, if the boundary conditions for E are similar to those for temperatures, and if the physical geometry of the problem is the same as that for the thermal problem, then the lines of constant electric potential are also the lines of constant temperature. Thus the electrical circuit board controls and regulates the thermal conductivity.

1.7.6 Layer the Deposited Protectives

Fibers have been covered by deposits of core glass silica ($SiCl_4$) and germanium ($GeCl_4$). Glass fibers can be reinforced by boron fibers in the vapor deposition process. The simplest boron hydride is borane (BH_3), which interacts with glass silica ($SiCl_4$) to form a silicon–boron bond.

Boron fibers are five times as strong and twice as stiff as steel. They are made via a CVD process in which boron vapors are deposited onto a fine tungsten or carbon filament [68]. Boron provides strength; stiffness is lightweight and possesses excellent compressive properties as well as buckling resistance.

Special Materials Inc. (formerly Textron) uses CVD for creating the boron layers. The process uses fine tungsten wire for the substrate and boron trichloride gas as the boron source [69].

The boron manufacturing process is precisely controlled and constantly monitored to assure consistent production of boron filaments with diameters of 4.0 and 5.6 mil (100 and 140 μm).

The mechanical properties of boron are represented in Table 1.4. It has a coefficient of thermal expansion of 2.5 ppm/°F (4.5 ppm/°C), density of 0.093 lb/in³ (2.57 g/cm²), and temperature performance of 350°F.

Combining the boron fiber with graphite prepreg, a high-performance material Hy-Bor has been produced with properties indicated in Table 1.4.

On the aluminum skin of the shuttle, it is proposed that layers of adhesive, silica fiber for heat resistance, and finally glass coating could be deposited, as shown in Figure 1.25.

The thermoprotective layers consist of silica fiber and glass coating. The thickness of the silica fiber is 2 to 3 in and the thickness of the glass coating is 0.5 to 1.0 in.

A vacuum "camera" for outside vapor deposition is shown in Figure 1.26.

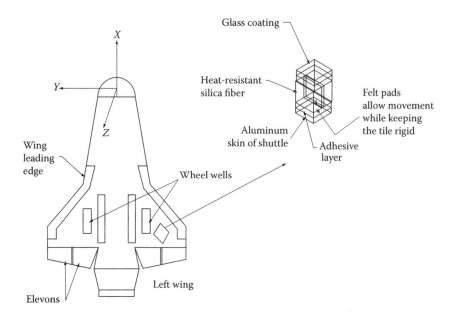

FIGURE 1.25
Deposition layers focus on the shuttle tiles.

The construction of a vacuum camera is very similar to the autoclaves used in curing aviation parts in that a large vacuum chamber is required. The nozzle profile repeats the shuttle bottom configuration and was installed on the "torch." The torch configuration compares with the torch used for the glass fiber deposition of preforms [63].

The components of deposition are transferred to the burn camera. The torch mounted on the carriage moves in the perpendicular coordinate y direction and the longitudinal axis x.

A liquid-oxygen tank is separated from the liquid-hydrogen tank located some distance from the vacuum camera. In the process of the burn, 500 kg of liquid oxygen use 100 kg of liquid hydrogen. These components are represented by fuel that is mixed with a soot component such as glass silica.

The silicon–boron layer is deposited on the aluminum bottom of the shuttle surface. The glass silica ($SiCl_4$) core layer and upper cladding layers have deposits of silica ($SiCl_4$) and oxygen (O_2).

Outside the TBCs the lattice structure shown consists of ceramic layers zirconia (ZrO_2) + scandia + yttria (Y_2O_3) (see Figure 1.27) [29,30].

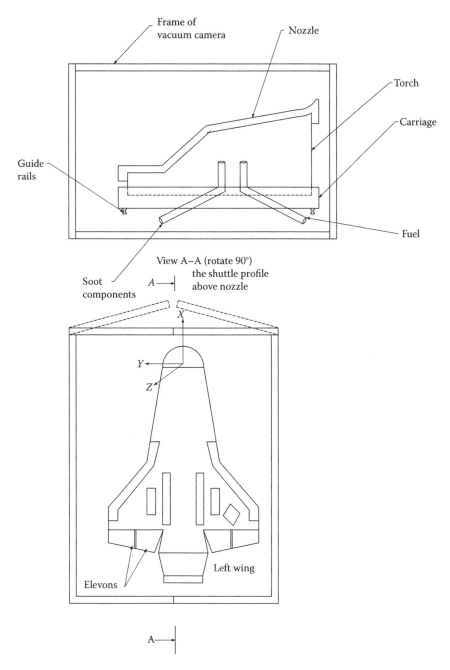

FIGURE 1.26
Vacuum camera for outside vapor deposition.

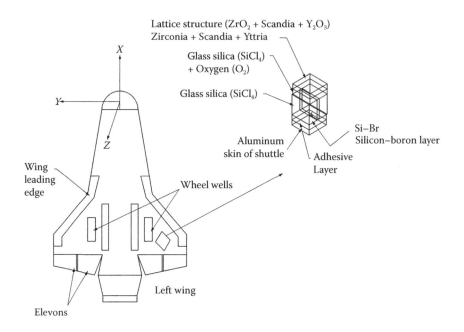

FIGURE 1.27
Deposition layers on the aluminum bottom of the surface of the shuttle.

1.7.7 Experimental Calculations

The lattice structural ceramic layers have a constant thermal conductivity and a uniform inner and outer surface temperature. At a given thickness of the area that is normal to the heat flow by conduction, we find $h \times L$, where L is the lattice cylinder length.

Following the work of Pitts and Sissom [48]:

$$T_2 - T_1 = \frac{q''}{khL} \ln \frac{H}{h}$$ (1.86)

where L is the length of lattice cylinder, H is the thickness of the ceramic layers, and h is the current coordinate of the ceramic layer. Therefore, the flux heat of internal energy transfer can be determined as:

$$q'' = \frac{khL(T_2 - T_1)}{\ln H/h}$$ (1.87)

and the relationship ($\ln H/h/khL$) is the thermal resistance of the single ceramic layer.

TABLE 1.8

Composite Properties for Thermal Stability

Property	Gr–/Ep (0°, 90°)	Gr/–Ep (0°)	C–C (0°, 90°)	Gr/–Mg (0°, 90°)	Gr–/Al (0°)
Absorptance	0.85	0.85	0.85	0.35	0.25
Emittance	0.85	0.85	0.85	0.35	0.25
Thermal conductivity through-thickness (Btu/ft-h/°F)	0.95	0.95	8.74	25.45	47.744
Specific heat (Btu/lb/°F)	0.23	0.23	0.279	0.294	0.250
Density (lb/in³)	0.62	0.62	0.065	0.071	0.091
Thickness (in)	0.04	0.04	0.04	0.04	0.04
Heat flux (Btu/ft² h)	442	442	442	442	442
Coefficient of thermal expansion	−0.68*	−0.57*	−0.68*	−0.233*	0.5*
(strain/°F)	10^{-6}	10^{-6}	10^{-6}	10^{-6}	10^{-6}

* Results has been determined as a test of five specimens.

We have the lattice structure with multiple layers of ceramic/boron composites (a different thermal conductivity in each layer), and the flux heat of internal energy transfer that is found for two layers:

$$q'' = \frac{H \times L(T_1 - T_3)}{(1/k_1)\ln(H_2/H_1) + (1/k_2)\ln(H_3/H_2)} \tag{1.88}$$

k_1, k_2 are the coefficients of thermal conductivity for different ceramic layers.

The composite properties for thermal stability are shown in Table 1.8 [70].

Materials considered were graphite fabric-reinforced epoxy (Gr–Ep) as a baseline. Composite, graphite fabric reinforced carbon–carbon (C–C), unidirectional graphite–magnesium (Gr–Mg), and unidirectional graphite–aluminum (Gr–Al) represent the next series of layers.

The thermal conductivity through-thickness values are shown in Table 1.8 (Btu indicates the British thermal unit) [71].

By flux plotting one can determine the heat transfer per unit length (q''/L).

If the inner surface of the aluminum body of the shuttle is at 300°F and the outer surface is at 1500°F, then the low thermal conductivity of the ceramic layers can protect the aluminum shuttle body. We consider that one layer thickness is equal to 0.04 in as shown in Table 1.8, and with a low coefficient of thermal conductivity 0.95, one can guarantee the permanent thermal stability of the lattice structure.

$$q''/L = H_1 k(T_2 - T_1) = 0.04 \times 0.95 \text{ Btu/h ft °F } (1500 - 300)°F = 45.6 \text{ Btu/h ft}$$

Since this is actually flux heat transfer in one direction through five layers, we can readily check the result by using Equation (1.86) for our system.

Hence, the heat transfer per unit length is:

$$\frac{q''}{L} = \frac{Hk(T_2 - T_1)}{\ln(H_2/H_1)} = \frac{(0.04 \times 5) \times 0.95 \times 1200}{\ln 2} = 330 \text{ Btu/h ft}$$

Radiation conductivity was not investigated in this work; however, a substantial amount of data for radiation heat transfer can be found in the work of Kundas et al. [73].

1.7.8 Conclusions

1. A CVD process has been developed, and is applicable to the thermo-protective layers of the space shuttle.

2. Low-thermal-conductivity ceramics have been proposed with their advantages as protective coatings and for the thermal stability.

3. Silicon–boron layers are stronger than carbon–boron layers, thus the silicon–boron bond strength is 1.5 times stronger than carbon–boron. The silicon–oxygen strength is much stronger than silicon–hydrogen.

4. Boron provides strength, stiffness, and is lightweight, possessing excellent compressive properties as well as buckling resistance to the bending moments.

5. The outside TBCs of the lattice structure consist of ceramic layers zirconia (ZrO_2) + scandia + yttria (Y_2O_3).

6. The flux heat of internal energy transfer was investigated in different directions. The low coefficient of thermal conductivity can guarantee the permanent thermal stability of the lattice structure, and therefore can protect future shuttles from failure.

7. It is proposed that the shuttle use a layered deposition on its aluminum bottom surface. The glass silica ($SiCl_4$) core layer and upper cladding layers have deposits of silica ($SiCl_4$) and oxygen (O_2) similar to the same procedure used for fiber communication deposition.

References

1. Lykins, C., and K. Watson, eds. 1995. *Integrated High Performance Turbine Technology (IHPTET) Brochure.* Wright Laboratory (WL/POT), Materials Directorate, Wright-Patterson Air Force Base, OH. (Information also available at http://www.pr.afrl.af.mil.)

2. Hill, R. J. 1993. The challenge of integrated high performance turbine engine technology (IHPTET). In *Proceedings of Eleventh International Symposium on Air Breathing Engines*, ed. F. S. Billig. American Institute of Aeronautics and Astronautics, September 19, 1993.

3. Zhu, S., M. Mizuno, Y. Kagawa, J. Cao, Y. Nagano, and H. Kaya. 1999. Creep and fatigue behavior in Hi-Nicalon-fiber–reinforced silicon carbide composites at high temperatures. *Journal of the American Ceramic Society* 82 (1): 117–128.

4. Chermant, J. L., G. Boitier, S. Darzena, G. Farizy, J. Vicens, and J. C. Sangleboeuf. 2002. The creep mechanism of ceramic matrix composites at low temperature and stress, by a material science approach. *Journal of the European Ceramic Society* 22: 2443–2460.

5. Morscher, G. N., G. Ojard, R. Miller, Y. Gowayed, U. Santhosh, J. Ahmad, and R. John. 2008. Tensile creep and fatigue of sylramic-iBN melt–infiltrated SiC matrix composites: retained properties, damage development, and failure mechanisms. *Composites Science and Technology* 68: 3305–3313.

6. Y. Golfman. 2007. Vapor-phase deposition for the thermoprotective layers for the space shuttle. *Journal of Advanced Materials (Special Edition)*.

7. High-performance composites: an overview. *International Edition*. November 2002.

8. Specialty Materials, Inc. Manufacturing of Boron SCS Silicon Carbide Fibers. http://www.specmaterials.com.

9. Johnson, D. P. 2001. Thermal cracking in scaled composite laminates. *Journal of Advanced Materials* 33 (1).

10. Bailey, J. E., P. T. Curtis, and A. Parvizi. 1979. On the transverse cracking and longitudinal splitting behavior of glass and carbon fiber reinforced epoxy. *Proceedings Royal Society London* A366: 599–623.

11. Golfman, Y. 2004. The fatigue strength prediction of aerospace components. *Journal of Advanced Materials* 36 (2): 39–43.

12. Talreja, R. 1987. *Fatigue of Composite Materials*. Technomic Publishing Co.

13. Kachanov, L. 1958. Rupture time under creep conditions. *Izvestiia Akademii Nauk SSR* 8: 16–31.

14. Rabotnov, M. 1969. *Creep Problem in Structural Members*. Amsterdam: North-Holland.

15. Barbero, E. J. 1999. *Introduction to Composite Materials Design*. Taylor & Francis.

16. Mallick, K. et al. 2003. Thermo-micromechanics of microcracking in a cryogenic pressure vessel. 44th Structures, Structural Dynamics, and Materials Conference, Norfolk, VA.

17. Hecht, J. 1999. Meeting the manufacturing challenge of optical fiber. *Laser Focus World*, March, 121–125.

18. Golfman, Y. 2001. Fiber draw automation control. *Journal of Advanced Materials* 34 (2).

19. Golfman, Y. 1994. Ultrasonic non-destructive method to determine modulus of elasticity of turbine blade. *SAMPE* 29 (4).

20. Golfman, Y. 1994. Effect of thermoelasticity for composite turbine disk. 26th International SAMPE Technological Conference, October 17–20, 1994.

21. Apparatus and Method for Monitoring Tension in a Moving Fiber by Fourier Transform Analysis. U.S. Patent Number 4,692,615, September 8, 1978.

22. University of Virginia Aerogel Research. http://fourier.mech.virginia.edu/microbx/home2.html.

23. Thermal Properties of Silica Aerogels. http://eande.lbl.gov/ECS/aerogels/satcond.htm, March 11, 2005.

24. Ma, X. Q., F. Borit, V. Guipont, and M. Jeandin. 2002. Thin alumina coating deposition by using controlled atmosphere plasma spray system. *Journal of Advanced Materials* 34 (4): 52–57.

25. Golfman, Y. 2001. Fiber draw automation control. *Journal of Advanced Materials* 34 (2): 35–40.

26. How Silica Aerogels Are Made. http://eande.lbl.gov/ECS/aerogels/saprep.htm.

27. Emelyanov, A. N. 1994. The thermal conductivity and diffusivity of transition metal carbides high temperatures. *High Temperature-High Pressure* 26 (6): 663–671.

28. Zhou, Y., Y. Wang, G. Song, T. Lei, and Z. Huang. 2003. Thermal diffusivity and thermal conductivity of ZrCp/W composites. *Journal of Advanced Materials* 35 (2): 41–45.

29. Mess, D. 2003. Low-conductivity thermal-barrier coatings. *Tech Briefs* 27 (6).

30. Mess, D. 2003. Scandia and yttria stabilized zirconia for thermal barriers. *Tech Briefs* 27 (10).

31. Golfman, Y. 1993. Ultrasound nondestructive method to determine modulus of elasticity of turbine blades. *SAMPE Journal* 29 (4): 31–35.

32. Golfman, Y. 1994. Effect of thermoelasticity for composite turbine disk. 26th International SAMPE Technical Conference, Atlanta, GA, pp. 10–15.

33. Golfman, Y. 2001. Nondestructive evaluation of aerospace components using ultrasound and thermography technologies. *Journal of Advanced Materials* 33 (4): 21–25.

34. Golfman, Y. 2002. Nondestructive evaluation of parts for hovercraft and ekranoplans. *Journal of Advanced Materials* 34 (4): 3–7.

35. Golfman, Y. 2003. Dynamic aspects of the lattice structures manufacturing from carbon/carbon. *Journal of Advanced Materials* 35 (2): 3–8.

36. Golfman, Y. 2004. The fatigue strength prediction of aerospace components using reinforced fiber/glass or graphite/epoxy. *Journal of Advanced Materials* 36 (2): 39–43.

37. Golfman, Y. 2004. The interlaminar shear stress analysis of composites sandwich/carbon fiber/epoxy structures. *Journal of Advanced Materials* 36 (2): 16–21.

38. Army Technology-TSS International. Explosion-Proof and Self-Sealing Fuel Tanks. www.army-technology.com/contractors/protection/tss. Accessed 23 June 2010.

39. Era Aviation, Inc. www.era-aviation.com/eranf/hs_gc_as_aux_fuel.stm. Accessed 27 February 2009.

40. Information Technology Associates. Self-Sealing-Fuel Tank Builder [(rubber goods) alternative fuel-cell assembler)]. http://www.occupationalinfo.org/75/752684046.html.

41. GKN Demonstrates Fuel Tank and Flotation System Capabilities. www.robertsonaviation.com/otherproducts.asp.

42. Abrate, S. 1998. *Impact on Composite Structures*. New York: Cambridge University Press, pp. 135–160.

43. Golfman, Y. 2001. Fiber draw automation control. *Journal of Advanced Materials* 34 (2): 35–40.

44. Hewlett-Packard Co. 1990. Effective Machinery Measurements Using Dynamic Signal Analysis. Application Note 243-1, Hewlett-Packard.

45. Golfman, Y. 2001. Nondestructive evaluation of aerospace components using ultrasound and thermography technologies. *Journal of Advanced Materials* 33 (4): 21–35.

46. Landay, L. D., and E. M. Lifshitz. 1959. *Theory of Elasticity.* London: Pergamon Press.

47. Momentive Performance Materials, Inc. 2007. Hot-Pressed Boron Nitride Shapes (QTZ-81507). http://www.momentive.com/momentiveInternetDoc/Internet/Static%20Files/Documents/4%20Color%20Brochures/81507.pdf.

48. Pitts, D., and L. Sissom. 1997. *Heat Transfer.* New York: McGraw-Hill.

49. Golfman, Y. 2007. Self-sealing fuel tank technology development. *Journal of Advanced Materials, Special Edition* 3: 29–33.

50. Dumoulin, J. 1988. *NSTS Shuttle Reference Manual.* http://science.ksc.nasa.gov/shuttle/technology/sts-newsref/.

51. Malone, J. 1996. Shuttle Super Lightweight Fuel Tank? Completes Test Series. Marshall Space Flight Center, Huntsville, AL.

52. Will Shrink-Wrap Fix Shuttle Fuel Tank? Science & Space. November 25, 2003. http://www.cnn.com.

53. Golfman, Y. 2007. Vapor-phase deposition for the thermoprotective layers for the space shuttle. *Journal of Advanced Materials* 3: 58–64.

54. Suplinskas, R. J., and A. W. Hauze. 1984. Boron Coated Silicon Carbide Filaments. U.S. Patent No. 4481257.

55. Pierce, A. D. 1981. *Acoustics: An Introduction to Its Physical Principles and Applications.* New York: McGraw-Hill Book Company.

56. Golfman, Y. 2003. Dynamic aspects of the lattice structures behavior in the manufacturing of carbon–epoxy composites. *Journal of Advanced Materials* 35 (2): 3–8.

57. Regel, B. R. 1972. Kinetic concept of strength. *Journal of Mechanics of Polymer.* Riga, Latvia, pp. 98–112.

58. Krivopal, B. A. 1968. Investigation of durability for polyamid-12. *Mechanics of Polymer.* Riga, Latvia, 1968.

59. Kinsler, L. E., A. R. Frey, A. B. Coppens, and J. V. Sanders. 1950. *Fundamentals of Acoustics.* New York: John Wiley & Sons.

60. Ma, X. Q., F. Borit, V. Guipont, and M. Jeandin. 2002. Thin alumina coating deposition by using controlled atmosphere plasma spray system. *Journal of Advanced Materials* 34 (4): 52–57.

61. Zhand, X. D., H. Zhang, R. J. Grylls, T. J. Lienert, C. Brice, H. L. Fraser, D. M. Keicher, and M. E. Schlienger. 2001. Laser-deposited advanced materials. *Journal of Advanced Materials* 33 (1): 17–23.

62. Lours, P., S. L. Roux, and G. Bernhart. 2001. Oxidation behavior of creep resistant oxide-dispersion–strengthened alloy PM2000. *Journal of Advanced Materials* 34 (2).

63. Golfman, Y. 2001. Fiber draw automation control. *Journal of Advanced Materials* 34 (2): 35–40.

64. Golfman, Y. 2003. Dynamic aspects of the lattice structures behaviour in the manufacturing of carbon–epoxy composites. *Journal of Advanced Materials* 35 (2): 3–8.

65. Thomas, S. E. 1991. *Organic Synthesis. The Roles of Boron and Silicon.* New York: Oxford University Press.

66. Landay, L. D., and Lifshitz, E. M. 1959. *Theory of Elasticity.* London: Pergamon Press.
67. Leakov, A. B. 1972. *Theory of Thermal Diffusivity.* Moscow.
68. High-performance composites: an overview. *Journal of International Edition.* November 2002.
69. Specialty Materials, Inc. Manufacturing of Boron SCS Silicon Carbide Fibers. http://www.specmaterials.com.
70. Rubin, L., D. Y. Chang, E. Y. Robinson, and C. Tseng. 1989. Advanced composites for a dimensionally stable space structure. 21st International SAMPE Technical Conference, September 25–28, 1989.
71. Gibilisco, S. 2002. *Physics Demystified.* New York: McGraw-Hill.
72. Kundas, S., V. Gurevich, A. Pyaschenko, and V. Okovity. 2000. Simulation and experimental studies of particles interaction with plasma jet in vacuum plasma spraying process. *Journal of Advanced Materials, SAMPE* 32 (3).

2

Impregnation Process

2.1 Impregnation Process for Prepregs, Braided Composites, and Low-Cost Hybrid Polymer and Carbon Fibers

2.1.1 Introduction

Prepreg, a combination of reinforcing fibers and matrix resin, is the starting material for many composite parts used in the aircraft, shipbuilding, and plastic industries. There are many reinforcing fibers: E-glass, S-glass, aramid, quartz, carbon, Spectra, Teflon, and many specialty and nonwoven reinforcements.

Resin systems include epoxy, polyester, phenolic, silicone, polybutadiene, and cyanate ester. The test method for the verification of the physical characteristics of a prepreg system includes resin content, gelation time, volatile content, and resin flow. High-strength polymer fibers are a critical component of lightweight personnel armor materials systems and of military systems such as inflatable beams for shelters (air beams), shelter fabrics, and cordage. Products such as Kevlar®, Spectra®, Twaron®, Zylon®, and Dyneema® are all examples of commercial high-strength polymeric fiber materials. These fibers are widely used in personnel armor components including vests, small arms protective insert plates, and helmets. Commercially available high-strength fibers are relatively expensive and are produced in much smaller quantities than more commodity-type fibers such as polyamide, polyester, or polyolefin fibers. The commodity fibers typically have mechanical properties (tensile strength, tensile modulus) on an order of magnitude lower than those of the high-performance fibers.

The continuous molding/pultrusion process is used for a wide range of military applications, including light armor, clothing, shelters, and airdrop systems.

The use of braiding as a fabrication process is becoming more prevalent in the composites industry for design flexibility, ease of fabrication, and economic consideration. Resin must be injected at a standard pressure or vacuum pressure on the braided preform. Standard pressure is the air pressure that will support a column of mercury 0.760 m high. Resins must have a low viscosity and must generate no volatiles upon curing. The several resins used are vinyl ester, polyesters, hot-melt bismaleimide, and liquid epoxies. The viscosity range has been considered from 200 to 600 cp and most injections are completed in 60 min. After the injection, the resin should be able to be cured 70% to 80% in 10 min in the heated mold.

Today, the resin vacuum pressure infusion process involves packing by hand, including the vacuum bag. Dry textile preforms are resin-impregnated, consolidated, and cured in a single step, eliminating costly prepreg tape manufacture and ply-by-ply layup. Under a high vacuum pressure, the laminate layers are saturated with vinyl ester or polyester resin. The quality control of the prepreg system and the impregnation of dry aviation parts include the same control characteristics plus the homogeneity and impregnation control. The homogeneity impregnation is only possible if an automation process is used.

This chapter [12] describes the control needed for an automation process for the impregnation of prepregs and dry braided parts.

The goal of this work was to evaluate a moving head resin at a hemispherical injection process for the impregnation of a "complex" shape in the braided preform versus a resin injection process.

The aim of Golfman's work [13] was to investigate hybrid fiber composite technology and apply novel processing techniques to commodity fibers to produce fibers with tensile strengths comparable to current commercial high-strength fibers (>20 g/denier), but at significantly lower cost (<$10/lb). The ultimate objective of this chapter is to improve the availability and cost of high-performance commodity fibers for continuous molding/pultrusion process for a wide range of military applications, including light armor, clothing, shelters, and airdrop systems.

The reinforcing fibers can be either directionally aligned or woven into a fabric while the matrix can be a thermosetting or a thermoplastic polymer. The resin was injected at 40 to 90 psi, followed by two gel stages: 300°F for 30 min followed by 90 min at 350°F [1]. In the case of the infusion process the thick laminate stack sucked 15 drums of resin dry in less than 7 h, and then was cured quietly under vacuum overnight [2].

In the case of thermosetting matrices, there are two main processes used to impregnate the matrix into the fiber bundles: hot-melt and solution-dip impregnation [3–5]. The solution-dip process is a solvated process in which the resin viscosity is lowered by the addition of solvents—usually acetone, methyl ethyl ketone, or xylene [6]. The solution-dip process, shown in Figure 2.1, is generally applied to woven fabrics [7].

The solution-dip process consists of first dipping the woven fabric in a resin bath where it is impregnated with a mixture of resin and solvent. The role of the solvent is to decrease the viscosity of the resin to a level suitable for impregnation.

The resin content of the prepreg is adjusted to the desired level by controlling the temperature and solvent content of the bath as well as the quantity of resin squeezed out in the nip rollers.

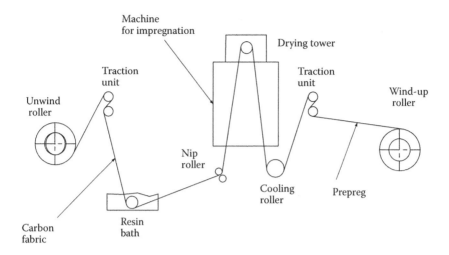

FIGURE 2.1
Schematic of a solution-dip prepreg operation.

After the nip rollers, the prepreg passes through a drying tower where a controlled amount of solvent is evaporated. Finally, the prepreg is cooled down and wound on a roll before being shipped and delivered. Then, the composite part manufacturer can cut the prepreg into a desired shape and orientation and combine it into a composite structure.

It is very important to determine the content of the binder, mass of the binder, and fiber for the hot-melt and solution-dip impregnation processes.

2.1.2 Automation Control for the Process of Prepreg Impregnation

The automation control for the process of prepreg impregnation is the method of material balance. This method avoids the mass influence interference factors for different electrophysical properties of fiber and resin, changing geometrical parameters for prepregs, and so forth. This method is measured by the quantity of binder taken out of a piece of fiber that has the numerical value of length.

The percent content of binder on the numerical impregnation of the piece of fiber should be determined as [8]:

$$X = \frac{P_b}{P_b + P_f} \times 100\% \qquad (2.1)$$

where P_b is the weight content of the binder (which includes resin, hardener, and solvent) and P_f is the mass of fiber piece.

If we know the weight content of the binder taken out of the numerical piece of fiber, we can determine the binder concentration:

$$P_b = P_{mv}k \tag{2.2}$$

where P_{mv} is the mass of varnish taken out of the numerical piece of fiber and k is the coefficient of varnish concentration. Varnish is a resin mixed with 10% acetone. Therefore, the mass of the fiber piece could be determined as:

$$P_f = gL \tag{2.3}$$

where g is the mass of linear meter of fiber and L is the length of the numerical piece of fiber.

By inputting Equations (2.2) and (2.3) into Equation (2.1), we get:

$$X = \frac{P_{mv}k}{gL + P_{mv}k} \times 100\% \tag{2.4}$$

If the binder concentration and mass of linear meter of fiber does not change in the process of impregnation, then the percentage contained in the binder can be determined as the mass of varnish taken out from the impregnation bath of the numerical piece of fiber, P_{mv}.

Based on this principle, the system of automatic control was developed and the percent contained in the binder automatically regulated by filling up the bath during the process of fiber impregnation.

The block schematic for automation control of the binder flow is shown in Figure 2.2.

The quantity of binder in the bath at all times is registered by the display mode of pos. 3. The signals transfer to the display mode by a differential transformer sensor in pos. 2, which measures the displacement of the impregnation bath.

At the end of cooling, a camera was installed to show the electrical length transformer (pos. 47), whose function consisted of giving signals to the counter distance (pos. 7) when it passed the numerical length of fiber. The signals pass to control the block (pos. 5) assembly using a relay schema.

When the quantity of binding material reaches a maximum in the impregnation bath, the contact of the display mode devices (pos. 3) and through the control block (pos. 5) switches off the valve (pos. 6), which is responsible for providing the binding. The control block (pos. 5) switches on the transformer length (pos. 4) and the counter distance (pos. 7).

When the numerical piece of fiber passes through the counter distance (pos. 7), a signal is given to the control block (pos. 5), which switches on the

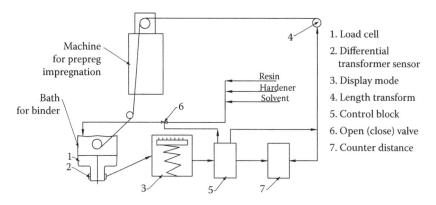

FIGURE 2.2
Block schematic for automation control of binder flow prepreg impregnation.

display mode and switches on the valve (pos. 6) to allow additional binder into the bath.

The binding material passes into the bath to increase the numerical mass when this mass reaches a maximum value of the sensor (pos. 2), which gives a signal to the display mode (pos. 3), and then the work following the same scheme is repeated.

The display mode registers a sharp peak and the height characterizes the mass of the binding. We examined the error to determine the percent content of binding.

We can see from Equation (2.1) that it is more convenient for a logarithmic expression for the relative error to measure the percent content of the binding. (Normally, the percent content of the binding equals 30).

$$\frac{\Delta X}{X} = \pm \frac{\partial X}{X} = \pm \partial \left| \ln X(P_b, P_f) \right| \tag{2.5}$$

The logarithmic expression can be shown as:

$$\ln X(P_b, P_f) = \ln \frac{P_b}{P_b + P_f} = \ln P_v - \ln(P_b + P_f) \tag{2.6}$$

Finally,

$$\frac{\partial X}{X} = \partial \left| P_b - \ln(P_b + P_f) \right| = \frac{\partial P_b P_f}{P_b(P_b + P_f)} - \frac{\partial P_f}{P_b + P_f} \tag{2.7}$$

We change the mathematical differential to a maximum relative error in order to measure the percent content of binder.

$$\frac{\Delta X}{X} = \pm \left| \frac{\Delta P_b P_f}{P_b(P_b + P_f)} + \frac{\Delta P_f}{P_b + P_f} \right| \qquad (2.8)$$

The error to measure the percent content of the binder can be evaluated as an absolute significant value from the measured value.

Now, we input Equation (2.1) into Equation (2.8):

$$\Delta X = \pm \left| \frac{\Delta P_b P_f + \Delta P_f P_b}{(P_b + P_f)^2} \right| 100 \qquad (2.9)$$

Here, ΔP_b, ΔL are the absolute errors in measuring the mass of the binder and fiber.

If we followed Equations (2.2) and (2.3), the absolute errors ΔP_b and ΔP_f would be:

$$\Delta P_b = (k \pm \Delta k)\Delta P_v; \quad \Delta P_f = (g \pm \Delta g)\Delta L \qquad (2.10)$$

where, ΔP_v and ΔL are the absolute errors of measuring the mass of the varnish and the numerical length of material, while Δk and Δg are the technological errors in the concentration of binder and the mass of length in the piece of fiber.

The absolute error to determine the percent content of binder below is, if we input Equations (2.1), (2.2), (2.3), (2.10) into Equation (2.9),

$$\Delta X = \pm \left| \frac{\dfrac{(k + \Delta k)\Delta P_v}{g - \Delta g} + \Delta L \dfrac{X}{1 - X}}{L\left(1 + \dfrac{X}{1 - X}\right)^2} \right| 100 \qquad (2.11)$$

It is a comparison of the error/variation between the "new" moving head impregnation process and the present infusion process.

2.1.3 Theoretical Investigation

If the properties for a hemispherical braided nose cap are changed under acting loads, then the temperature of the stress–strain relationship must change in a matrix form [9].

For an orthotropic material there are ten stiffness constants; the first six constants Q_{11}, Q_{22}, Q_{33}, Q_{12}, Q_{21}, Q_{66} were designated in Equation (2.12). For orthotropic materials we have 10 stiffness constants:

$$Q_{11} = \frac{E_1}{1 - \mu_{12}\mu_{21}}; \quad Q_{12} = \frac{\mu_{21}E_1}{1 - \mu_{21}\mu_{12}}$$

$$Q_{21} = \frac{\mu_{21}E_2}{1 - \mu_{12}\mu_{21}}; \quad Q_{22} = \frac{E_2}{1 - \mu_{21}\mu_{12}}$$

$$Q_{13} = \frac{E_1}{1 - \mu_{12}\mu_{21}}(\alpha_1 + \mu_{21}\alpha_2)$$ (2.12)

$$Q_{23} = \frac{E_2}{1 - \mu_{32}\mu_{23}}(\mu_{32}\alpha_1 + \alpha_2)$$

which Prof. V. Z. Parton [14] called the Duhamel–Neumann law. Here, E_1, E_2 are moduli of normal elasticity in warp and fill directions, and μ_{12}, μ_{21}, μ_{23}, μ_{32} are Poisson's ratio of material. The first symbol designates the direction of force, and the second symbol designates the direction of transverse deformation. α_1, α_2 are coefficients of thermal expansion in warp and fill directions.

Therefore, we operate only under the thermal stresses if the mechanical deformation is equal to zero [10]. Correlation between the thermal stresses σ_{ij} and the stiffness constants Q_{ij} can be shown as:

$$\sigma_{ij} = -Q_{ij} * a_j * T$$ (2.13)

For a hemispherical nose with an orthotropic structure, the thermal radial stresses are determined by Parton and Perlin [14]. Here, E_1, E_2 are the modulus of normal elasticity in radial and tangential directions, and μ_{12}, μ_{21} is Poisson's ratio of material; the first symbol designates the direction of force, and the second symbol designates the direction of transverse deformation; α_1, α_2 are the coefficients of thermal expansion in radial and tangential direction, and T is the temperature gradient.

We use ultrasonic measurements in the process of impregnation control. In Golfman's work [10], an ultrasonic nondestructive method to determine the modulus of elasticity of turbine blades was described. The equations for the modulus of elasticity in tensor form for the solid parts was shown in Equation (1.26)

$$E_{ij} = C_{ij}^2(1 - \mu_{ij}\mu_{ji})\rho$$

where C_{ij} is the velocity of propagation of the ultrasonic wave in a solid body, μ_{ij} is Poisson's ratio, and ρ is the density of material (1.998 g/cm³). For a hemispherical nose with an orthotropic structure, the thermal and tangential radial stresses were found by Golfman [12].

When the ultrasound wave passes through the solid and liquid body simultaneously, we combine the sum of two moduli: the modulus of elasticity for solid E_1 and moduli of viscosity E_2.

$$\sum E = E_1 + E_2 = C_1^2(1 - \mu_{12}\mu_{12})\rho + C_2^2\lambda\mu_{12} \tag{2.14}$$

where C_2 is the velocity of propagation of the ultrasound wave in the liquid body, λ is the diffusion coefficient, and μ_{12} is the density of the mixture components.

The compression strength acting on the mixture can be determined as:

$$\sigma = \frac{F}{A} = \frac{\rho_2 Vm}{\pi R_m^2} \tag{2.15}$$

Here, f is the compression force acting in the mixture, means the weight of the mixture ($F = \rho_2 V_m$), V_m is the volume of the mixture, A is the area of the mixture ($A = \pi R_m^2$), and R_m is the middle radius of the mixer.

Therefore, the volume of the mixture correlates with the velocity propagation ultrasound wave in a solid and liquid body.

2.1.4 Dry Braided Impregnation of Aviation Parts

It is very difficult to support an equal degree of impregnation in the transfer molding process. The automation process was designed for the impregnation of a nose cap. A block schematic for automation control of the dry braided impregnation nose cap is represented in Figure 2.3.

In the mixer (pos. 1) we transferred the resin, hardener, and solvent. All three components were stored in the tank (pos. 17, 18, 19) and transferred to the mixer (pos. 1) by centrifugal pumps (pos. 21, 22, 23). The binder flow to impregnate the nozzle is seen in (pos. 20). The ultrasound sensor from shipping (pos. 3) to receiving (pos. 5) passes through the mixture. When the volume of the binder decreased the signal of the ultrasound from the ultrasound sensor we went to the computer (pos. 12), and the microprocessor immediately sent a signal to valve 13 and the valve switchoff. Simultaneously, the microprocessor sent the signals to the switch on the valve's stations (14, 15, 16) and the components were filled by the volume in the mixer pos. 1. The ultrasound sensor again sent a command to the computer and the computer opened valve 13 and impregnation began.

The dry braided nose cap (pos. 9) was installed on the rotation mandrel (pos. 10). The nose cap was able to rotate with a flexible speed as a result

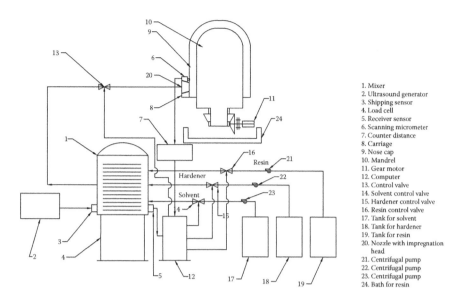

FIGURE 2.3
Block schematic for automation control of impregnation for the dry braided nose cap.

of miter-bevel gears and the gear motor (pos. 11). To control impregnation length we used a laser-scanning micrometer (pos. 6), which was installed on the carriage (pos. 8). The laser-scanning micrometer was also installed on the carriage (pos. 8). A laser-scanning micrometer has a photo detector that passes signals to the computer so that we can visually see the geometrical configuration of the nose cap.

Figure 2.4 shows the automation control of the nose cap impregnation process.

The pipe of the binding carriage (pos. 1) connects with the impregnation nozzle (pos. 3), which is installed on the carriage (pos. 2), which carries electrical wire together with the laser-scanning micrometer (pos. 4). The special guide support (pos. 6) is assembled with the frame (pos. 7).

The flexible shaft (pos. 5) driven by the gear motor (pos. 8) is there in order for the carriage pos. 2 to be able to travel and repeat the trajectory of the outside configuration of the nose cap.

The mandrel (pos. 12) rotates with the nose cap driven by the miter-bevel gears and the gear motor (pos. 11). Companies such as Whitlock or Colotronic Systems produce mixers that can blend an accuracy from two to six components. The accuracy of the weighting systems is ±0.1%.

The rest of the resin flow into the resin bath is located under the nose cap, (pos. 12), shown in Figure 2.4.

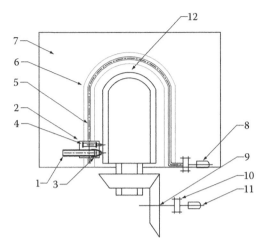

1. Pipe for binding carriage
2. Carriage for nozzle
   and laser transportation
3. Nozzle
4. Laser scanning
   micrometer
5. Flexible shaft
6. Guide support
7. Frame
8. Gear motor
9. Miter-bevel gear
10. Coupling
11. Gear motor
12. Nose cap

FIGURE 2.4
Automation control nose cap impregnation.

2.1.5 Automatic Control of the Dry Braided Nose Cap

Whitney et al. [9] refers to the acoustic transducers directly installed on the surface of the nose cap. For automation and quality control in aviation and marine structures, the ultrasound defectoscope with immersion transducers is used. In this case, the best acoustic environment used was water.

To determine the delaminating, flaws, voids, and porosity, we used a shelter method where the acoustic sound penetrated through the body of the composites. Figure 2.5 shows the automatic ultrasound control for braided nose cap.

The nose cap is in (pos. 2); the fabricated composite is placed in the bath (pos. 1). The bath is filled with water. Immersion transmitting transducers in (pos. 5) are installed in the top of the frame (pos. 3). Immersion receiving transducers (pos. 6) are installed in the bottom of the frame (pos. 3). When we switch on the synchronizer (pos. 7), electrical signals are transmitted to the generator, which sends electrical signals to the immersion transmitting transducers. Impulses affect (pos. 4) immersion transmitting transducers, which transfer the impulses to elastic oscillations depicting the focal zone. Immersion receiving transducers transfer the elastic oscillations to electrical signals and reflect a scanning picture onto the computer screen (pos. 8). If a panel has defects, the ultrasound beam declines and the receiving transducers cannot receive signals. We use a fiber optic cable (pos. 9) for transmitting optic signals to the computer. The DC electrical motor (pos. 4) is supported to move the motion frame with the transducers to force and back the automatic scanning of the composite panel.

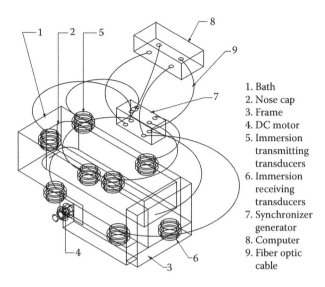

1. Bath
2. Nose cap
3. Frame
4. DC motor
5. Immersion
   transmitting
   transducers
6. Immersion
   receiving
   transducers
7. Synchronizer
   generator
8. Computer
9. Fiber optic
   cable

FIGURE 2.5
Automatic ultrasound control for braided nose cap.

2.1.6 Experimental Results

We determined the absolute error to measure the percent content of the binder for the reinforcing E-glass fiber impregnating epoxy varnish. For the epoxy varnish we know $g = 360 \pm 30$ g, $k = 0.52 \pm 0.02$, and $X = 30 \pm 2\%$. The automation control binder flow system permits us to measure and register the mass of the bath with the binder, which has an error ± 5 g, so $\Delta P_v = 10$ g. The length transformer counting the numerical piece of length with an error of ≤ 50 mm with an overall length is 2.5 m. Figure 2.6 shows the correlation between the absolute error to measure the percent content of binder and the numerical length of the reinforcing E-glass fiber counted by the length transformer.

We can see that the measurable allowable limits of the binder concentration and the mass of length of E-glass fiber do not have a significant influence on the error to measure content binder. The percent of content concentration binder is reduced when there is an increase in the length of the E-fiber piece. However, if we increase the length of the numerical E-fiber, the time of measurement also increases because the length transformer is worked slowly. Therefore, it is very important to calculate the optimal length of reinforcing E-glass fiber.

Equation (2.11) shows the configuration of the hyperbola.

$$\Delta X = \frac{A}{L} \tag{2.16}$$

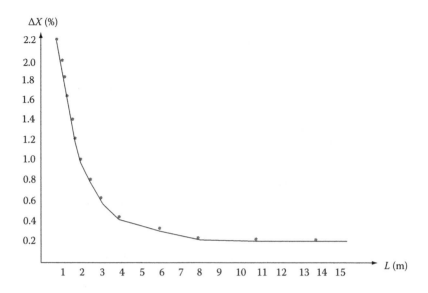

FIGURE 2.6
Correlation between error measuring percentage content of the binder and the numerical length of the E-glass fiber.

$$A = \left| \frac{\dfrac{(k + \Delta k)\Delta P_v}{g - \Delta g} + \Delta L \dfrac{X}{1 - X}}{\left(1 + \dfrac{X}{1 - X}\right)^2} \right| 100 \qquad (2.17)$$

We see that for the optimal length of the numerical piece we chose the absolute pick of the hyperbola L', because the increase in length L in the condition of $L > L'$ does not increase ΔX significantly.

The equation for the hyperbola is:

$$\Delta X * L = (L')^2, \text{ where } L' = A^{1/2}$$

$$L' = \left| \frac{\left\{ \dfrac{(k + \Delta k)\Delta P_v}{g - \Delta g} + \Delta L \dfrac{X}{1 - X} \right\}}{1 + \dfrac{X}{1 - X}} \right| \quad \text{(in meters)} \qquad (2.18)$$

Therefore, for a length that is more than 8 m in error, the measuring percentage content of the binder is close to 0.2%.

2.1.7 Conclusions

1. The automation control for the process of prepreg impregnation was used at the solid-dip prepreg stations.
2. We determined the absolute error to measure the percentage of the content of binding for reinforcing E-glass fiber impregnating epoxy varnish.
3. We calculated the critical numerical length L' with previous values for the epoxy varnish: $g = 360 \pm 30$ g, $k = 0.52 \pm 0.02$, $X = 30 \pm 2\%$. The result was $L' = 4.3$ m, $\Delta X = \pm 0.43\%$. This result fits very well with the experimental data.
4. The method of material balance allowed us to control the percentage of content for the binding and the prepreg's quality. For a length that is more than 8 m in error, the measuring percentage content of the binder is close to 0.2%.
5. The dry braided impregnation process for aviation parts was developed and this process was implemented using a nose cap.
6. The automation control for the dry braided nose cap was also established.

2.2 High-Strength, Low-Cost Polymer Fibers Hybrid with Carbon Fibers in Continuous Molding/Pultrusion Processes

2.2.1 Introduction

KaZaK Composites (KCI) created technology for pultrusion of proprietary profiles and deck plates for military systems [15]. KCI successfully designed a stanchion using carbon fibers for the U.S. Navy; pultruded stanchions are self-righting after impact, returning to their original state.

Straight position with a near-zero permanent deformation after a 45° overload, a steel stanchion is returned to the straight position after a bend of 45°, because he work in a range of elastic deformations.

Polyurethane is a matrix material for composites with a high strain to failure and superior impact damage resistance. Polyurethane is a commodity high-strength fiber with a low cost comparable to current commercial high-strength fibers.

Dr. Jung investigated the Lewis acid–base complication of nylon 6,6 complex. In this process hydrogen bonds in nylon 6,6 polymer are broken suppressing crystalline. This amorphous polymer is a high-strength fiber and exhibits dramatically high modulus, tensile strength, and tenacities [16]. This process uses the dry-jet wet spinning forms.

Prof. Hara developed a novel liquid crystalline polymer (LCP) that incorporates ionic groups in the polymer. When ionic groups are introduced into the polymer, lateral interaction occurs between polymer chains. This interaction results in a material with significant increase in mechanical properties (compression and tensile) [17].

Dr. Mukhopadhyay modified the processing parameters (spin line stress, spinning temperature, cooling procedure, take-up velocity, and postspinning parameters) of the molding/pultrusion process. These modifications resulted in an increase in the following mechanical properties of the fibers (modulus, tensile strength, and tenacity) [18].

Mr. Cowan improved the processing parameters (highly reliable high-speed spinning process) with the addition of nucleating additives. The resulting fibers exhibit low shrinkage and better strength properties [19].

2.2.2 Technological Process

We have analyzed the tensile strength of commodity fibers and have concentrated on liquid polymers such as polypropylene and polyester.

We found commercial companies manufacturing polypropylene and polyesters with a high tenacity of approximately 10 g/denier. We combined melting, hot impregnation, cooling, and modification of the pultruded process, including six different approaches. We selected KaZaK Composites, LLC, Woburn, MA as a base facility for testing.

A dry polyester fiber (Figure 2.7) (pos. 1) was melted in furnace (pos. 2), and pulled to bath (pos. 3) where it was impregnated by novel liquid crystalline polymer resin (Prof. Hara's approach) and rapidly cooled in cooling camera (pos. 4).

At rapid cooling, the melt of a semicrystallite polymer contains α-crystallites and β form. A β form occurs to β crystallites if we add special nucleates. The content of β crystallites depends on the conditions of crystallization and on the kind of nucleator and its concentrations.

Cooling procedure. The polymer is cooled by air or a liquid after leaving the die-spinner (pos. 5). Very high rates of cooling may be obtained. The rate of cooling in air is far in excess of 15°C per minute, aided by quenching in a liquid.

The high rate of cooling prevents excessive crystallization of the polymer, which affects the subsequent drawing of the spun filaments.

The melt temperature in the furnace in pos. 2 is 175°C.

The fiber is cooled in a cooling camera below room temperature and stretched using a pulley (see Figure 2.7).

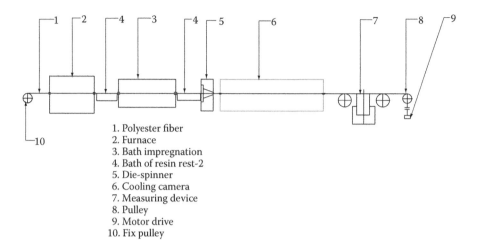

1. Polyester fiber
2. Furnace
3. Bath impregnation
4. Bath of resin rest-2
5. Die-spinner
6. Cooling camera
7. Measuring device
8. Pulley
9. Motor drive
10. Fix pulley

FIGURE 2.7
Stretch and impegnation of the polyester fiber.

We measure tenacity using a measuring device with a load cell at pos. 7.

A stretch or pultrusion fiber gives us a more homogeneous and crystalline structure. Fiber stiffness depends on the fiber diameter.

Reducing fiber diameter close to *I* mk will increase tensile modulus, tenacity, and strength to close theoretical parameters. We call this the effect of nanocomposites. KCI uses a pultruded process for manufactured carbon fiber structural elements.

Dry fiber filaments were impregnated by liquid epoxy and then pultruded through the heat system.

A big problem for KCI was not having a cooling system, which is absolutely necessary for reducing internal stresses. Hydraulic cylinders create pressure in the process of stretching fiber.

We used KCI's automatic pultrusion line for commodity fiber modification.

In Figure 2.8, we show a die-spinner drawing, which can reduce fiber diameter.

In Figure 2.9 we show pultruded installation for commodity fibers. On the frame (pos. 1), we installed six rolls with dry commodity fibers (pos. 2) (three rolls of polyester and three rolls of polypropylene).

Dry fiber was pultruded through heat electrical camera (pos. 3) and bath impregnation (pos. 4 and 6). Between this bath, a bath of rest resin (pos. 5) was installed. In bath (pos. 4), we inputted nucleating compound following the approach described by Mukhopadhyay et al. [18].

In bath (pos. 6) we inputted an ionics group following Prof. Hara's approach [17]. Heat electrical camera (pos. 3) was used as support for curing this filament with impregnated fibers. The melting fibers impregnated by liquid polymer resins and diameter fiber were reduced by die-spinner (pos. 7).

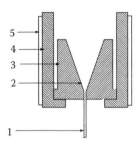

1. Die-drawn fiber
2. Biocomponent polymers
3. Die
4. Aluminum cylinder
5. Bond heater

FIGURE 2.8
Die-spinner drawing.

Fiber diameter reduced from 200 to 10 mk. Cooling camera (pos. 8) fixed this diameter by reducing internal stresses.

Commodity fibers by pressure 10 psi of hydraulic cylinders (pos. 9 and 10) made solid and after measure tenacity piled on pulley (pos. 12). Electromotor (pos. 14) and reducer (pos. 13) regulated speed and productivity.

Tensile strength of commodity fiber increased by impregnation of liquid polymer process, heat/cooling system, and the pultruded process.

In Figure 2.10, we show the spinner. The holes drilled at a 45° angle changed the diameter from 200 to 10 mk.

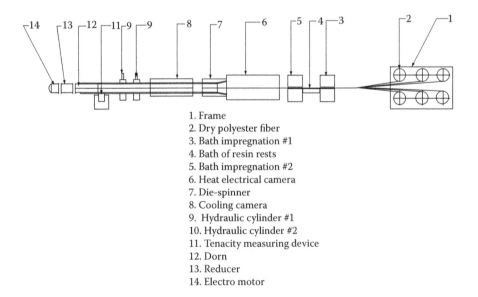

1. Frame
2. Dry polyester fiber
3. Bath impregnation #1
4. Bath of resin rests
5. Bath impregnation #2
6. Heat electrical camera
7. Die-spinner
8. Cooling camera
9. Hydraulic cylinder #1
10. Hydraulic cylinder #2
11. Tenacity measuring device
12. Dorn
13. Reducer
14. Electro motor

FIGURE 2.9
Pultruded installation for multilayered commodity fibers.

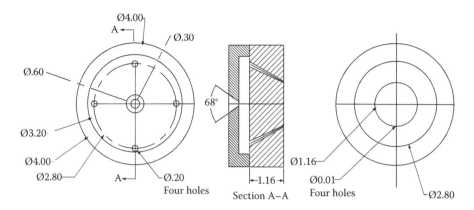

FIGURE 2.10
Aluminum spinner.

Fiber cooling by hot air occurs in the drawing area and we measure tenacity using tension device and load cell. The nucleating compounds are dibenzylidene sorbitol (DBS)-based compound; sodium benzoate, sodium, and lithium phosphate salt (e.g., sodium 2,2-methylene-bis-(4,6-di-*tert*-butylphenyl), otherwise known as NA-11, NA-21 [19].

We analyze the tensile strength, tenacity, and modulus of elasticity of commercial high-strength fiber materials and commodity polymers in Table 2.1.

Carbon fiber is a brittle material without a yield point, and it does not strain when hardened, which means that the ultimate strength and breaking strength are the same [22–25]. So replacing brittle carbon fiber on plasticity commodity fiber is a basic task of this proposal. Polyesters such as PBT have a modulus of elasticity 1000 to 4500 ksi, which is much higher than PP modulus 210 to 260 ksi [26,27,28]. However, tensile strength and tenacity of PP is two times less than PBT. PP tensile strength variation is 21,800 to 40,600 psi, tenacity variation is 10.88 to 20.2, and low modulus of elasticity is 210 to 260 ksi.

The tensile strength of PP is very close to DuPont Kevlar properties: tensile strength = 495,000–525,000 psi; tenacity = 23.6. The modulus of elasticity of PP is less than that of polyester and Kevlar [20].

The high denier polyester fibers are present in the Thinsulate batts to increase the low bulk and bulk recovery provided to the batt by the microfibers alone. For use in winter sports outerwear garments, these various insulating materials are often combined with a layer of film of porous poly (tetrafluoroethylene) polymer of the type disclosed in U.S. Patent No. 4,187,390.

Although polyesters such as polyethylene terephthalate (PET) and polyamids such as nylons are generally more expensive to manufacture than

TABLE 2.1

Tensile Strength of Commercial High-Strength Fiber Materials and Commodity Polymers

Material	Tensile Strength Ultimate (psi)	Tenacity Grams/Denier	Modulus of Elasticity (ksi)
Commercial high-strength fiber materials	841,000	42	
Zylon			
PBO-AS poly(P-phenylene-2,6-benzobisoxazole) fiber			
Kevlar 149 fiber, diameter 12 μm DuPont Kevlar	49,500–525,000	23.6	16,300
Polyamids Nylon with Kevlar (D638) Boedeker Plastics, Inc.	16,000	7.9	1,300
Zytel 101 Nylon 6,6 chips DuPont Co.	12,000	6.0	203–450
Aromatic Polyesters PEEK polyetheretherketone, Aramid fiber-filled	11,000–12,000	6.9	493–595
PPS (polyphenylene sulfide)	10,000–18,000	9.0	319–798
PPS + carbon fiber 50%	14,100	12.7	1,200–5,000
HMW polyethylene	5,800	3.0	725
thermoset polyurethane liquid	580–7,250		63.8–580
PBT-polybutylene terephthalate	12,000	6.0	1,000–4,500
Carbon fiber-filled acrylonitrile butadiene styrene (ABS)	24,900	12.4	
Polyolefine, polypropylene fiber grade Grade count = 8	21,800–40,600	10.88–20.2	210–260

fibers such as polypropylene (see Table 2.2), using polyester fiber recycling exchanges dramatically reduces the price.

Shell Chemical is producing a new type of polyester fiber. The fiber is poly (trimethylene terephthalate) (PTT) and bears the trade name Corterra®. PTT fibers have many similarities to, but some important differences from, the more common polyester, poly (ethylene terephthalate) (PET) fibers. The vast majority of polyester textile fibers are PET. Its sister polymer, poly (butylenes terephthalate) (PBT) has a tenacity of 12.4 (see Table 2.1). PTT is synthesized by polycondensation of trimethylene glycol with either a terephthalic acid or diethyl terephthalate. Trimethylene glycol is now commercially producible through the hydroformulation of ethylene oxide, allowing for the economic

TABLE 2.2

Conventional Polymers	Cost ($/lb)	Source, Web Site
Polypropylene pellets	0.50	agroplastics.com
Polyester fiber recycling	0.20	recycle.net
Polyethylene pellets	40	asia.recycle.net
PPS pellets	60–120	peachbelt.polychange.com
PEEK pellets	200	gengen23tripod.com
Nylon 6,6	92	plasticbrokers.com

production of PTT. The added number of methylene units affects the physical and chemical structure of PTT. PTT is easily heat-set and can be spun in PTT/PET bicomponent fibers [21].

2.2.3 Conclusions

1. A novel LCP incorporates ionics groups in the polymer. The lateral interactions occur between polymer chains, which significantly increase the mechanical properties (compression and tension).

2. Polyurethane matrix on an LPC avoids volatility; polybutylene terephthalate PBT matrix has less water absorption (0.03% to 0.05%), glass fiber content of 0% to 40%, elongation of 1.2% to 3.5%, and higher tensile strength than polyurethane.

3. Pultruded hybrid fiber composites develop and design the stretch and impregnation polypropylene/polyester fiber technology and multilayer process.

We added a cooling system to reduce the internal stresses.

References

1. Stover, D. 1994. Braiding and RTM succeed in aircraft primary structures. *High Performance Composites* 24–27.
2. Lazarus, P. 1997. Reporting from the resin infusion front. *Professional Boat Builder* 44: 30–35.
3. Ahn, K. J., and J. C. Seferis. 1993. Prepreg process analysis. *Polymer Composites* 14 (4): 346.
4. Ahn, K. J., and J. C. Seferis. 1993. Prepreg process science and engineering. *Polymer Engineering and Science* 33 (18): 1177.

5. Lee, W. J., J. C. Seferis, and D.C. Bonner. 1986. Prepreg processing science. *SAMPE Quarterly* 17 (2): 58.

6. Hayes, B. S., and J. C. Seferis. 1998. Self-adhesive honeycomb prepreg systems for secondary structural applications. *Polymer Composites* 19 (1): 54.

7. Buehler, F. U., J. C. Seferis, and S. Zeng. Consistency evaluation of a qualified glass fiber prepreg system. *Journal of Advanced Materials* 34 (2): 41.

8. Golfman, Y., and B. Buriakin. 1972. The automation process fiber impregnation by epoxy resin. *The Shipbuilding Technology Journal* 5 (85): 85–88.

9. Whitney, J. M., I. M. Daniel, and R. B. Pipes. 1982. *Experimental Mechanics of Fiber Reinforced Composite Materials.* Englewood Cliffs, NJ: Prentice-Hall.

10. Golfman, Y. 1994. Effect of thermoelasticity for composite turbine disk. 26th International SAMPE Technical Conference, Atlanta, GA.

11. Golfman, Y. 1993. Ultrasonic nondestructive method to determine modulus of elasticity of turbine blades. *SAMPE Journal* 29 (4): 31–35.

12. Golfman, Y. 2007. Impregnation process for prepregs and braided composites. *Journal of Advanced Materials,* Special Edition 3: 65–71.

13. Golfman, Y. 2009. High-Strength Low-Cost Polymer Fibers Hybrid with Carbon Fibers in Continuous Molding/Pultrusion Process. *Jounal of Advanced Materials* 41 (1): 35–39.

14. Parton, V. Z., and P. I. Perlin. 1984. Mathematical methods of the theory of elasticity. *Magazine World vols. 1 and 2.* Moscow: Mir Publishers.

15. Fanucci, J. President of KaZaK Composites, Inc. kazakcomposites.com.

16. Kotek, R., A. Tonelli, and N. Vasanthan. High Modulus Aliphatic Nylon Fibers via Lewis-Acid Complexation. NTC Project N05-NS05, http://www.ntcresearch.org.

17. Hara, M. Novel Polymeric Materials with Superior Mechanical Properties. Ionic Interaction, http://www.stormingmedia.us/32/3219/A321973.html.

18. Mukhopadhyay, S., B. L. Deopura, and R. Alagrusamy. 2004. Production and properties of high modulus-high tenacity polypropylene filaments. *Journal of Industrial Textiles* 33 (4): 245–268.

19. U.S. Patent # 7,041,368, May 9, 2006. High Speed Spinning Procedures for the Manufacture of High Denier Polypropylene Fibers. Milliken & Company.

20. Polyester Fibers. http://www.fbi.gov/hq/lab/fsc/backissu/july2001/houck.htm. U.S. Commercial Polyester Fiber Production, DuPont Company, 1953.

21. Pandit, S. B., and V. M. Nadkarni. 1993. Toughening of unsaturated polyesters by reactive liquid polymers. National Chemical Laboratory, India. *Engineering Chemical Research* 32: 3089–3099.

22. Urban, M. Aramids, Material Research Science Engineering Center, University of South Mississippi. www.psic.ws/macrog/aramid.him.

23. Bogun, M. et al. 2006. Influence of the As-Spun Draw Ratio on the Structure and Properties of PAN Fibers Including Montmorillonite. Facility of Materials Engineering & Ceramics, Krakow, Poland.

24. Carbon Fiber Manufacturing Composite Structures. http://www.zoltek.com/aboutus/news/34/.

25. Manufacturing Plastics & Resin. Ruth Ellen Carey Communications. www.internal-auditor.com.

26. Suzuki, A. E. 1997. Preparation of high modulus nylon 4,6 fibers by high-temperature zone-drawing polymer. *Polymer* 38 (12): 3085–3089.

27. Suzuki, A., and M. Isshihara. 2002. Application of CO_2 laser heating zone drawing and zone annealing to nylon 6 fibers. *Journal of Applied Polymer Science* 83 (8): 1711–1716.
28. Zachariades, E., and T. Kanamoto. 1988. New model for the high modulus and strength performance of ultradrawn polyethylenes. *Journal of Applied Polymer Science* 35: 1265–1281.

3

Strength Criteria and Dynamic Stability

3.1 Develop a Validated Design and Life Prediction Methodology for Polymeric Matrix Composite

3.1.1 Introduction

The currently used design methodology for polymeric matrix composites (PMCs) is based on the assumption of linear elastic stress–strain behavior of the composite. In certain composites such as carbon–epoxy and carbon–polyimide, and for certain lay-ups, laminate stress–strain curves are found to be highly nonlinear, especially at elevated temperatures. Exposure to oxygen and moisture can further enhance this nonlinearity due to constituent material property changes and increased micromechanical damage. In such materials and environments, use of linear stress analysis methods can result in inaccurate results, and thereby less than satisfactory designs.

Similarly, current PMC component design methods are found to be insufficiently accurate in predicting time-dependent deformation and damage behavior, thereby providing inadequate estimates of component life and durability. A validated living methodology would result in a significant reduction in component development testing.

The purpose of this research is to develop validated nonlinear analysis methods that would lead to a more accurate design and life prediction methodology for PMC components.

Polymer matrix composites used in high-temperature applications, such as frames for aircraft structures, turbine engines, and engine exhaust washed structures, are known to have limited life due to environmental degradation. For example, high temperature, pressure, and the presence of moisture limit the life of some polyamide composite components to only 100 h of service for worst-case operational conditions.

The thermo-oxidative behavior of the composite is significantly different from that of the fiber and matrix constituents as the composite microstructure, including the fiber–matrix interfaces, introduces anisotropy in the diffusion behavior [2].

The fiber, matrix, and interface regions that constitute the composite domain fundamentally influence durability and degradation mechanisms in composites [3].

Studies were conducted to examine the influence of the thermo-oxidative resistance characteristics of the fiber and matrix resin on the thermal stability of composites made from these materials. Celion 6000 graphite fiber, PMR-15 matrix resin, and Celion 6000/PMR-15 unidirectional composites were isothermally aged in air-circulating ovens at 288, 316, 329, and 343°C.

Microscopy (SEM) studies indicate extreme oxidative erosion of the graphite fiber occurs at elevated temperatures in the presence of the polyimide matrix [4].

The effect of sub-T_g environmental aging on the durability of two high-performance polymeric composites has been investigated. The material systems under study were a thermoplastic-toughened cyanate ester resin (Fiberite 954-2) and a semicrystalline thermoplastic resin (Fiberite ITX), and their respective carbon fiber composites (CFCs), IM8/954-2 and IM8/ITX.

Specimens were aged for periods of up to 9 months in environmental chambers at 150°C and in one of three different gas environments: nitrogen, a reduced air pressure of 13.8×10^{-3} MPa (2 psi air), or atmospheric ambient air (14.7 psi air). The glass transition temperatures, T_g, of the two resin systems were monitored as a function of aging time and environment. The changes in T_g showed effects of both physical aging and chemical degradation; the latter appeared to be sensitive to the oxygen concentration in the aging environment. Flexure tests were performed on eight-ply unidirectional (90°) IM8/954-2 and IM8/ITX composites, aged up to 6 months in the three gas environments at 150°C. The samples showed a 30% to 40% loss in the bending strength after aging. These strength reductions were sensitive to the oxygen concentrations in the aging environment. Stress–strain tests were also conducted on the same composites to measure the ultimate properties of the materials before and after aging in the three different environments at 150°C. The results showed a decrease of 40% to 60% in the ultimate strain to failure with aging. The modulus of both composite systems on the other hand increased by up to 20% after aging for 6 months, possibly as a consequence of the physical aging phenomena. In both systems, the greatest reduction in "useful" mechanical properties occurred in the ambient air environment, whereas the least reduction occurred in nitrogen. Weight loss in the plain resin and composite samples was monitored as a function of aging time and environment. Typically, all of the samples showed 1% to 2% weight loss after 9 months of aging at 150°C, and the composite samples lost much more weight (on a polymer basis) than unreinforced resin specimens over the same aging period. The weight loss data as well as all the above-mentioned observations were indicative of an oxidation process in the composites [5]. The impact damage resistance and tolerance of CFCs,

IM8/954-2, and IM8/ITX, was investigated by Parvatareddy et al. [6]. Impact tests were conducted at impact velocities of 6.3, 10.1, 19.8 and 25.4 m/s by using a gas-gun.

The damage area was evaluated by C-scan and x-radiography techniques.

In both the unidirectional and cross-ply composites, damage increased progressively with aging time. Tension after impact strength tests were also conducted on the cross-ply composites. Strength values fell by as much as 70% to 75% of original tensile strength in both material systems and were dependent on variables such as aging time, aging environment, and impact velocity.

Extensive effort is currently being expended to demonstrate the feasibility of using high-performance polymer–matrix composites as engine structural materials over the expected operating lifetime of the aircraft, which can extend from 18,000 to 30,000 h, which was studied by Bowles [7].

To accomplish this goal, it is necessary to pursue the development of thermal and mechanical durability models for graphite fiber–reinforced polymer–matrix composites.

Numerous investigations have been reported regarding the thermo-oxidative stability (TOS) of the polyimide PMR-15 [1–5]. A significant amount of this work has been directed at edge and geometry effects, reinforcement fiber influences, and empirical modeling of high-temperature weight loss behavior. It is yet to be determined if the information obtained from the PMR-15 composite tests is applicable to other matrix–matrix composites.

The condensation-curing polymer Avimid N is another advanced composite material often considered for structural applications at high temperatures. Avimid N has better TOS than PMR-15, but the latter is more easily processed. The differences between the microcracking resistance and mechanical properties' durability of the two different polyimides were unexpected a priori. The Avimid N composites experienced extensive microcracking during high-temperature aging. It was probably this microcracking, and not the internal void content, that was the primary cause of the lower compression properties for this condensation-curing polymer. This type of microdamage has been observed in PMR-15 composites with the identical fabric reinforcement when they were aged for more than 10,000 h at 204 and 260°C.

Graphite sheet coating was used for improved oxidation stability of carbon fiber reinforced PMR-15 composites [8].

Expanded graphite was compressed into graphite sheets and used as a coating for carbon fiber reinforced PMR-15 composites. BET analysis of the graphite indicated an increase in graphite pore size on compression; however, the material was proven to be an effective barrier to oxygen when prepreged with PMR-15 resin. Oxygen permeability of the PMR-15/graphite was an order of magnitude lower than the compressed graphite sheet. By providing a barrier to oxygen permeation, the rate of oxidative degradation of PMR-15 resin was decreased. As a result, the composite TOS increased by up to 25%. The addition of a graphite sheet as a top ply on the composites yielded little change in the material's flexural strength or interlaminar shear strength.

3.1.1.1 Concept of Nonlinear Composite Orthotropic Models

In modern design of aircraft structures where requirements for optimal calculation must be met, assumption of nonlinear elastic stress–strain behavior of the components is very important.

In current practice, extensive full-scale testing is used mainly because the reliability of the mathematical model has not been established to the satisfaction of engineers who have to make or approve design decisions. Improved reliability of mathematical models will make it possible to reduce the scope of experimental programs and the time required for developing information on which design decisions can be based with confidence. Also, extensive testing programs can improve the safety of design at substantial cost, but cannot be used for the optimization of design with respect to weight and durability. Only reliable mathematical models can do that.

The problem of simulating the elastostatic response of fastened structural connections is discussed [9]. The interaction between the fasteners and the contact plates was simulated by distributed springs. Friction was treated by the additional of external tractions in an iterative process. A nonlinear relation between the transferred force and the relative displacements represented each fastener.

In several studies [10–12], there was concern with the accuracy and reliability of the mathematical model. Two types of error have to be considered [2]:

1. The difference between the exact solution of the mathematical problem formulated to represent a physical system and the actual response or behavior of the physical system are called errors of idealization.

2. The difference between the exact solution of the mathematical problem formulated to represent a physical system or process and its numerical approximation are called errors of discreditation.

It is important to verify that the mathematical/numerical solution is close to the experimental solution; for example, the two aircraft frames (plates) made from carbon fiber–epoxy or carbon fiber–polyamide connect by fasteners, as shown in Figure 3.1.

Classical engineering calculation stresses distribution into two orthotropic plates consisting of calculated bending, torsion moments, and cutting forces. Two orthotropic plates are fixed around the counter, and bend in plane x, y by equal distributed load. In case of an ellipse hole we designate the geometrical side as a and the diameter of the fastener as b. The gap between the hole and the fasteners is designated by b_0.

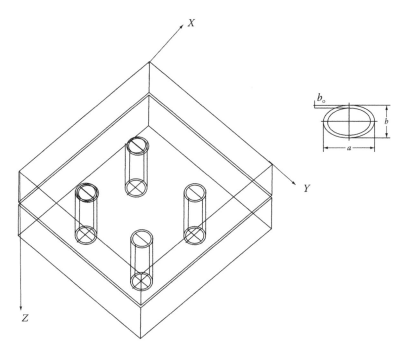

FIGURE 3.1
Two plates fixed by fasteners.

Bending, torsion moments, and cutting forces fill are found from the following equation:

$$M_x = \int\limits_{-h/2}^{h/2} \sigma_x z \partial z; \quad M_y = \int\limits_{-h/2}^{h/2} \sigma_y z \partial z \qquad (3.1)$$

$$H_{xy} = \int\limits_{-h/2}^{h/2} \tau_{xy} z \partial z; \quad N_x = \int\limits_{-h/2}^{h/2} \tau_{zx} z \partial z; \quad N_y = \int\limits_{-h/2}^{h/2} \tau_{zy} z \partial z,$$

where M_x, M_y are the bending moments, H_{xy} is the torsion moment, N_x, N_y are the cutting forces, and h is a thickness of both plates.
   The equilibrium equation is:

$$\frac{\partial N_x}{\partial x} + \frac{\partial N_y}{\partial y} + q = 0 \qquad (3.2)$$

We assume that all moments are constant and found in the work of Lekhnitskii [13] as:

$$M_x = -\left(D_{11} \frac{\partial^2 \omega}{\partial x^2} + D_{12} \frac{\partial^2 \omega}{\partial y^2} + 2D_{16} \frac{\partial^2 \omega}{\partial x \partial y} \right);$$

$$M_y = -\left(D_{12} \frac{\partial^2 \omega}{\partial x^2} + D_{22} \frac{\partial^2 \omega}{\partial y^2} \, 2D_{26} \frac{\partial^2 \omega}{\partial x \partial y} \right); \qquad (3.3)$$

$$H_{xy} = -\left(D_{16} \frac{\partial^2 \omega}{\partial x^2} + D_{26} \frac{\partial^2 \omega}{\partial y^2} + 2D_{66} \frac{\partial^2 \omega}{\partial x \partial y} \right);$$

$$N_x = -\left(D_{11} \frac{\partial^3 \omega}{\partial x^3} + 3D_{16} \frac{\partial^3 \omega}{\partial x^2 \partial y} + (D_{12} + D_{66}) \frac{\partial^3 \omega}{\partial x \partial y} + 2D_{26} \frac{\partial^3 \omega}{\partial y^3} \right);$$

$$N_y = -\left(D_{16} \frac{\partial^3 \omega}{\partial x^3} + (D_{12} + 2D_{66}) \frac{\partial^3 \omega}{\partial x \partial y^2} + 3D_{26} \frac{\partial^3 \omega}{\partial x \partial y} + D_{26} \frac{\partial^3 \omega}{\partial y^3} \right);$$

Here,

$$D_1 = E_1 \frac{h^3}{12}; \quad D_2 = E_2 \frac{h^3}{12}; \quad D_3 = D_1 \mu_{21} + \frac{G^{12} h^3}{6}; \qquad (3.4)$$

$$D_{12} + 2D_{66} = D_3; D_{16} = D_{26} = 0;$$

where
  D_1, D_2 = Bend stiffness for orthotropic material in x, y, xy directions
  D_3 = Torsion stiffness for orthotropic material
  E_1, E_2 = Tensile/compression modulus for orthotropic material
  G_{12} = Shear modulus for orthotropic material

We input cutting force $N_x N_y$ in equilibrium Equation (3.2) and obtain a differential equation for bending orthotropic plates:

$$D_1 \frac{\partial^4 \omega}{\partial x^4} + 2D_3 \frac{\partial^4 \omega}{\partial x^2 \partial y^2} + D_{22} \frac{\partial^4 \omega}{\partial y^4} = q \qquad (3.5)$$

Equation (3.5) was solved precisely via relative flexure ω by Lekhnitskii [5].

$$\omega = \frac{qa^4}{64D'} \left(1 - \frac{x^2}{a^2} - \frac{y^2}{b^2} \right)^2 \qquad (3.6)$$

where

    q = Load acting perpendicular to plane x, y

    a = Hole diameter; $D' = 1/8(3D_1 + 2D_3c^2 + 3D_2c^4)$

    $c = a/b$

We spread this flexure for fasteners installed between two plates (see Figure 3.1).

Concentration stress occurs when flexure is at maximum.

$$\omega_{max} = -\frac{qa^4}{64D'} \qquad (3.7)$$

We except the boundary conditions: $x_0 = 0$, $x_1 = a$; $y_0 = 0$, $y_1 = b$

$$\frac{\partial \omega}{\partial x} = 0; \quad \frac{\partial \omega}{\partial y} = 0; \quad \frac{\partial \omega}{\partial x_1} = a; \quad \frac{\partial \omega}{\partial y_1} = b; \qquad (3.8)$$

The geometrical profile of the ellipse hole with fasteners can be expressed by the following:

$$\left(1 - \frac{x^2}{a^2} - \frac{y^2}{b^2}\right)^2 \qquad (3.9)$$

After differential Equation (3.6) by

$$\frac{\partial \omega^4}{\partial x^2} \quad \text{and} \quad \frac{\partial \omega^4}{\partial y^2}$$

we get displacements Δx and Δy.

$$\Delta x = \frac{qa^6}{128D'}\left(\frac{4}{b^2} - \frac{24}{b^4}\right);$$

$$\Delta = \frac{qa^4b^2}{128D'}\left(1 - \frac{4}{a^4} - \frac{24}{a^4}\right) \qquad (3.10)$$

In this case, the field of deformation is never consent with the field of stresses. We assume that connection between displacements and deformations has a nonlinear character. Nonlinear deformations in plane x, y directions and shear directions are:

$$\varepsilon_x^2 = \frac{\Delta x^2}{x^2}; \quad \varepsilon_y^2 = \frac{\Delta y^2}{y^2}; \quad \varepsilon_{xy}^2 = \frac{\Delta x^2 \Delta y^2}{x^2 y^2} \qquad (3.11)$$

Now we can find tensile/compression and shear stresses and thermal-oxidation stresses.

$$\sigma_x = E_x \varepsilon_x^2 + E_x \alpha \Delta T \tag{3.12}$$

$$\tau_{xy} = G_{xy} \varepsilon_{xy}^2 + G_{xy} \alpha_{xy} \Delta T$$

where
  E_x = Tensile/compression modulus of elasticity
  G_{xy} = Shear modulus of elasticity (we get nonlinear deformation from Equation 3.11)
  α = Coefficient of thermal expansion
  ΔT = Gradient of temperature

Obviously, thermal normal/shear stresses will be increased when increasing aging temperatures such as 288, 316, 329, and 343°C.

Thermal stresses analysis, shown in Figure 3.2, has influence during the thermo-oxidation process.

Two strength criteria of orthotropic plates are the tools that control the environmental conditions [13]:

$$\frac{\sigma_x^2 + \alpha \tau_{xy}^2}{(\sigma_x^2 + \tau_{xy}^2)^{1/2}} \leq \sigma_{xy}^s \tag{3.13}$$

$$\frac{\sigma_x^2 + \beta \tau_{xy}^2}{(\sigma_x^2 + \tau_{xy}^2)^{1/2}} \leq \tau_{xy}^s \tag{3.14}$$

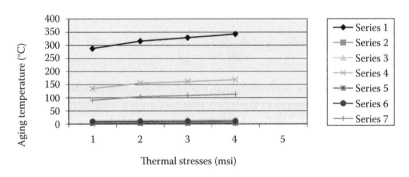

FIGURE 3.2
Thermal stress analysis.

where σ_x, τ_{xy} are how the stresses normal and shear act in plane xy. The strength relation coefficients α and β change depending on reinforced volume fractions, carbon fiber–epoxy, or carbon fiber polyimide in anisotropic composites.

$$\alpha = \sigma_{xy}^s / \tau_{xy}^s \quad \beta = \tau_{xy}^s / \sigma_{xy}^s$$

The results of stress combinations in plane are equal or less than strength σ_{xy}^s, τ_{xy}^s. Here, σ_{xy}^s, τ_{xy}^s is a strength of material test by experimental specimens.

Two orthotropic composite plates are shown in Figure 3.1. In the case of the ellipse hole, we designate the geometrical side as a and the diameter of the fastener as b. The gap between the hole and the fasteners is designated by b_o.

Criteria in Equations (3.13) and (3.14) are very good tools to check the strength in the interface field when act interlaminar normal and shear stresses ($\sigma_z \tau_{zx}$).

3.1.2 Experimental Investigation

We select diameters for the fasteners and the hole, which have ellipse geometry. Dimension variations were selected from 50 to 100 mm.

Displacements are varied:

$$\Delta x = 9.2 \times 10^{-2} - 17.9 \times 10^{-2}; \quad \Delta y = 7.7 \times 10^{-2} - 26.4 \times 10^{-2}$$

Deformation in plane x,y and shear deformation is also varied.

$$\varepsilon_x = 1.23 \times 10^{-3} - 2.12 \times 10^{-3}; \quad \varepsilon_{xy} = 1.35 \times 10^{-6} - 8.07 \times 10^{-6};$$

$$\varepsilon_x^2 = 1.51 \times 10^{-3} - 4.49 \times 10^{-3}; \quad \varepsilon_{xy}^2 = 1.82 \times 10^{-6} - 65.12 \times 10^{-6}$$

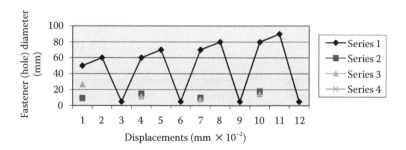

FIGURE 3.3
Correlation between fastener diameter (hole) and displacements.

TABLE 3.1

Modulus of Elasticity and Coefficient of Thermal Expansion for the Prepreg Laminates Based on IM7 Carbon Fibers

Material	IM7/PEEK	IM7/PPS	IM7/Boron/PEEK	IM7/Boron/PPS
Modulus of elasticity, E_z (GPa/msi)	165/24.2	165.7/24.2	188.3/27.5	188.3/27.5
Modulus of shear elasticity, G_{zx} (GPa/msi)	117.3/17.2	117.3/17.2	126.5/18.5	126.5/18.5
Coefficient of thermal expansion, CTE (ppm/°F)	0.02	0.06	0.57	0.57

Linear correlation between fastener (hole) diameter and displacements are established as shown in Figure 3.3.

The modulus of elasticity and coefficient of thermal expansion for the prepreg laminates based on the IM7 carbon fibers used for aircraft parts are shown in Table 3.1.

Carbon fiber IM7 consists of 60% fiber and 40% polyimide resin volume fractions; in the case of add 10% boron, 50% carbon fiber, and 40% polyimide resin volume fractions. Thermal stress components acting perpendicular to plane *x, y* calculated for the prepreg laminates of the IM7 carbon fibers are shown in Table 3.2.

Shear stress components dramatically depend on the temperature degradation process (see Table 3.3).

3.1.3 Conclusions

1. A nonlinear model for PMC components design based on physical thermo-oxidation process. The mathematical model has been reflected the influence a thermal oxidation process and thermal stress components has anisotropic character.

TABLE 3.2

Thermal Stress Components for the Prepreg Laminates Based on IM7 Carbon Fibers

Temperature (°C)	IM7/PEEK (MPa/psi)	IM7/PPS (MPa/psi)	IM7/Boron/PEEK (MPa/psi)	IM7/Boron/PPS (MPa/psi)
288	0.03/4.4	0.09/13.0	0.92/134.6	0.92/134.6
	0.03/4.4	0.09/13.0	0.92/134.6	0.92/134.6
316	0.03/4.8	0.09/14.4	1.06/154.6	1.06/154.6
329	0.03/5.0	0.09/14.9	1.11/161.0	1.11/161.0
343	0.03/5.18	0.10/15.56	1.15/167.8	1.15/167.8

TABLE 3.3

Thermal Shear Stress Components for the Prepreg Laminates Based on IM7 Carbon Fibers

Temperature (°C)	IM7/PEEK (MPa/psi)	IM7/PPS (MPa/psi)	IM7/Boron/PEEK (MPa/psi)	IM7/Boron/PPS (MPa/psi)
288	0.02/3.0	0.064/9.3	0.624/90.5	0.624/90.5
316	0.02/3.4	0.073/10.1	0.717/104.0	0.717/104.0
329	0.02/3.5	0.073/10.6	0.746/108.3	0.746/108.3
343	0.02/3.6	0.075/11.0	0.779/113.0	0.779/113.0

2. Results of stress calculations were verified by tools based on two criteria control strength parameters. The numerical solutions are very close to the experimental solutions. This method gives a green light for using the logarithm software for automation parameters control.

3.2 History of Design and Life Prediction Methodology for PMC

The currently used design methodology for fiber reinforced polymers (FRPs) is based on the assumption of linear elastic stress–strain behavior of the polymers.

In certain composites, such as carbon–liquid polymers, and for certain lay-ups, laminate stress–strain curves are found to be highly nonlinear, especially at elevated temperature. Inaccurate results stress components can be an influence of strength predictions and micromechanical damage.

Lightweight and high-strength FRPs that are currently used in the aviation industry pose a challenge to designers for establishing a reliable failure criterion for FRP in case of biaxial and triaxial stress conditions.

A fourth-order polynomial strength criterion provides better approximation to experimental data than a second-order criterion, and also explains the nature of failure mechanisms of anisotropic reinforced polymers if we accurately input stress components.

A computer algorithm has been written to investigate the effect of different loading conditions on the failure of explosion, deep-space radiation, or ultraviolet lights. Results are provided for various out of plain ply orientations.

Thermoplastic polymers have anisotropic properties and wide applications; however, rational design 0°, +45°, −45°, −90° will be closed to orthotropic properties.

FRP appears attractive because the time for the technological curing process is less than for carbon–epoxy composites.

Accurate prediction strength is so important; however many designers do not understand which criteria must be selected. A comparison among several recent criteria for the failure analysis composite was conducted by Icardi and Fererro [15]. To assess their accuracy, the authors use finite element analysis and various composite materials, different concepts, boundary conditions, and thickness ratio. The implementation of different criteria is based on different assumptions, but authors operating with finite element analysis do not easily understand basic assumptions. In the case of thick laminates the prediction of 2-D and 3-D criteria do not agree. 3-D effects to take the place when we have free edges in close to geometric discontinuous spaces.

Some possible principles can be used in formulating strength criteria:

1. The strength criteria must be invariant with respect to coordinate transformation.

2. The strength criteria must satisfy Drucker's postulate; that is, the strength surface is a plot of the limiting values of strength in a nine-dimensional stress and space must be convex.

3. The general criteria must be transform failure criteria with components that are directly responsible for thresholds of failure. These components will be determined on the test data.

3.2.1 Basic Features of Natural Strength

Strength criteria of failure and plasticity were investigated by Mises in 1928 [16]. Mises suggested that functional plasticity for anisotropy crystals looked like a second-order polynomial, and he named it "potential of plasticity." He assumed that the potential of plasticity does not depend on invariant stress tensors. This is based on the assumption that hydrostatic pressure does not influence the beginning of crystal fluidity. This assumption makes a correlation between constituencies of tensor plasticity and modulus of elasticity. Hill [17,18] investigated this correlation.

The mathematical polynomial proposes that the tensile/compression strengths are equal, and shear strength does not depend on the bending and/or twisting directions.

We designate the following:

$X_\pm, Y_\pm, Z_\pm$ are the tensile/compression strengths in Cartesian coordinates x, y, z.

$S_{12}^\pm, S_{13}^\pm, S_{23}^\pm$ are the shear strengths acting in plane xy and out-of-plane xz, yz.

$\sigma_{11}\sigma_{22}\sigma_{33}$ are the normal stresses acting in x, y, z directions.

$\tau_{12}, \tau_{13}, \tau_{23}$ are the shear stresses acting in plane xy and out-of-plane xz, yz.

In 1971, Tsai and Wu [19] developed the criteria of Hill–Mises and designated tensile and compression and shear strengths with positive and negative signs.

$$
\left(\frac{1}{X_+}-\frac{1}{X_-}\right)\sigma_{11}+\left(\frac{1}{Y_+}-\frac{1}{Y_-}\right)\sigma_{22}+\left(\frac{1}{Z_+}-\frac{1}{Z_-}\right)\sigma_{33}+\frac{\sigma_{11}^2}{X_+X_-}+\frac{\sigma_{22}^2}{Y_+Y_-}+\frac{\sigma_{33}^2}{Z_+Z_-}+\frac{\tau_{12}^2}{S_{12}^+S_{12}^-}
$$

$$
+\frac{\tau_{13}^2}{S_{13}^+S_{13}^-}+\frac{\tau_{23}^2}{S_{23}^+S_{23}^-}+2F_{12}\sigma_{11}\sigma_{22}+2F_{13}\sigma_{11}\sigma_{33}+2F_{23}\sigma_{22}\sigma_{33}=1
$$

$$(3.15)$$

The term F_{ij}, which represents the stress interaction, is independent from the material properties.

Azzi and Tsai [20] adapted Hill's criterion to thin laminates.

In the latest version of Hashin's criterion [21], the stress components σ_{11}, σ_{12}, and σ_{13} are assumed to be responsible for fiber failure.

Stress components σ_{22}, σ_{33}, σ_{12}, σ_{13}, and σ_{23} are responsible for matrix failure.

Stress components σ_{11}, σ_{22}, and σ_{12} are acting in plane xy.

Stress components σ_{22} and σ_{33} are acting in the out-of-plane in directions y, z.

Tensile failure in fiber test is:

$$
\left(\frac{\sigma_{11}}{X_+}\right)^2+\frac{1}{S_{12=13}^2}\left(\tau_{12}^2+\tau_{13}^2\right)=1 \tag{3.16}
$$

For compression test $\sigma_{11}=-X(\sigma_{11}<0)$.

The matrix failure in the traction tensile test is approximated as:

$$
\left|\frac{\sigma_{22}+\sigma_{33}}{Y_+}\right|^2+\frac{1}{S_{23}^2}\left(\tau_{23}^2-\sigma_2\sigma_3\right)+\left(\frac{\tau_{xy}}{S_{12=13}}\right)^2+\left(\frac{\tau_{xz}}{S_{12=13}}\right)^2=1 \tag{3.17}
$$

$$
(\sigma_{11}+\sigma_{22})>0
$$

where Y_+ is a transverse tensile strength, S_{12} is the shear strength on the plane xy, and S_{13}, S_{23} are the out-of-plane shear strengths xz and yz.

Hashin [21] assumed that the shear strength in the xy plane, S_{12}, is equal to the interlaminar shear strength S_{13}. The stress components $(\sigma_{11}+\sigma_{33})>0$.

The matrix failure in the traction compression test is approximated as:

$$
\frac{1}{Y_-}\left(\frac{Y_-^2}{4S_{23}^2}-1\right)(\sigma_{22}+\sigma_{33})+\frac{1}{4S_{23}^2}(\sigma_{22}+\sigma_{33})^2+\frac{1}{S_{23}^2}\left(\tau_{23}^2-\sigma_{22}\sigma_{33}\right)+\frac{1}{S_{12=13}^2}\left(\tau_{12}^2+\tau_{13}^2\right)=1
$$

$$(3.18)$$

In the case of matrix failure in compression, the stress components $(\sigma_{22} + \sigma_{33}) < 0$.

This deals with the experimental evidence [21] that when the matrix fails in the presence of the transverse isotropic pressure (i.e., $\sigma_{22} = \sigma_{33}$), this pressure can reach values much larger than the actual compressive failure stress. However, it is not physically clear how to incorporate this effect in other failure modes [16]. The contribution of σ_x does not appear in the matrix failure modes of Equations (3.17) and (3.18), since any possible plane of failure must be parallel to the fibers. The linear interaction term is not considered in the tensile mode. Doubts arise for this, and in general, for the interaction between stresses and strength as it results from the interaction between tensor invariants.

The presences of the out-of-plane shear strength S_{23} and S_{13} in Equation (3.16) for matrix failure requires the presences of shear stress components τ_{23} and τ_{13}.

Tensile stresses σ_{11}, σ_{12}, and σ_{13} are assumed to be responsible for fiber failure; however, in Equation (3.16), σ_{12} and σ_{13} are absent.

3.3 Strength Criteria for Anisotropic Materials

3.3.1 Introduction

Lightweight and high-strength anisotropic composites such as carbon–carbon, graphite–epoxy, and so forth, under different combinations of applied stress components (biaxial and triaxial stress conditions) pose a challenge to designers for establishing a reliable failure criterion. Stimulated stress components in a composite turbine disk under the influence of centrifugal loads and airstream pressure is a typical example. In this section we propose to use a fourth-order tensor polynomial failure criterion for investigating the failure mechanisms in turbine disk.

A fourth-order polynomial strength criterion provides better approximation to experimental data than a second-order criterion, and also explains the nature of the failure mechanisms of anisotropic materials.

A computer algorithm has been written to investigate the effect of different loading conditions on the failure of a turbine disk. Results are provided for various out-of-plane ply orientations.

Strength predictions of anisotropic materials require a failure criterion according to all the stress components under uniaxial, biaxial, and triaxial stress conditions. The strength of composites can be predicted using a second-order polynomial [22]. The strength criterion of the second order is not capable of handling airstream load, particularly for strong anisotropic materials such as carbon–carbon and graphite–epoxy. For evaluating triaxial and biaxial stress conditions in wood and fiberglass, Ashkenazi [23–25] and Goldenblat and Kopnov [26,27] used fourth-order polynomial strength criterion. Fourth-order polynomial gives better approximation to experimental data than second-order polynomials. This criterion is also capable of handling airstream pressure. In the present work, a fourth-order polynomial is used to estimate the strength of a turbine disk.

3.3.2 Strength Theory

The ability to predict the strength of high-performance composite materials under complex loading conditions is a necessary ingredient for rational design. Criteria that can be used to predict strength arise from two radically different yet complementary approaches to the problem; namely, empirical theories and micromechanics. Some possible principles that can be used in formulating a strength criterion were given by Goldenblat and Kopnov [26,27]. These principles are:

1. The strength criterion must be invariant with respect to coordinate transformation.
2. The strength criterion must satisfy Drucker's postulate, that is, the strength surface (a plot of the limiting values of strength in a nine-dimensional stress space must be convex).
3. The criterion should be as simple as possible.

From the studies of Malmeister [28], Goldenblat and Kopnov [26,27], and Ashkenazi [23,24], the criterion of strength for composite materials is given as:

$$\left(\sum a_{ik}\sigma_{ik}\right)^{\alpha} + \left(\sum a_{pqrs}\sigma_{pq}\sigma_{rs}\right)^{\beta} + \left(\sum a_{ikpqrs}\sigma_{ik}\sigma_{pq}\sigma_{rs}\right)^{\gamma} + \ldots \leq 1 \qquad (3.19)$$

where σ_{ik} are components of tensor stress and a_{ik}, a_{pqrs}, and a_{ikpqrs} are components of tensor strength with different valences satisfying the following conditions of symmetry: $a_{ik} = a_{ki}$, $a_{pqrs} = a_{rspq}$, $a_{pqrs} = a_{pqsr}$.

$$\left(\sum a_{ik}\sigma_{ik}\right) + \left(\sum a_{pqrs}\sigma_{pq}\sigma_{rs}\right)^{1/2} \ldots \leq 1 \qquad (3.20)$$

Goldenblat and Kopnov took $\alpha = 1.0$, $\beta = 1/2$, and $\gamma = 1/3$, but they limited for two terms. An equation of strength in triaxial stress conditions in tensor form was given by Ashkenazi [23,24] as

$$a_{ikem}\sigma_{ik}\sigma_{ik} - \left[\frac{(\sigma_{ik}\delta_{ik})^2 + \sigma_{ik}\sigma_{ik}}{2}\right]^{1/2} = 0$$

$$\text{if } i = k \qquad (3.21)$$

where $\delta_{ik} = 0$ if $i \neq k$. This criterion can be used separately for tensile and compressive loads.

Expanding Equation (3.21), the criterion of strength for triaxial stress conditions is obtained as:

$$\frac{\sigma_x^2 + c\sigma_y^2 + b\sigma_z^2 + d\tau_{zx}^2 + p\tau_{zx}^2 + s\sigma_x\sigma_y + t\sigma_y\sigma_z + f\sigma_z\sigma_x}{\left(\sigma_x^2 + \sigma_y^2 + \sigma_z^2 + \tau_{xy}^2 + \tau_{yz}^2 + \tau_{zx}^2 + \sigma_x\sigma_y + \sigma_y\sigma_z + \sigma_z\sigma_x\right)^{1/2}} \leq [\sigma_{bx}] \qquad (3.22)$$

where

$$c = \frac{X}{Y}; \ b = \frac{X}{Z}; \ d = \frac{X}{S_{12}}; \ p = \frac{X}{S_{23}}; \ r = \frac{X}{S_{13}}$$

$$s = \frac{4X}{S_{12}^{45}} - c - d - 1; \ t = \frac{4X}{S_{23}^{45}} - c - b - p; \ f = \frac{4X}{S_{13}^{45}} - b - r - 1 \qquad (3.23)$$

All coefficients are necessary and test laminates spaces, where

X, Y, Z = Tensile (compression) strength in x, y, z directions
X^{45}, Y^{45}, Z^{45} = Normal strengths to act in diagonal directions
S_{12}, S_{13}, S_{23} = Shear strengths to act in plane xy and out of plane xz, yz
$S_{12}^{45}, S_{13}^{45}, S_{23}^{45}$ = Shear strengths to act in diagonal directions under angle 45° in plane xy and interlaminar planes xz, yz
$\sigma_x, \sigma_y, \sigma_z, \tau_{xz}, \tau_{yz}, \tau_{xy}, \tau_{zx}$ = Normal and shear variable stresses, respectively, depending on loading history

The coefficients c, b, d, p, r, s, t, and f are relative relations of strength and are dependent on the quality of materials and are determined experimentally. Coefficients are actually variable and depend on coinciding the fiber direction with loading directions and technological factors. K_0 is the factor of safety, which is the ratio of strength and the resultant stress of material.

For determining the strength in different directions, specimens are tested in both tension and compression zones separately. Equal physical dimensions for stress components and the rank of tensors provide flexibility of sign convention for ply orientation. All parameters in strength in the criterion change with respect to the rotation of the axis of symmetry.

The symmetrical tensor for strength of orthotropic materials is given in Table 3.4.

For example, in the case of triaxial compression loads, when only normal stresses are acting and the shear stresses are equal to zero, the hydrostatic pressure can be computed [24,29] using the equation given below:

$$P = \frac{\sqrt{6}}{4\left(\dfrac{1}{X_{12}^{45}} + \dfrac{1}{X_{31}^{45}} + \dfrac{1}{X_{32}^{45}}\right) - \left(\dfrac{1}{X} + \dfrac{1}{Y} + \dfrac{1}{Z} + \dfrac{1}{S_{12}} + \dfrac{1}{S_{31}} + \dfrac{1}{S_{23}}\right)} \qquad (3.24)$$

TABLE 3.4

Symmetrical Tensor of Strength for Orthotropic Material

	11	22	33	12	23	31
11	$\dfrac{1}{X}$	$\dfrac{1}{2}\left(\dfrac{4}{2X^{45}} - \dfrac{1}{X} - \dfrac{1}{Y}\dfrac{1}{S_{12}}\right)_{xy}$	$\dfrac{1}{2}\left(\dfrac{4}{2X^{45}} - \dfrac{1}{X} - \dfrac{1}{Y}\dfrac{1}{S_{13}}\right)_{zx}$	0	0	0
22	$\dfrac{1}{2}\left(\dfrac{4}{2X^{45}} - \dfrac{1}{X} - \dfrac{1}{Y}\dfrac{1}{S_{21}}\right)_{xy}$	$\dfrac{1}{Y}$	$\dfrac{1}{2}\left(\dfrac{4}{2X^{45}} - \dfrac{1}{X} - \dfrac{1}{Y}\dfrac{1}{S_{23}}\right)_{yz}$	0	0	0
33	$\dfrac{1}{2}\left(\dfrac{4}{2X^{45}} - \dfrac{1}{X} - \dfrac{1}{Y}\dfrac{1}{S_{32}}\right)_{zx}$	$\dfrac{1}{2}\left(\dfrac{4}{2X^{45}} - \dfrac{1}{X} - \dfrac{1}{Y}\dfrac{1}{S_{32}}\right)_{yz}$	$\dfrac{1}{Z}$	0	0	0
12	0	0	0	$\dfrac{1}{S_{12}}$	0	0
23	0	0	0	0	$\dfrac{1}{S_{23}}$	0
31	0	0	0	0	0	$\dfrac{1}{S_{31}}$

All parameters of strength can be determined experimentally. The results are obtained by using the strength criteria and can be compared with that of the results that are gated experimentally on the laminates and structures with different layers.

For nonzero normal stress components σ_x and σ_y acting on the material, we can find the in-place shear stress τ_{xy} using strength criterion Equation (3.22).

$$\tau_{xy} = \left|\frac{-2Dd + 1 + \left[(1-2Dd)^2 - 4d^2(D^2 - D_1)\right]^{1/2}\,^{1/2}}{2d^2}\right| \tag{3.25}$$

where

$$D = \sigma_x^2 + c\sigma_y^2 + e\sigma_x\sigma_y$$

$$D_1 = \sigma_x^2 + \sigma_y^2 + \sigma_x\sigma_x$$

$$d = \frac{X}{S_{12}};\quad c = \frac{X}{Y};\quad a = \frac{X}{X_{12}^{45}}$$

$$e = 4a - c - d - 1;$$

In-plane shear stresses τ_{xy} and interlaminar shear stress components τ_{zy} and τ_{zx} appear when loads do not concur with the axis of symmetry of elasticity. Similarly, interlaminar shear stress in the ZY plane and ZX plane can be obtained and are given by Equations (3.26) and (3.27).

$$\tau_{xy} = \left| \frac{-2D_1 d_1 + 1 + \left[(1 - 2D_1 d_1)^2 - 4d_1^2 (D_1^2 - D_{12}) \right]^{1/2}}{2d_1^2} \right|^{1/2} \tag{3.26}$$

where

$$D_1 = \sigma_z^2 + c\sigma_y^2 + e\sigma_z \sigma_y$$

$$D_{12} = \sigma_z^2 + \sigma_y^2 + \sigma_z \sigma_y$$

$$d_1 = \frac{X}{S_{32}}; \quad c_1 = \frac{X}{Y}; \quad a_1 = \frac{X}{X_{32}^{45}}$$

$$e = 4a_1 - c_1 - d_1 - 1$$

and

$$\tau_{xx} = \left| \frac{-2D_2 d_2 + 1 + \left[(1 - 2D_2 d_2)^2 - 4d_2^2 (D_2^2 - D_{13}) \right]^{1/2}}{2d_2^2} \right|^{1/2} \tag{3.27}$$

where

$$D_2 = \sigma_z^2 + c\sigma_x^2 + e\sigma_z \sigma_x$$

$$D_{13} = \sigma_z^2 + \sigma_x^2 + \sigma_z \sigma_x$$

$$d_2 = \frac{X}{S_{31}}; \quad c_2 = \frac{X}{Y}; \quad a_2 = \frac{X}{X_{31}^{45}};$$

$$e = 4a_2 - c_2 - d_2 - 1$$

Normal stresses σ_x, σ_y, and σ_z would be calculated using the equations of the theory of elasticity. Substituting these values in Equations (3.25) through (3.27), the shear stresses are calculated.

3.3.2.1 Stresses in a Rotating Disk

Stresses in a rotating disk can be calculated using the following equation [30]:

$$\sigma_x = \sigma_r = \frac{3 + \mu_{r\theta}}{8} \rho \omega^2 \left(b_1^2 + a_1^2 - \frac{a_1^2 b_1^2}{r^2} - r^2 \right)$$

$$\sigma_y = \sigma_\theta = \frac{3 + \mu_{r\theta}}{8} \rho \omega^2 \left(b_1^2 + a_1^2 - \frac{a_1^2 b_1^2}{r^2} - \frac{1 + 3 + \mu_{r\theta}}{3 + \mu_{r\theta}} r^2 \right) \tag{3.28}$$

The material has Poisson's ratio, $\mu_{r\theta} = 0.036$ and density $\rho = 0.1497 \times 10^{-3}$ lb/in³. The dimensions are $a_1 = 0.4$ in, $b_1 = 4.0$ in, and $r = 3.0$ in, as shown in Figure 3.4.

The disk is rotating at 3627 rad/s. Normal stresses σ_x and σ_y are calculated using Equation (3.28). The results of the stress analysis of the disk gives $\sigma_x = 35.4/5138.497$ MPa/psi, and $\sigma_y = 67.8/9835.13$ MPa/psi. The shear stresses τ_{xy} are obtained using Equation (3.25) and are equal to 0.82/118.57 MPa/psi.

In Tables 3.5 and 3.6, it can be seen that the stress levels in the disk increase with the increase in the angular velocity (see Figure 3.5).

This method can be used to develop requirements of the material properties and composite ply orientation. Parameters of strength for a representative graphite material are given in Table 3.7.

The factor of safety K for angular velocity ω is given by:

$$K = \frac{\omega_{max}}{\omega_r} = 1.35$$

For angular velocity of 4800 rad/s, the in-plane shear stress τ_{xy} is about 211.77 psi, and $\sigma_y = 17{,}950.52$ psi, which is almost equal to the parameter of the material in axial y direction $\sigma_{by} = 18{,}000$ psi. If the parameter of strength in the y direction is increased to 36,000 psi, the angular velocity can be increased up to 7000 rad/s and the factor of safety is:

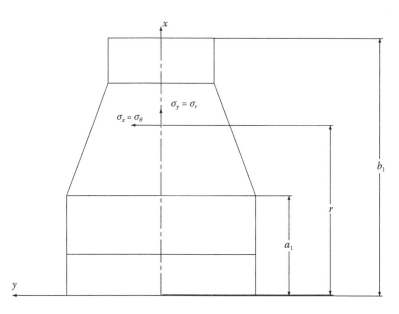

FIGURE 3.4
Rotating disk.

TABLE 3.5

Significances of Stress Components (σ_x, σ_y, τ_{xy}) in a Turbine Depend on Rotation Velocity

ω	σ_x (MPa/psi)	σ_y (MPa/psi)	τ_{xy} (MPa/psi)
3627.00	35.4/5138.50	67.8/9835.13	0.82/118.57
3700.00	36.8/5347.43	70.6/10,235.01	0.85/123.16
3800.00	38.9/5640.38	74.5/10,795.73	0.89/129.60
3900.00	40.9/5941.15	78.4/11,371.40	0.94/136.22
4000.00	43.1/6249.73	82.5/11,962.03	0.98/143.00
4100.00	45.3/6566.12	86.7/12,567.61	1.03/149.96
4200.00	47.5/6890.33	90.9/13,188.14	1.08/157.08
4300.00	49.8/7222.36	95.3/13,823.62	1.13/164.38
4400.00	52.5/7562.18	99.8/14,474.05	1.18/171.85
4500.00	54.5/7909.82	104.4/15,139.44	1.23/179.49
4600.00	57.0/8265.27	109.1/15,819.78	1.29/187.30
4700.00	59.5/8628.54	114/16,516.08	1.35/195.28
4800.00	62.06/8999.61	118.8/17,225.32	1.40/203.44
4900.00	64.7/9378.50	123.8/17,950.52	1.46/211.77
5000.00	67.3/9765.21	128.9/18,690.67	1.52/220.26
5100.00	70.0/10,159.72	134.1/19,445.77	1.58/228.93
5200.00	72.8/10,562.05	139.4/20,215.83	1.64/237.77
5300.00	75.6/10,972.19	149.8/21,000.84	1.7/246.78
5400.00	78.5/11,390.14	150.3/21,800.80	1.76/255.97
5500.00	81.5/11,815.90	156/22,615.71	1.83/265.32

Note: Limit σ_{by} = 124 MPa/18.000 psi.

$$K = \frac{\omega_{max}}{\omega_r} = 1.93$$

The strength criteria given in Equation (3.21) can be designated as a curve or a surface. Strength curve for nonzero stress components σ_z and τ_{zy} is

$$\sigma_z^4 + 2d\sigma_z^2\tau_z^2 - \sigma_z^2 + d^2\tau_{zy}^4 - \tau_{zy}^2 = 0 \tag{3.29}$$

where

$$d = \frac{X}{S_{31}}$$

TABLE 3.6

Significances of Stress Components (σ_x, σ_y, τ_{xy}) in a Turbine Disk Depend on Rotation Velocity

ω	σ_x (MPa/psi)	σ_y (MPa/psi)	τ_{xy} (MPa/psi)
5600.00	84.5/12,249.47	161.6/23,445.58	1.9/274.85
5700.00	87.5/12,690.86	167.5/24,290.39	1.96/284.54
5800.00	90.6/13,140.06	173.4/25,150.16	2.03/294.41
5900.00	93.8/13,597.07	179.5/26,024.89	2.10/304.45
6000.00	96.9/14,061.90	185.6/26,914.56	2.17/314.67
6100.00	100/14,534.53	191.8/27,819.19	2.24/325.05
6200.00	103.5/15,014.98	198.2/28,738.77	2.30/335.61
6300.00	106.9/15,503.24	204.6/29,673.31	2.39/346.33
6400.00	110.3/15,999.31	211.2/30,622.79	2.46/357.23
6500.00	113.8/16,503.20	217.8/31,587.23	2.54/368.30
6600.00	117.3/17,014.89	224.6/32,566.62	2.62/379.54
6700.00	120.9/17,534.40	231.5/33,560.97	2.7/390.95
6800.00	124.5/18,061.72	238.4/34,570.26	2.77/402.54
6900.00	128.25/18,596.86	245.5/35,594.51	2.85/414.20
7000.00	131.9/19,139.80	252.6/36,633.71	2.94/426.22

The strength surface for nonzero stress components σ_z, σ_y, and τ_{zy} is;

$$\sigma_Z^2 + c\sigma_y^2 + e\sigma_z\sigma_y + d\tau_{zy} - \left(\sigma_Z^2 + \sigma_y^2 + \sigma_z\sigma_y + \tau_{zy}\right)^{1/2} = 0 \qquad (3.30)$$

where

$$d = \frac{Z}{S_{32}}; \quad c = \frac{Z}{Y}; \quad e = 4a - c - d - 1; \quad a = \frac{Z}{Z_{32}^{45}};$$

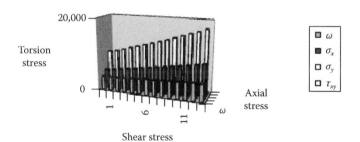

FIGURE 3.5

Correlation between velocity of rotation and values has stress components on turbine disk.

TABLE 3.7

Parameters of Strength for Graphite–Epoxy Composite

	Normal Acting x,y,z (MPa/psi)	Strength in Axis, ×10³ (MPa/psi)		In-Plane Shear Strength, ×10³ (MPa/psi)	Interlaminar Shear Strength, ×10³ (MPa/psi)	
Significance	σ_{bx}	σ_{by}	σ_{bz}	τ_{bxy}	τ_{byz}	τ_{bz}
Compression	0.34/50	0.8/12	0.34/50	0.35/5	0.06/10	0.06/10
Tension	0.68/100	0.12/18	0.35	0.03/5	0.06/10	0.06/10

	Normal Acting Diagonal, ×10³ (MPa/psi)	Strength in 45° Direction (MPa/psi)				
Significance	σ_{bxy}^{45}	σ_{bzy}^{45}	σ_{bzx}^{45}			
Compression	0.206/30	0.206/30	0.34/50			
Tension	0.406/59	0.08/12	0.362/52.5			

where Z, Y, and Z_{32}^{45} are the parameters of strength in the x, y, z directions in tension or compression and S_{32} is the interlaminar shear strength.

For biaxial stress conditions in the YZ plane (i.e., $\sigma_y = 0.0$), Equation (3.21) is modified as:

$$\frac{\sigma_Z^2 + a\tau_{zy}^2}{\left(\sigma_Z^2 + \tau_{zy}^2\right)^{1/2}} \leq \left|S_{32}^R\right| \tag{3.31}$$

where S_{32}^R is the resultant shear strength acting in the interlaminar layers. The strength criterion for the tension zone is given as:

$$\frac{\sigma_Z^2 + a\tau_{zy}^2}{(\sigma_Z^2 + \tau_{zy}^2)^{1/2}} \leq \left|Z^R\right| \tag{3.32}$$

where Z^R is the resultant normal strength acting perpendicular to layers.

It has been assumed that in the compression zone τ_{zy} is responsible for delamination of laminate layers. In the tension zone σ_z in combination with τ_{zy} is responsible for delamination of laminate layers.

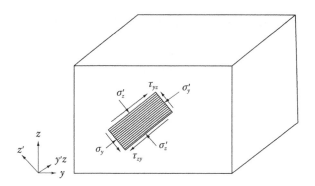

FIGURE 3.6
Compression and shear stress direction for ply rotation.

For the compression zone, shear stress is given by:

$$\tau_{zy} = \frac{\sqrt{1 - 2D\sigma_Z^2 \pm \left[(2d\sigma_Z^2 - 1)^2 - 4d(\sigma_Z^2 - \sigma_Z^4) \right]^{1/2}}}{2d} \tag{3.33a}$$

$$\sigma_z = \frac{\sqrt{1 - 2D\tau_Z^2 \pm \left[(1 - 2d\tau_Z^2)^2 - 4\tau_{zy}^2 (d^2\tau_Z^2 - 1) \right]^{1/2}}}{2d} \tag{3.33b}$$

Both cases are shown in Figures 3.6 and 3.7.

In biaxial stress conditions (see Figure 3.6), compression strength in the *y* direction is greater than compression strength in the *z* direction. In the *y*

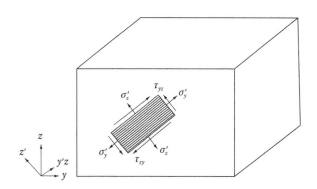

FIGURE 3.7
Tension and shear stress direction for ply rotation.

direction compression stress coincides with the fiber direction, and in the z direction compression stress coincides with resin.

Tensile strength (see Figure 3.7) in the y direction is greater than the tension strength in the z direction, because fiber orientation coincides with tension stress in the y direction and work, and in the z direction work resin properties are less than fiber properties.

Figure 3.8 shows the variation of interlaminar shear stress τ_{zy} as a function of normal stress σ_z. The curve is convex and therefore satisfies Drucker's condition only when σ_z is equal to or greater than 1.

Now by using the above calculations, optimum directions of perpendicular laminate layers in the ZY plane can be selected. A change in direction of airstream load in arbitrary coordinate system results indicates in normal stress σ_z and interlaminar shear stresses τ_{zy} and τ_{zx}.

Stress components σ_x, σ_y, and σ_z can be calculated using the sign convention for ply orientation by the following relations.

- In the XY plane:

$$\sigma_x = \sigma_{byx}^\alpha \sin^2 \alpha; \quad \sigma_y = \sigma_{byx}^\alpha \cos^2 \alpha; \tag{3.34a}$$

- In the XZ plane:

$$\sigma_z = \sigma_{bxz}^\alpha \sin^2 \alpha; \quad \sigma_x = \sigma_{bxz}^\alpha \cos^2 \alpha \tag{3.34b}$$

- In the ZY plane:

$$\sigma_z = \sigma_{byz}^\alpha \sin^2 \alpha; \quad \sigma_y = \sigma_{byz}^\alpha \cos^2 \alpha \tag{3.34c}$$

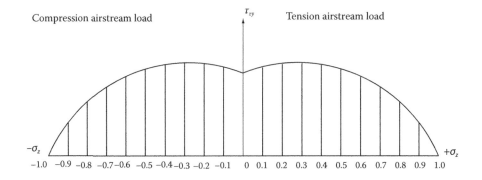

Interlaminar shear stress τ_{zy} calculated to use Equation (3.33a)

FIGURE 3.8
Interlaminar shear stress analysis.

For σ_{bx}^{α} and σ_{by}^{α} as a function of α are given by the equations developed in 1946 by Rabinovich [31].

- In the XY plane:

$$\sigma_{byx}^{\alpha} = \frac{\sigma_{bx}}{\cos^4 \alpha + b \sin^4 2\alpha + c \sin^4 \alpha} \tag{3.35}$$

where

$$c = \frac{X}{Y}; \quad a = \frac{X}{X_{12}^{45}}; \quad b = a - \frac{1+c}{4}$$

- In the XZ plane:

$$\sigma_{bxz}^{\alpha} = \frac{\sigma_{bz}}{\cos^4 \alpha + b_1 \sin^4 2\alpha + c_1 \sin^4 \alpha} \tag{3.36}$$

where

$$c_1 = \frac{X}{Z}; \quad a_1 = \frac{X}{\sigma_{13}^{45}}; \quad b_1 = a_1 - \frac{1+c_1}{4}$$

- In the ZY plane:

$$\sigma_{byz}^{\alpha} = \frac{\sigma_{by}}{\cos^4 \alpha + b_2 \sin^4 2\alpha + c_2 \sin^4 \alpha} \tag{3.37}$$

where

$$c_2 = \frac{Z}{Y}; \quad a_2 = \frac{Z}{Z_{32}^{45}}; \quad b_2 = a_2 - \frac{1+c_2}{4}$$

Parameters for carbon–carbon are shown in Table 3.8.

TABLE 3.8

Parameters are Related Strength for Carbon–Carbon

	a	c	b	a	c	b	a	c	b
Compression	0.4	0.24	0.09	1.0	1.0	0.5	0.4	0.24	0.09
Tension	0.305	0.18	0.01	1.9	2.0	−3.3	1.5	3.6	0.35

In the research, a fourth-order criterion for predicting the static strength of composite materials is given for different loading conditions. Influence of time, temperature, moisture, and scales factor are introduced parametrically by Parton [34], Golfman et al. [29,35] and Kerchtein [32].

3.3.3 Conclusions

Equations for surface and curve for strength criteria in triaxial and biaxial stress conditions maintain stresses in the second and fourth degrees.

1. Fourth-order polynomial strength criteria not only give us the option for much better approximation to experimental data than second-order criteria (Tsai Wu, Chamis, Hoffman, and Hill in [22]), but also explains the nature of the phenomenon of strong anisotropic materials.

2. The surface strength for strong anisotropic material may maintain convex and concave plots due to different character critical conditions on the plots. Pappo and Ivenson [33] discussed this paradox using the failure criteria of the second order.

3. The fourth-order polynomial is simpler for engineering applications. Computer programs for biaxial and triaxial stress conditions in different combinations of normal and shear stresses acting on the composite material with the influence of hydrostatic pressure are very useful in practical applications.

Appendix

This research was conducted to investigate the effect of different acting loading conditions on the failure of turbine disks for gas and steam turbines.

The instrument of investigation was fourth-order polynomial strength criteria. These criteria, based on analytical modeling and experimental validation of failure mechanisms, is manufactured from strong composite materials such as carbon–carbon and graphite–epoxy. Fourth-order polynomial strength criteria provide better approximations to experimental data than second-order criteria, and also precisely explain the nature of failure mechanisms fabricated from strong anisotropic materials. Designers can use this computer algorithm as a tool for selected aircraft dimensions and submarine construction when airstream load and hydrostatic pressure must be taken into consideration.

3.4 Theoretical Prediction of the Forces and Stress Components of Braided Composites

3.4.1 Introduction

The use of braiding as a fabrication process is becoming more prevalent in the composites industry due to design flexibility, ease of fabrication, and economic considerations [36]. The theoretical model that predicts the tensile elastic properties of triaxially braided composites has been developed [37]. The braid is divided into its three components (uniaxial fibers and off axis fibers), and the stiffness of each component is analyzed separately. The stiffness of each of the three parts of the braid are then combined to obtain the overall stiffness of the structure. The aim of this section is to show the theoretical prediction of the forces and stress components acting in braided construction.

The use of braiding as a fabrication method presents design problems from the dry braided shape with the next move to injected resin for impregnation of fabric. Although there is no interaction between plies when analyzing 2-D laminates, each layer of a triaxial braid has three components that interact with each other. The longitudinal fibers and the $\pm\theta$ braid fibers are interlocked, and thus cannot be considered independent components. It is this characteristic of the braid that presents the challenge in the design process. At the present time there are braiding machines and the possibility to design triaxially braided composite structures.

A number of models have been developed that predict the effective properties of multidirectional structural composites, many of which are based on the analysis of a representative unit cell of the braided structure. These models include the fabric geometry model [38] and the model for spatially oriented fiber composites [39].

3.4.2 Development of the Model

A model has been developed to analyze a triaxial braid, with no reinforcing fibers between the layers of braid. The braid is considered as having three separate components, or three plies, all coexisting in the same space. This means that each ply has a thickness equal to that of the full layer of braid. These three components are longitudinal fibers, braid fiber A, and braid fiber B. Although only the type of fiber, such as glass, is normally used for the off axis fibers, using two sets of braid fibers allows the model to analyze hybrid braids, such as glass–carbon. The braid fibers are oriented at some angle relative to the longitudinal fibers, usually referred to as the braid angle or helix angle. The longitudinal fibers are therefore considered to be oriented at $0°$ with respect to the $x_{(1)}$ direction (see Figure 3.9).

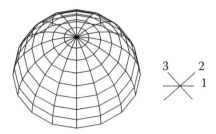

FIGURE 3.9
Braid fiber model.

All fibers are interwoven and do not lie flat next to each other, and therefore appear as a fiber kink. In our study the effect of fiber kink will be neglected in the analysis of the braid.

The "effective matrix" will include the modulus of elasticity in the direction of braid fiber A–E_{11} and in the direction of braid fiber B–E_{22}

$$E_{11}(n_{EFF}) = E_1(f)_n V(f)_n + E_1(m)(1 - V(f)_n) \tag{3.38}$$

$$E_{22}(n_{EFF}) = E_2(f)_n E_2(m)/E_2(m)V(f)_n + E_2(f)_n(1 - V(f)_n) \tag{3.39}$$

$$G_{12}(n_{EFF}) = G_{12}(f)_n G_{12}(m)/G_{12}(m)V(f)_n + G_{12}(f)_n(1 - V(f)_n) \tag{3.40}$$

$$\mu_{12}(n_{EFF}) = v_{12}(f)_n V(f)_n + v_{12}(m)(1 - V(f)_n) \tag{3.41}$$

$$\mu_{21}(n_{EFF}) = v_{12}(f)_n E_{22}/E_{11}(n_{EFF}) \tag{3.42}$$

where

n = Ply number in the effective matrix
$E_1(f)$ = Young's modulus of the secondary fibers in the "1" direction
$E_2(f)$ = Young's modulus of the secondary fibers in the "2" direction
$G_{12}(f)$ = Shear modulus of the secondary fibers
$\mu_{12}(f)$ = Primary Poisson's ratio for the secondary fibers
$V(f)$ = Volume fraction of the secondary fibers present in the ply
$E(m)$ = Young's modulus of the matrix material
$G_{12}(m)$ = Shear modulus of the matrix material
$\mu_{12}(m)$ = Poisson's ratio of the matrix material
$E_{11}(n_{EFF})$ = Young's modulus for ply n of the effective matrix in the "1" direction
$E_{22}(n_{EFF})$ = Young's modulus for ply n of the effective matrix in the "2" direction

$G_{12}(n_{EFF})$ = Shear modulus for ply n of the effective matrix
$\mu_{12}(n_{EFF})$ = Primary Poisson's ratio for ply n of effective matrix
$\mu_{21}(n_{EFF})$ = Secondary Poisson's ratio for ply n of effective matrix

Once the properties for each ply in the effective matrix have been determined, the stiffness matrix for each ply is calculated:

$$|Q_n| = \begin{vmatrix} Q_{11} & Q_{12} & Q_{16} \\ Q_{21} & Q_{22} & Q_{26} \\ Q_{61} & Q_{62} & Q_{66} \end{vmatrix}_n \tag{3.43}$$

where

$$(Q_{11})_n = E_{11}(n_{EFF})/\left|(1 - \mu_{12}(n_{EFF})\mu_{21}(n_{EFF})\right| \tag{3.44}$$

$$(Q_{12})_n = -\mu_{12}(n_{EFF})E_{22}(n_{EFF})/\left|(1 - \mu_{12}(n_{EFF})\mu_{21}(n_{EFF})\right| \tag{3.45}$$

$$(Q_{22})_n = E_{22}(n_{EFF})/\left|(1 - \mu_{12}(n_{EFF})\mu_{21}(n_{EFF})\right| \tag{3.46}$$

$$(Q_{66})_n = G_{12}(N_{EFF}) \tag{3.47}$$

$$(Q_{12}) = (Q_{21}) \tag{3.48}$$

and

$$Q_{16} = Q_{61} = Q_{26} = Q_{62} = 0$$

To determine the properties of the effective matrix that correspond to the direction of the primary fibers, the transformed stiffness matrix must be determined for each ply in the effective matrix:

$$(Q^*)_n = (T)^{-1}(Q)_n(T) \tag{3.49}$$

where (T) in matrix form could be shown as:

$$(T) = \begin{vmatrix} \cos^2\theta & \sin^2\theta & 2\sin\theta\cos\theta \\ \sin^2\theta & \cos^2\theta & -2\sin\theta\cos\theta \\ \sin\theta\cos\theta & \sin\theta\cos\theta & \cos 2\theta - \sin 2\theta \end{vmatrix} \tag{3.50}$$

For orthotropic material there are ten primary stiffness constants: Q_{11}, Q_{22}, Q_{33}, Q_{12}, Q_{21}, Q_{13}, Q_{31}, Q_{44}, Q_{55}, and Q_{66} (see Equation 2.12).

One study [40] discusses and describes an ultrasonic nondestructive method to determine modulus of elasticity of turbine blades. The equations for the modulus of elasticity were done in tensor form.

An ultrasonic nondestructive method to determine the modulus of elasticity of a nose cap was described by Golfman [41].

The resultant forces acting on a laminate cross section can be obtained by integrating the corresponding stress through the laminate thickness h:

$$
\begin{Bmatrix} N_x \\ N_y \\ N_{xy} \end{Bmatrix} = \int_{-h/2}^{h/2} \begin{Bmatrix} \sigma_x \\ \sigma_y \\ \sigma_z \end{Bmatrix} dz \tag{3.51}
$$

If the transformed stiffness matrix is assumed to be constant through the thickness of each ply, and it is assumed that the midplane strains are the same for all plies, then the resultants force can be directly related to the midplane strains. Replacing the continuous integral by the summation of integrals, which represents the contribution of each layer in the laminate, leads to the following:

$$
\begin{Bmatrix} N_x \\ N_y \\ N_{xy} \end{Bmatrix} = \sum_k \begin{bmatrix} Q_{11}^* & Q_{12}^* & Q_{16}^* \\ Q_{21}^* & Q_{22}^* & Q_{26}^* \\ Q_{61}^* & Q_{62}^* & Q_{66}^* \end{bmatrix} \int_{-h/2}^{h/2} dz \begin{Bmatrix} \varepsilon_x \\ \varepsilon_y \\ \gamma_{xy} \end{Bmatrix} \tag{3.52}
$$

When the properties for an orthotropic nose cap are changed under acting loads and temperature, the stress–strain relations must change in a matrix form [42].

$$
\sigma_{ij} = Q_{ij}^* \left(\varepsilon_{ij} - \alpha_{ij}^* T \right) \tag{3.53}
$$

where
Q_{ij}^* = Primary stiffness constants
ε_{ij} = Known as deformation
α_{ij} = Coefficient of temperature expansion
T = Temperature gradient

The differential equation of heat conductivity without the exothermic reaction of curing of the nose cap is as follows:

$$\frac{dT}{dt} = \beta \left(\frac{d^2T}{dr^2} + \frac{1}{r} * \frac{dT}{dr} \right)$$

(3.54)

where

t = Time of curing
R, r = Outside and middle radius of the nose cap
β = Coefficient of thermal conductivity

In selecting the boundary conditions: $T(r,0) = 0$; $T(R,t) = bt$; and b is the velocity of curing (cooling) process. The first approach to the solution is:

$$T(r,t) = \frac{bR^2}{\beta} * \left[\frac{\lambda t}{R^2} - \frac{1}{4}\left(1 - \frac{r^2}{R} \right) \right]$$

(3.55)

The gradient of temperature T in the period of curing (cooling) can be responsible for geometrical parameters of the nose cap, thermal conductivity of the epoxy resins, and the velocity of curing b.

The Kochi equation correlated deformation with displacements in tensor form and could be shown as:

$$\varepsilon_{ik} = \frac{1}{2}\left(\frac{\partial u_i}{\partial x_k} + \frac{\partial u_k}{\partial x_i} \right)$$

(3.56)

Therefore:

$$\varepsilon_x = \frac{\partial u}{\partial x}; \ \varepsilon_y = \frac{\partial v}{\partial y}; \ \varepsilon_z = \frac{\partial w}{\partial z}; \ \gamma_{xy} = \frac{\partial v}{\partial x} + \frac{\partial u}{\partial y};$$

$$\gamma_{xz} = \frac{\partial w}{\partial x} + \frac{\partial v}{\partial z}; \ \gamma_{yz} = \frac{\partial v}{\partial z} + \frac{\partial w}{\partial y};$$

(3.57)

where u, v, and w are the displacements in primary x, and braid A-y and braid B-z directions; ε_x, ε_y, ε_z are the deformations in primary x, and braid A-y, and braid B-z directions; γ_{xy}, γ_{xz}, γ_{yz} are the angle deformations, where the first index is the indicated force direction and the second index indicates the deformation direction. Now every point has 3 degrees of freedom and displacement can be described as a polynomial equation.

$$u = Q_{11}^* x^3 + Q_{12}^* y^3 + Q_{13}^* z^3 + Q_{16}^* xyz$$

$$v = Q_{21}^* x^3 + Q_{22}^* y^3 + Q_{23}^* z^3 + Q_{26}^* xyz \qquad (3.58)$$

$$w = Q_{31}^* x^3 + Q_{32}^* y^3 + Q_{33}^* z^3 + Q_{36}^* xyz$$

For every point on the surface of the nose cap, we can find linear and angle deformation:

$$\varepsilon_x = \frac{\partial u}{\partial x} = 3Q_{11}^* x^2 + Q_{16}^* yz$$

$$\varepsilon_y = \frac{\partial v}{\partial y} = 3Q_{22}^* y^2 + Q_{26}^* xz$$

$$\varepsilon_z = \frac{\partial w}{\partial z} = 3Q_{33}^* z^2 + Q_{36}^* xy$$

$$\gamma_{xy} = \frac{\partial v}{\partial x} + \frac{\partial u}{\partial y} = 3Q_{21}^* x^2 + 3Q_{12}^* y^2 + Q_{26}^* yz + Q_{16}^* xz$$

$$\gamma_{xz} = \frac{\partial w}{\partial x} + \frac{\partial v}{\partial z} = 3Q_{31}^* x^2 + 3Q_{23}^* z^2 + Q_{36}^* yz + Q_{26}^* xy$$

$$\gamma_{yz} = \frac{\partial v}{\partial z} + \frac{\partial w}{\partial y} = 3Q_{23}^* x^2 + 3Q_{32}^* z^2 + Q_{26}^* yz + Q_{36}^* xy$$

(3.59)

There is stiffness in Q_{11}^*, Q_{12}^*, Q_{21}^*, Q_{23}^*, Q_{13}^*, Q_{66}^*, which can be found using Equations (1.3) and (2.12) and $Q_{16}^* = Q_{61}^* = Q_{26}^* = Q_{36}^* = Q_{63}^* = 0$.

Therefore, the linear and angle deformations can be found using the following equations:

$$\varepsilon_x = 3Q_{11}^* x^2; \quad \varepsilon_y = 3Q_{22}^* y^2; \quad \varepsilon_z = 3Q_{33}^* z^2 \qquad (3.60)$$

$$\gamma_{xy} = 3Q_{21}^* x^2 + 3Q_{12}^* y^2; \quad \gamma_{xz} = 3Q_{31}^* x^2 3Q_{12}^* y^2; \quad \gamma_{yz} = 3Q_{23}^* z^2 + 3Q_{32}^* y^2 \qquad (3.61)$$

where

x, y, z= Coordinates on the surface of the nose cap
x   = Linear displacement in the longitudinal direction
y   = Linear displacement in braid fiber A direction
z   = Linear displacement in braid fiber B direction

So for the deformation prediction we can use finite element analysis using the polynomial Equations (3.60) and (3.61).

3.4.3 Experimental Results

Several braid configurations were evaluated using the model, including a carbon braid with no longitudinal reinforcement and a carbon–Kevlar hybrid braid with carbon fiber longitudinal reinforcement. The effect of the braid angle on the apparent longitudinal and transverse stiffness of the carbon braid would be expected. Table 3.9 shows the properties of elasticity for carbon–carbon composites.

Modulus E_{11}, E_{22}, and E_{33} are found using nondestructive ultrasound methods when the braid angle is 50°. The experimental axial stiffness of a hybrid braid on a tube was braided using P25 carbon and Kevlar 49 with a value of 3.48×10^{10} N/m².

The model predicted an axial stiffness of 3.79×10^{10} N/m² as shown in the work of Redman and Douglas [37].

3.4.4 Conclusions

1. A theoretical model has been developed to predict the forces and stress components of braided composites.

2. The main goal was to predict the primary stiffness of braid construction using a nondestructive ultrasound method and the linear and angle deformations that occurred during static and dynamic loading.

3. The micromechanical structure of braided composites was studied and linear deformations were found to be capable of accepting displacements using polynomial equations.

4. We used finite element analysis in our determinations.

5. Primary stiffness and linear/angle deformation were predicted.

6. The program was designed to be used as a parametric tool so that the design engineer could quickly determine the proper combination of materials and braid angles to obtain the forces and stresses. Efforts are continuing and ongoing to predict the strength of the final structure.

TABLE 3.9

Properties of Elasticity for Carbon/Carbon Composite

	E_{11}	E_{22}	E_{45}	G_{21}
Properties of elasticity on the patterns	3.56×10^{10}	2.59×10^{10}	2.24×10^{10}	0.84×10^{10}
Properties of elasticity on the nose cap	3.14×10^{10}	2.5×10^{10}	2.2×10^{10}	0.81×10^{10}
	μ_{12}	μ_{21}	μ_{45}	
Properties of elasticity on the patterns	0.818×10^{10}	0.818×10^{10}	0.68×10^{10}	
Properties of elasticity on the nose cap	0.81×10^{10}	0.818×10^{10}	0.68×10^{10}	

Note: Values of characteristics are in N/m².

3.5 Nonlinear Correlation between Modulus of Elasticity and Strength

A good approximation between the modulus of elasticity as a tensor of the fourth-order polynomial, and strength, also a tensor of the fourth-order, has the geometrical identification surfaces and gives us the assumption that there exists a correlation link between stiffness and strength in the 3-D modulus.

$$X_{\pm} = E_{11}\left(\varepsilon_{11}^{2} - \delta_{11}^{2}\Delta T\right)$$

$$Y_{\pm} = E_{22}\left(\varepsilon_{22}^{2} - \delta_{22}^{2}\Delta T\right)$$

$$Z_{\pm} = E_{33}\left(\varepsilon_{33}^{2} - \delta_{33}^{2}\Delta T\right)$$

$$S_{12} = G_{12}\left(\gamma_{12}^{2} - \delta_{12}^{2}\Delta T\right)$$

$$S_{13} = G_{13}\left(\gamma_{13}^{2} - \delta_{13}^{2}\Delta T\right)$$

$$S_{23} = G_{23}\left(\gamma_{23}^{2} - \delta_{23}^{2}\Delta T\right)$$

$$(3.62)$$

where

$\varepsilon_{11}, \varepsilon_{22}, \varepsilon_{33}$ = Linear deformations

$\gamma_{12}, \gamma_{13}, \gamma_{23}$ = Angle of deformations

$\delta_{11}, \delta_{22}, \delta_{33}, \delta_{12}, \delta_{13}, \delta_{23}$ = Coefficients of thermal expansion (CTE)

ΔT = Temperature gradient

Nonlinear deformation ε_{ik} consists of elasticity and plasticity parts. This tensor form of strain as a second invariant can be described as:

$$\varepsilon_{ik}^2 = \varepsilon_{11}^2 + \varepsilon_{22}^2 + \varepsilon_{33}^2 + 2\gamma_{12}^2 + 2\gamma_{13}^2 + 2\gamma_{23}^2$$

$$i \neq k = 0; \quad I = k = 1$$

$$\varepsilon_{11}^2 = \frac{1}{1^2(\Delta x_1^2 + 2\Delta x_1 \Delta x_2 + \Delta x_2^2)}$$

$$\varepsilon_{22}^2 = \frac{1}{b^2(\Delta y_1^2 + 2\Delta x_1 \Delta x_2 + \Delta x_2^2)}$$

$$\varepsilon_{33}^2 = \frac{1}{h^2(\Delta z_1^2 + 2\Delta x_1 \Delta x_2 + \Delta x_2^2)}$$

$$\gamma_{12}^2 = \frac{1}{1^2(\Delta \alpha_1^2 + \Delta \alpha_1^2 \Delta \alpha_2^2 + \Delta \alpha_2^2)} \tag{3.63}$$

$$\gamma_{13}^2 = \frac{1}{b^2(\Delta \beta_1^2 + \Delta \beta_1^2 \Delta \beta_2^2 + \Delta \beta_2^2)}$$

$$\gamma_{23}^2 = \frac{1}{h^2(\Delta \chi_1^2 + \Delta \chi_1^2 \Delta \chi_2^2 + \Delta \chi_2^2)}$$

where l, b, and h are designated as the length, width, and high dimensions of sample laminate, respectively; Δx_1, Δx_2, Δy_1, Δy_2, Δz_1, Δz_2 are decrements of linear displacements; $\Delta \alpha_1$, $\Delta \alpha_2$, $\Delta \beta_1$, $\Delta \beta_2$, $\Delta \chi_1$, $\Delta \chi_2$ are the decrements of angle displacements.

The field of strains in the strong anisotropic materials has never been considered with the field of stresses. The methodology for measuring strain under similar loading forces using embedded fiber optic strain sensors was done by Murioz and Lopez Anido [43].

Nodal quantities represent decrements of linear and angle displacements and are used in finite element analysis.

3.5.1 Experimental Results and Test Data of Thermoplastic Polymers

We designate $\varepsilon_{ik} - \delta_{ik}^2 \Delta T$ equal coefficients K_{ij}. We assign Equation (3.62) as:

$$X_{\pm} = K_{11}E_{11}$$

$$Y_{\pm} = K_{22}E_{22}$$

$$Z_{\pm} = K_{33}E_{33}$$

$$S_{12} = K_{12}G_{12}$$

$$S_{13} = K_{13}G_{13}$$

$$S_{23} = K_{23}G_{23}$$

(3.64)

where coefficients K_{11}, K_{22}, K_{33}, K_{12}, K_{13}, K_{23} are the relations between the strength and modulus of elasticity and determined by experiments.

Physical and mechanical property test data were provided by Phoenixx TPC [44]. We used test data from Thermoplastic Polymers. Phoenixx TPC prepregs were laminated by a pultrusion process. Thermo-Lite™ prepreg series 1 consisted of carbon fiber IM7 and polyetheretherketone (PEEK), polyamide. Thermo-Lite prepreg series 2 consisted of carbon fiber IM7 and polyphenylene sulfide (PPS). Thermo-Lite prepreg series 3 consisted of carbon fiber K63712 and polyetheretherketone (PEEK), polyamide. Thermo-Lite prepreg series 4 consisted of carbon fiber K63712 and PPS (Figure 3.10).

The coefficient K_{ij} has discrete characters. The coefficient of thermal expansion for the prepreg laminates based on the IM7 and K63712 carbon fibers is the basis of calculus coefficients K_{ij} (see Tables 3.10 and 3.11).

The variation of strain for double plate joint was investigated by Beck [44]. Variation is 0.0236 to 0.044 from initial crack 6.8 mm to final crack 12.7 mm

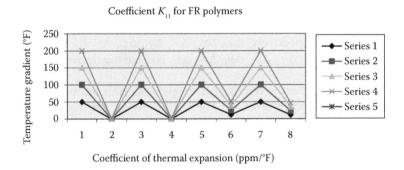

FIGURE 3.10
Coefficient of thermal expansion thermoplastic polymers.

TABLE 3.10

Modulus of Elasticity and Coefficient of Thermal Expansion for Prepreg Laminates Based on IM7 Carbon Fibers

Material	IM7/PEEK	IM7/PPS	IM7/Boron/ PEEK	IM7/Boron/ PPS
Modulus of elasticity, E	165.6	165.6	188.5	188.5
(GPa/msi)	24.2	24.2	27.5	27.5
Coefficient of thermal		0.06	0.57	0.57
expansion, CTE (ppm/°F)			130	

Note: IM7–60, boron fibers–40% volume fractions.

on the base 287 mm. In Beck's study [44], fiber-optic strain sensors embedded in FRP investigated the correlation between cracks and strain.

Tables 3.12, 3.13, and 3.14 represent the mechanical property data for Thermo-Lite Series 1, 2, and 3 and relations coefficient K_{11}.

3.5.2 Conclusions

1. Analysis of fourth tensor rank criteria shows that portion of shear stress for simple combination with tension (compression) stress is only 30% to 40%.

2. A good approximation between modulus of elasticity, as a tensor of the fourth-order polynomial, and strength, also a tensor of the fourth-order, has the geometrical identification surfaces and gives the assumption that there exists a correlation link between stiffness and strength in 3-D modulus.

TABLE 3.11

Modulus of Elasticity and Coefficient of Thermal Expansion for Prepreg Laminates Based on K63712 Carbon Fibers

Material	K63712/ PEEK[a]	K63712/PPS[a]	K63712/ Boron/ PEEK[b]	K63712/Boron/ PPS[c]
Modulus of elasticity, E	383.4	383.4	383.4	383.4
(GPa/msi)	56.0	56.0	56.0	56.0
Coefficient of thermal	−0.50	−0.48	0.00	0.00
expansion, CTE (ppm/°F)				

[a] K63712 –60, Matrix –40% volume fractions.
[b] 16% Boron fibers volume fraction.

TABLE 3.12

Mechanical Property Data, Thermo-Lite Series 1

Laminate Property	Fiber Orientation	Test Data	Relation Strength/ Modulus
Tensile strength	0°	1.623/235	0.015
(MPa/ksi)	90°	0.046/11.2	0.086
Tensile modulus	0°	106.8/15.6	
(GPa/ksi)	90°	0.001/1.3	
Compression strength (MPa/ksi)	0°	0.932/136	0.0093
Compression modulus (GPa/ksi)	0°	10/14.7 131	
Flexural strength (MPa/ksi)	0°	1.56/227	0.014

3. Coefficients K_{11}, K_{22}, K_{33}, K_{12}, K_{13}, K_{23} are the relations between strength and modulus of elasticity determined by experiments of a data test of Thermo-Lite thermoplastic composite materials.

4. This nonlinear correlation link has physical phenomena and explains failure analysis in biaxial and triaxial stress conditions.

5. Designers can use numerical finite element analysis and select optimal aircraft dimensions using fourth-order polynomial criteria. These criteria include airstream loads and hydrostatic pressure.

TABLE 3.13

Mechanical Property Data, Thermo-Lite Series 2

Laminate Property	Fiber Orientation	Test Data	Relation Strength/ Modulus
Tensile strength	0°	1.97/285	0.0146
(MPa/ksi)	90°	0.086/12.5	0.083
Tensile modulus	0°	103.4/19.5	
(GPa/ksi)	90°	1.02/1.5	
Compression strength (MPa/ksi)	0°	1.25/185	
Compression modulus (GPa/ksi)	0°	117/17.1	0.018
Flexural strength (MPa/ksi)	0°	1.56/275	0.0159
Flexural modulus (GPa/ksi)	0°	117.1/17.2	

TABLE 3.14

Mechanical Property Data for Thermo-Lite Series 3

Laminate Property	Fiber Orientation	Test Data	Relation Strength/ Modulus
Tensile strength	0°	2.08/302	0.016
(MPa/ksi)	90°	0.048/7.1	0.055
Tensile modulus	0°	128.6/18.8	
(GPa/ksi)	90°	1/1.3	
Compression strength	0°	1.27/165	0.0094
(MPa/ksi)			
Compression modulus	0°	119.7/17.5	
(GPa/ksi)			
Flexural strength	0°	1.70/247	0.0148
(MPa/ksi)			

3.6 Dynamic Stability Aspects for Hybrid Structural Elements for Civil Aircraft

3.6.1 Introduction

The lightweight hybrid structural composites in the aerospace industry will reduce weight and oil expenses. The core operations of companies like Apple and Millicon & Company [45] are manufacture and assembly of carbon nanotube (CNT) composite elements, aluminum frames with carbon fiber components, and improved engineering capabilities to support new product development.

Recently a new structural thermoplastic like PPS was found to reduce the weight of construction in wings aluminum frames by 20%.

Carbon fiber fuselage manufactured by automated fiber placement machines have also been investigated.

Fortron PPS in composites has replaced aluminum in Airbus Fokker wing leading-edge nose parts [45]. This is the first time the aircraft industry has used thermoplastic composites outside the cabin as a structural element in wings. Fortron PPS is half the weight of aluminum, which results in fuel savings and increased flight range. Fortron PPS is tougher, stronger, and more ductile than other similar materials, and maintains its properties over a very broad range of temperatures, up to 240°C and well below –40°C.

The Airbus A340-500/600 series with a leading edge nose on the wings manufactured by Netherlands-based Fokker Special Products opens the way for hybrid aluminum, CNTs, and structural Fortron PPS.

Premium Aerotec is the most important aerostructures supplier for the new Airbus A350 XWB long-haul aircraft, whose fuselage is largely made

FIGURE 3.11
View of the Airbus A340-600.

of CFC materials. This cutting edge CFC technology is essential for production of the highly complex fuselage structure for the A350 XWB, with its lightweight design. It involves creating the outer skin with a fiber placement machine and curing it in an autoclave [46]. The autoclave (pressurized oven), 25 m long and 8 m in diameter, was delivered to the plant in Augsburg.

A view of the Airbus A340-600 is shown in Figure 3.11.

The composite reduces the weight of the wing leading edge nose parts by 20%, makes fabrication faster and easier, improves impact resistance, and resists extreme temperatures and chemically aggressive liquids such as hydraulic fluids, fuel, and deicing agents. The wing leading edge nose parts are shown in Figure 3.12.

Faster fabrication of package layers on the wood models spread this technology on the flaps or ailerons and the keel beam.

Our proposal is to design structural wings as an aluminum frame with CNTs and PPS resin. Hybrid panels are shown in Figure 3.13. Aluminum frame (pos. 1) and aluminum ribs (pos. 2) would be sustained by the carbon nanotubes/thermoplastics PPS panels (pos. 3, 4, 5, 6, 7).

Carbon nanotubes are an extremely strong material for wing structures. Spinning of carbon nanotubes directly from the vapor phase is a more efficient process. Machine-fabricated hybrid panels are discussed by Hinrichsen and Bautista [49].

3.6.2 Continuous Laminate Process

Continuous laminate layers solidify when temperatures reach 265°C. Use of this continuous process avoids hydraulic presses, autoclaves, and vacuum bags. Moreover, thermoplastics do not expel volatile organic compounds (VOCs). This method can solidify big panels, and also costs less than other systems. PPS resin has been delivered in carbon fiber laminate (see Figure 3.14).

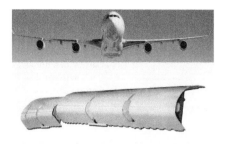

FIGURE 3.12
Wing leading edge nose parts.

PPS resin was heated in a reactor (extruder) until reaching a melting temperature of 340°C. Then, following the crystallization kinetic studies described by Nohara et al. [50], we cooled the resin to 265°C and held for 2 min.

Then the resin was driven by electro motor and vacuum pump through piping to a brush head, which was fed to the laminate package. The brush head

$$Cl \longrightarrow \bigcirc \longrightarrow Cl + NO_2 S \longrightarrow \left[\bigcirc \right] - S \right] + 2NOCl$$

FIGURE 3.13
Polyphenyline sulfide.

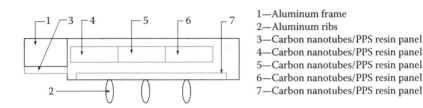

1—Aluminum frame
2—Aluminum ribs
3—Carbon nanotubes/PPS resin panel
4—Carbon nanotubes/PPS resin panel
5—Carbon nanotubes/PPS resin panel
6—Carbon nanotubes/PPS resin panel
7—Carbon nanotubes/PPS resin panel

FIGURE 3.14
Structural wing with PPS panels.

has fillers that are used as channels for the resin. The brush head is installed at the x, y table and feeds to the textile laminate package (Figure 3.15).

3.6.3 Fuselage and Wing Vibrations

Turbulence factors such as air flow in the case of lightning or hurricanes could be a basic cause of wing vibrations. In this case, if the natural frequencies of wings coincide with the force frequencies of wings, a parametric resonance may occur, resulting in disaster.

A carbon fiber fuselage manufactured by an automated fiber placement machine with CNC control looks like a solid shell with the loads distributed by bending moments and cutting forces. We investigate the natural frequencies for flexible wings and attempt to determine a more flexible wing consisting of an aluminum frame and carbon fiber PPS skin.

Natural frequencies, f, can be found by solving this differential equation of force order relative deflections in x, z coordinates [51].

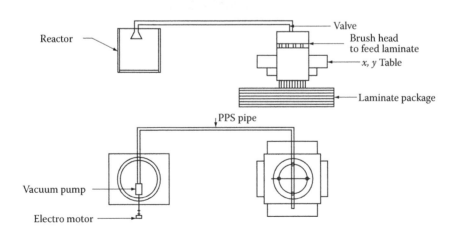

FIGURE 3.15
Automatic system for transfer of PPS resin.

$$\frac{\partial^2 \omega}{\partial t^2} \times \frac{g}{h\eta} \left(D_{1-} \frac{\partial^2 \omega}{\partial x^2} + 2D_3 \frac{\partial^2 \omega}{\partial x^2 \partial z^2} + D_{2-} \frac{\partial^4 \omega}{\partial z^2} \right) = 0 \qquad (3.65)$$

where

ω = Deflection of wings as a result of bending moments and cutting forces

T = Times of flexure

g = Density of CFCs (g = 1.85 g/cm³)

η = Acceleration due to gravity (9.81 m/s²)

h = Height of wings (see Figure 3.16)

L = Length of wings

Here, D_{ij} is the stiffness of the wings parameters:

$$D_1 = \frac{E_1 h^3}{12}; \; D_3 = D_1 + 2D_6; \; D_6 = \frac{G_{13} x h^3}{12}; \; D_2 = \frac{E_3 h^3}{12} \qquad (3.66)$$

Here, E_1, E_3, G_{13} are the normal and shear modulus of elasticity in the x and z directions, and $\mu_{13}\mu_{31}$ are Poisson's ratio. The first letter in the subscript of μ represents the direction of force applied and the second letter represents the transverse direction of deformation.

Model wings consisting of hybrid aluminum with CNT and PPS skin shell manufacturing from carbon fiber epoxy are shown in Figure 3.16.

In the case of the free vibration of wings, we use the boundary conditions:

- If $x = 0$; $x = L$; $\omega = 0$ (where L is the length of wings),

$$\frac{\partial^2 \omega}{\partial x^2} + \mu_{21} \frac{\partial^2 \omega}{\partial z^2} = 0 \qquad (3.67)$$

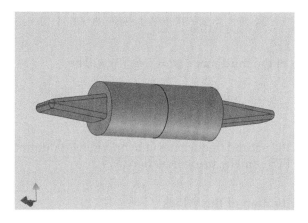

FIGURE 3.16
Model wings are hybrid aluminum with CNT and PPS skin shell manufacturing from carbon fiber epoxy.

- If $z = 0$; $z = h$; $\omega = 0$,

$$\frac{\partial^2 \omega}{\partial x^2} + \mu_{12} \frac{\partial^2 \omega}{\partial z^2} = 0 \tag{3.68}$$

These boundary conditions are known as functions of deflections

$$f_{mn} = \sin \frac{m\pi x}{L} \sin \frac{n\pi z}{h} \tag{3.69}$$

By inputting Equation (3.69) into Equation (3.65) to designate $k = L/h$. K is the present geometrical parameters (relationship of the length of thickness panel wings).

Now, we determine natural frequencies f_{mn} as:

$$f_{mn} = \frac{\pi^2}{L^2} \left(\frac{g}{h\eta} \right)^{1/2} \left[D_1 \left(\frac{m}{k} \right)^4 + 2D_3 n^2 \left(\frac{m}{k} \right)^2 + D_2 n^4 \right]^{1/2} \tag{3.70}$$

where m and n are semiconductors waves.

The frequency of the basic mode ($m = 1$; $n = 1$) will be:

$$f_{mn} = \frac{\pi^2}{L^2} \left(\frac{g}{h\eta} \right)^{1/2} \left[D_1 + 2D_3 k^2 + D_2 k^4 \right]^{1/2} \tag{3.71}$$

The frequency of the second mode ($m = 2$, $n = 2$) will be:

$$f_{mn} = \frac{4\pi^2}{L^2} \left(\frac{g}{h\eta} \right)^{1/2} \left[D_1 + 2D_3 k^2 + D_2 k^4 \right]^{1/2} \tag{3.72}$$

The frequency of the third mode ($m = 3$, $n = 3$) will be:

$$f_{mn} = \frac{9\pi^2}{L^2} \left(\frac{g}{h\eta} \right)^{1/2} \left[D_1 + 2D_3 k^2 + D_2 k^4 \right]^{1/2} \tag{3.73}$$

We determine the natural frequencies of hybrid wings (aluminum frame and thermoplastic PPS skin) by using Equation (3.74).

3.6.4 Force Vibration of the Wings

The loss of dynamic stability is a result of parametric resonance in the wings, when it coincides with the free and force frequencies. The wings are compressed in vertical direction 3 by force p_{mn}, which we can assign as:

$$p_{mn} = p_0 \sin\theta t - p_0 \cos\theta t \tag{3.74}$$

We need to solve the Matue equation relative to deflection:

$$\frac{\partial^2 f}{\partial t^2} + f_{min}^2 \left(1 - \frac{P_0}{P_{min}} \cos\theta t\right) f(t) = 0 \tag{3.75}$$

Here, f_{mn} are the frequencies of free vibrations determined by the formulas in Equations (3.70) through (3.73). P_{mn} is a critical value of compression force determined by Imbirchumun and A. Chichiturum [52]. The critical value of compressed force could be determined using Equation (3.75).

$$p_{mn} = \frac{\pi^2}{L^2} \left(\frac{g}{h\eta}\right)^{1/2} \left[D_1\left(\frac{m}{k}\right) + 2D_3 n^2 \left(\frac{m}{k}\right)^2 + D_2 n^4\right]^{1/2} \tag{3.76}$$

We select coefficient λ_{mn} as:

$$\lambda_{mn} = \frac{p_0}{2p_{min}} \tag{3.77}$$

Equation (3.75) should be assigned as:

$$\frac{\partial^2 f}{\partial t^2} + f_{min}^2 (1 - 2\lambda \cos\theta) f(t) = 0 \tag{3.78}$$

In the work of Golfman [51], the boundary conditions for stability losses were determined. For the basic tone of the force, the frequencies are:

$$\theta^* = 2f_{min}(1 \pm \lambda_{mn}) \tag{3.79}$$

where θ^* is a critical value of the frequency for the force load.
The second tone of the force frequency is:

$$\theta^* = f_{min}\left[\left(1 + 1/3\lambda_{mn}^2\right)\right]^{1/2} = 0 \tag{3.80}$$

The third tone of the force frequency is:

$$\theta^* = 2/3 f_{min}\left[\left(1 - \frac{9\lambda_{min}^2}{8 \pm 9\lambda_{min}}\right)\right]^{1/2} = 0 \tag{3.81}$$

3.6.5 Experimental Investigation

The moduli of normal and shear elasticity for a hybrid aluminum frame and thermoplastic skin was selected as a relation of 80% aluminum, 20% PPS resin, and carbon fiber plain satin weave. For example,

$$E_1 = 0.8E_{al} + 0.2E \text{ pps/cf or } G_{12} = G_{al} + G \text{ pps/cf} \tag{3.82}$$

$E_1 = E_2 = 248.6 \times 10^6$ GPA (2.48 $\times 10^6$ kg/cm^2; $G_{12} = 103.6 \times 10^6$ GPA (1.04 $\times 10^6$) kg/cm^2. Carbon fibers made with PPS resin are described in [47,48]. The challenge of reducing aluminum weight of airframe by adding carbon nanotubes is represented in [49].

Aluminum engineering properties data are presented in Table 3.15 [53].

Thickness (H) of the wings varied from 7.5 to 10.0 cm, and the length of wing selected was equal to 800 to 1000 cm. Geometrical parameter $k = L/H$

TABLE 3.15

Properties of Aluminum Oxide

94% Aluminum Oxide			
	Units of Measure	**SI/Metric**	**Imperial**
Mechanical			
Density	g/cm^3 (lb/ft^3)	3.69	230.4
Porosity	% (%)	0	0
Color	–	White	–
Flexural strength	MPa (lb/in^2 ×10^3)	330	47
Elastic modulus	GPa (lb/in^2 ×10^6)	300	43.5
Shear modulus	GPa (lb/in^2 ×10^6)	124	18
Bulk modulus	GPa (lb/in^2 ×10^6)	165	24
Poisson's ratio	–	0.21	0.21
Compressive strength	MPa (lb/in^2 ×10^3)	2100	304.5
Hardness	kg/mm^2	1175	–
Fracture toughness, K	MPa m$^{1/2}$	3.5	–
Maximum use temperature (no load)	°C (°F)	1700	3090
Thermal			
Thermal conductivity	W/m K (Btu in/ft^2 h °F)	18	125
Coefficient of thermal expansion	10^{-6}/°C (10^{-6}/°F)	8.1	4.5
Specific heat	J/kg K (Btu/lb °F)	880	0.21
Electrical			
Dielectric strength	ac—kV/mm (V/mil)	16.7	418
Dielectric constant	@ 1 MHz	9.1	9.1
Dissipation factor	@ 1 kHz	0.0007	0.0007
Loss tangent	@ 1 kHz	–	–
Volume resistivity	Ω cm	>10^{14}	–

TABLE 3.16

Three Modes for Natural Frequencies

Coefficient $K = L/h$	Sym Bending Stiffness, D' (kg cm)	Three Mode of Vibrations, f_1, f_2, f_3 (Hz)
4	23.26×10^4	3.05
		12.18
		27.4
8.3	25.4×10^4	3.36
		13.45
		30.24
16.6	46×10^4	6.09
		24.36
		54.81
33.3	92.56×10^4	12.31
		49.25
		110.79

varied from 8.3 to 33.3. Parameter k and bending stiffness have a critical influence on three modes. Bending stiffness $D_1 = D_2 = 206 \times 10^6$ kg cm, $D_6 = 86.6 \times 10^6$ kg cm, and $D_3 = D_1 + 2D_6 = 379.2 \times 10^6$ kg cm.

We designate sym $D' = (D_1 + 2D_3k^2 + D_2k^4)^{1/2}$. The fiber density that was selected was $g = 1.77 \times 10^4$ kg/cm³. The acceleration due to gravity was $\eta = 981$ cm/s². Frequencies of free vibrations for the three modes are presented in Table 3.16.

We calculate the force vibrations of wings using Equations (3.79, 3.80, and 3.81) (see Table 3.17). The variable coefficient λ_{mn} for practical calculus is equal to 1.5.

TABLE 3.17

Three Modes for Force Frequencies

Coefficient $K = L/h$	Three Modes of Natural Frequencies, f_1, f_2, f_3 (Hz)	Force Frequencies, Equation (3.79) (Hz)	Force Frequencies, Equation (3.80) (Hz)	Force Frequencies, Equation (3.81) (Hz)
4	3.05	15.25/−3.05	20.17/−4.03	9.42/0.8
	12.18	60.9/−12.18	80.5/−16.1	37.6/0.32
	27.4	137/−27.4	181.5/−36.25	84.6/0.7
8.3	3.36	16.8/−3.36	222/−4.45	10.15/0.008
	13.45	67.25/−13.45	88.9/−17.8	41.5/0.35
	30.24	156.2/−31.24	206.6/−41.3	93.36/0.798
16.6	6.09	30.45/−6.09	40.28/−8.05	18.8/0.16
	24.36	121.8/−24.36	161.1/−32.2	75.2/0.064
	54.81	274/−55.0	362.5/−72.76	169.3/1.44
33.3	12.31	61.6/−12.3	81.4/−16.28	38.0/.32
	49.25	246.25/−49.25	325.7/−65.15	152/1.3
	110.79	559/−111.8	740/−147.9	342.2/2.92

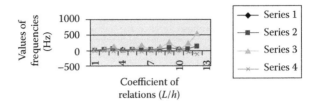

FIGURE 3.17
Natural and force frequencies (Equation 3.79).

Figures 3.17, 3.18, and 3.19 show that the values of natural and force frequencies calculated using Equations (3.79), (3.80), and (3.81) change depending on relations L/h.

3.6.6 Conclusions

1. Hybrid structural elements for the wings and fuselage of an Airbus A340-600 was investigated. A proposal has been established to use an aluminum frame and CNTs with PPS resin as the structural elements of the wings.

2. A hybrid aluminum frame with carbon nanotubes reduces the weight of the wings and bending stiffness by 20%. As a result, there was less natural vibration and a reduction in the risk of parametric resonance in risky situations such as lightning and hurricanes.

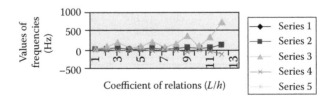

FIGURE 3.18
Natural and force frequencies (Equation 3.80).

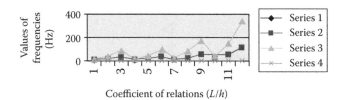

FIGURE 3.19
Natural and force frequencies (Equation 3.81).

3.7 Dynamic Aspects of the Lattice Structures Behavior in the Manufacturing of Carbon–Epoxy Composites

3.7.1 Introduction

Lattice structures that are made of carbon fiber reinforced plastics (CFRPs) have been used for interstage structures in launch vehicles. These lattice structures have been found under different combinations of applied compressive loads and bending moments, changeable temperature, and moisture.

The purpose of this research was to predict critical failure loads under stress variable parameters and strength characteristics of carbon fiber. Additionally the mechanical vibration was calculated and reduced due to maintenance of the appropriate electrical circuit mode used.

Carbon–epoxy composite materials have been used extensively in upper-stage structures of satellite vehicles to improve payload performance and reduce costs. The payload is improved by the lightweight but high compressive strength of carbon–epoxy composite structures. For example, the interstage structure connecting the third-stage rocket to the payload of the Japanese H-Z launch vehicle is a triangular-lattice cylinder that is made of CFRPs.

The lattice structure is referred to as the payload attachment fitting. The success of lattice cylinders for this type of application is essentially a result of their relatively high strength/weight ratio as compared to that of semi-monocoque cylinders, and the capability to accommodate the mounting of equipment such as electric and pyrotechnic devices. The efficiency of a CFRP lattice cylinder increases as the number of fibers in each of its layers that are aligned with the longitudinal layer axis increases. In the study of Hou and Gramoll [54], a new filament winding fabrication method for lattice shells was disclosed and had a significant potential for improving the performance of lattice cylinders that are used for spacecraft applications (see Figure 3.20).

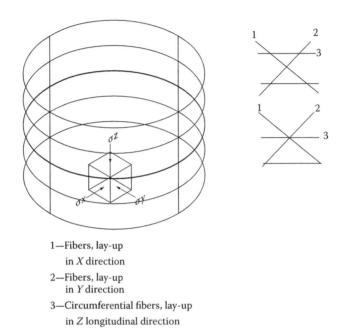

1—Fibers, lay-up
in X direction

2—Fibers, lay-up
in Y direction

3—Circumferential fibers, lay-up
in Z longitudinal direction

FIGURE 3.20
Lattice helical structure.

3.7.2 Theoretical Investigation

In this work we used experimental strength parameters of two families of lattice cylinders: triangular and hexagonal lattices. During launch, vehicles with the lattice cylinder triangular and hexagonal structures have been loaded by axial forces and bending moments. The specimens are subjected to normal axial and longitudinal stress and interlaminar shear stresses.

In a past study [55], we introduced strength criteria for anisotropic materials for the triaxial stress conditions in tensor form. These criteria can be used separately for tensile and compression loads.

Expanding Equation (3.84), the criterion of strength for triaxial stress conditions is obtained as the polynomial of the fourth order (see Section 3.2).

Normal variable stresses are σ_x, σ_y, σ_z, and shear stresses τ_{xy}, τ_{yz}, τ_{zx} will occur during a change of outside loads. Coefficients of strength c, b, d, p, r, s, t, and f are relative strengths that are also variable, and depending on the quality of materials are determined experimentally. k_0 is the safety factor for aviation construction and can be taken between 2 or 3. Dynamic aspects of application of these criteria have consisted of quick-change relations between normal σ_x, σ_y, σ_z, and shear stresses τ_{xy}, τ_{yz}, τ_{zx}, which are automatically programmed and manage to reduce these stresses and reduce vibration by the maintenance of the appropriate electrical circuits in opposite directions.

3.7.3 Practical Calculation of Failure Load

In triaxial stress conditions, when axial normal stresses σ_x equal σ_y and longitudinal stresses σ_z are equal and shear stresses $\tau_{xy}, \tau_{yz}, \tau_{zx}$ are zero, we get a value of hydrostatic pressure p:

$$p = \frac{\delta(6.0)^{1/2} |X|}{R(1+c+b+s+t+f)} \tag{3.83}$$

In Equation (3.83), we then substitute σ_x from Equation (3.84):

$$\sigma_x = \sigma_y = \sigma_z = \frac{pR}{\delta} \tag{3.84}$$

where
p = Hydrostatic pressure
R = Middle radius of lattice structure
δ = Middle thickness of lattice structure

The failure load will be determined as the relationship of hydrostatic pressure to square p/S.

By substituting coefficients c, b, s, t, and f for strength parameter σ_b, we get:

$$p = \frac{\delta(6.0)^{1/2} |X|}{R^4 \left\{ \left[\dfrac{X}{X_{xy}^{45}} + \dfrac{2X}{Y_{yz}^{45}} + \dfrac{2X}{Z_{xy}^{45}} \right] - \left[\dfrac{X}{Y} + \dfrac{X}{Z} + \dfrac{X}{S_{xy}} + \dfrac{X}{S_{zx}} - 2 \right] \right\}} \tag{3.85}$$

The value of hydrostatic pressure predictions for two carbon–epoxy composites are shown in Figure 3.21.

3.7.4 Value of Hydrostatic Pressure Predictions for Carbon–Epoxy Composites

Lattice structures work under mechanical vibrations. The appropriate equation of motion following Newton's second law becomes:

$$m\frac{\partial^2 x}{\partial z^2} + c\frac{\partial x}{\partial z} + Q_{11}x = p \sin \Omega t \tag{3.86}$$

where
m = Mass of lattice cylinder
c = Critical damping coefficient
Q_{11} = Stiffness of the lattice cylinder

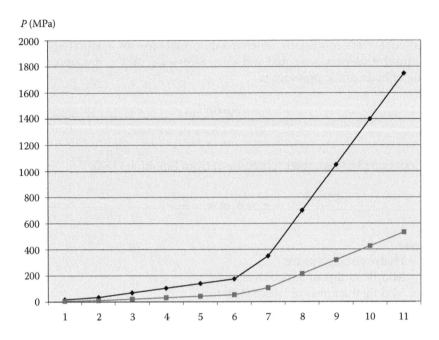

FIGURE 3.21
Value of hydrostatic pressure predictions for carbon–epoxy composites.

p = Hydrostatic pressure from Equation (3.87)
Ω = Forcing frequency
t = Time of wave propagation

The particular solution that applies to the steady-state vibration of the system should be a harmonic function of time such as [56]:

$$x = A\sin(\Omega t - \phi) \tag{3.87}$$

where A and ϕ are constant.
    Substituting x in Equation (3.87), we get:

$$-m\Omega^2 t^2 A\sin(\Omega t - \phi) + c\Omega t A\cos(\Omega t - \phi) + Q_{11}A\sin(\Omega t - \phi) = p\sin\Omega t \tag{3.88}$$

Submitting two boundary conditions:

$$\Omega t - \phi = 0; \quad \text{or} \quad \Omega t - \phi = \pi/2 \quad \text{results in:}$$

$$c\Omega t A = p\sin\Omega t \tag{3.89}$$

$$-m\Omega^2 t^2 A + Q_{11}A = p\sin\Omega t$$

Thus, the magnitude of amplitude changes from A_{1m} to A_{2m}:

$$A_{1m} = \frac{p \sin \Omega t}{c \Omega t}; \quad A_{2m} = \frac{p \sin \Omega t}{-m\Omega^2 t^2 + Q_{11}} \tag{3.90}$$

The electrical circuit for compensation vibration is shown in Figure 3.22 and the critical damping coefficient c for the carbon–epoxy composite and the lattice cylinder can be determined as the relationship between the potential energy W, and the energy lost during one deformation cycle, dW.

$$c = \frac{dW}{W} = m\omega\lambda = m2\pi f^{1-v}\lambda; \tag{3.91}$$

where
  m = Mass of lattice cylinder
  ω = Natural circular frequency; $\omega = 2\pi f^{1-v}$
  λ = Coefficient of internal friction
  f = Frequency of the cycle of variation of the deformation
  v = Exponent dependent on frequency f

According to Bok, $v = 0$, whereas according to Fokht, $v = 1$ [57]. Fokht's hypothesis concerning the proportionality of the nonelastic stress to the frequency is not confirmed by experiment, whereas the Bok hypothesis is in

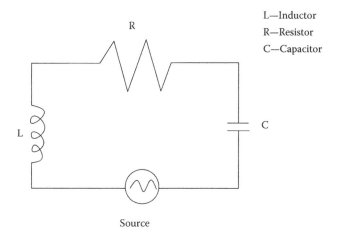

L—Inductor
R—Resistor
C—Capacitor

R

L

C

Source

FIGURE 3.22
Electrical circuit for compensation vibration.

TABLE 3.18

Modulus of Elasticity for Fiberglass in Warp-x and Fill-y Directions

Value (kg/cm²)	Fabrics with Phenol	Fabrics with Epoxy Resin	Fabrics with Epoxy Resin
E_x 10⁶	17.9	36.5	46
E_y 10⁶	13.1	26	16
E_z 10⁶	4.3	12.4	11.2
G_{xy} 10⁶	2.8	8.2	5.6
G_{yz} 10⁶	2.4	6.8	4.3
G_{zx} 10⁶	2.4	6.8	3.3
Poisson's ratio, xy	0.15	0.13	0.27

better agreement with experimental results at least in a rather wide range of frequencies.

Table 3.18 presents the modulus of elasticity for fiberglass in the warp x and fill y directions [58]. In the work of Golfman et al. [58], the critical damping coefficient c (see Table 3.19) for fiberglass for different angles relative to a warp/fill directions was performed. The critical damping coefficient was determined in the process of determining the free vibration of the patterns. The coefficient of internal friction λ was given in the process of testing the fiberglass for durability [58]. The ability of ultrasonic waves to travel in a web direction over a minimum time was established by Golfman [59].

The velocity of ultrasonic wave propagation was determined as:

$$V_0 = L/t \ 10^3 = Lf \ 10^3 \tag{3.92}$$

TABLE 3.19

Vibration Characteristics of Fiberglass

Relations of Fiber in Warp, Fill Directions	Angle (°)	Modulus of Elasticity, $E \times 10^6$ (kg/cm²)	Durability, σ 10^2 (kg/cm²)	Critical Damping Coefficient
1:10	0	46	22	1
	45	18	3.31	4.5
	90	16	2.83	3.7
1:01	0	46	13.6	1.4
	45	18		4.2
1:10	0	35.6	29.2	1.5
	45	21	9.7	9.75
	90	26	6.6	5.2
1:01	0	35.6	9.9	2.4
	45	21	6.9	7.1

where

L = Length between two acoustic heads

t = Time it takes for ultrasonic oscillations to reach from one head to the other

f = Frequency of ultrasonic wave propagation

Dynamic conditions may cause delay in the frequency of ultrasonic waves, because internal heat effects create resistance.

$$V_d = Lf\,10^3\lambda \tag{3.93}$$

The velocity of ultrasonic waves in lattice structures will be:

$$V = V_0 - V_d = Lf\,10^3(1 - \lambda) \tag{3.94}$$

In an earlier study [59], we determined the modulus of elasticity under angle α as:

$$E\alpha = V_\alpha^2\rho(1 - \mu_{1\alpha}\mu_{2\alpha}) \tag{3.95}$$

Substituting Equation (3.94) into Equation (3.95), we obtain the dynamic modulus of elasticity:

$$E\alpha^d = V_\alpha^2\rho(1 - \mu_{1\alpha}\mu_{2\alpha}) = L^2 f^2 10^6 (1 - \lambda)^2 \rho(1 - \mu_{1\alpha}\mu_{2\alpha}) \tag{3.96}$$

3.7.5 Electrical Analogies of Mechanical Vibration

The vibrations in the mechanical system can be reduced if we maintain an appropriate electrical circuit. The source of alternating voltage $E = E_0\sin\Omega t$ generates a current $I = dq/dt$ in the circuit, where q denotes the electric charge. The drop in electric potential (voltage) around the circuit is $L(dI/dt)$ across the inductor, the resistor R, and q/C across the capacitor. The sum of the voltage drops around the circuit equals the applied voltage according to Kirchoff's second law.

$$L\frac{\partial^2 q}{\partial z^2} + R\frac{\partial q}{\partial z} + \frac{1}{C}q = E_0 \sin\Omega t \tag{3.97}$$

where L is the inductor, R is the resistor, and C is the capacitor. Equation (3.99) is identical in mechanical form since L, R, and C are constants.

The mechanical–electrical analogs can be established by comparing these two equations [60]. The particular solution that applies to the steady-state vibration of the system should be a harmonic function of time, such as:

$$q = A\sin(\Omega t - \phi) \tag{3.98}$$

We find the first and second derivatives:

$$\partial q/\partial z \text{ and } \partial^2 q/\partial z^2$$

$$\partial q/\partial z = A\Omega t \cos(\Omega t - \phi) \tag{3.99}$$

$$\partial^2 q/\partial z^2 = -A\Omega^2 t^2 \sin(\Omega t - \phi)$$

Submitting Equations (3.100) and (3.101) into Equation (3.99), we get the amplitude from the electrical circuit.

$$A_e = \frac{E_0 \sin \Omega t}{L\Omega^2 t^2 \sin(\Omega t - \phi) - R\Omega t \cos(\Omega t - \phi) + 1/C \sin(\Omega t - \phi)} \tag{3.100}$$

Submitting two boundary conditions:

$$\Omega t - \phi = 0 \text{ or } \Omega t - \phi = \pi/2$$

in Equation (3.102), we get:

$$A_{e1} = \frac{E_0 \sin \Omega t}{-R\Omega t}; \quad A_{e2} = \frac{E_0 \sin \Omega t}{L\Omega^2 t^2 + 1/C} \tag{3.101}$$

In the case of the transfer of mechanical energy into electrical energy

$$A_{m1} = A_{e1}; \quad A_{m2} = A_{e2}$$

The source of alternative voltage will be found as:

$$E_{01} = -\frac{pR}{c}; \quad E_{02} = \frac{p(L\Omega^2 t^2 + 1/C)}{-m\Omega^2 t^2 + Q_{11}} = \frac{p(L\Omega^2/f^2 + 1/C)}{-m\Omega^2/f^2 + Q_{11}} \tag{3.102}$$

In Equation (3.102), we substitute time t for natural frequency f and as a result we find a material properties structure: mass of lattice cylinder (m), critical damping coefficient (c), stiffness of lattice cylinder (Q_{11}), and vibration characteristics: forcing frequency (Ω) and natural frequency (f).

Now again following Equation (3.102), a value for the hydrostatic pressure (p) correlates with the alternative voltage (E_{01}, E_{02}) and the vibration of lattice cylinders will be reduced by managing the inductance (L), resistance (R), and reciprocal of capacitance ($1/C$).

Natural frequency f can be found by solving the differential equation for an orthotropic lattice cylinder:

$$\frac{\partial^2 f}{\partial t^2} + \frac{g}{h\eta}\left(D_1 \frac{\partial^4 f}{\partial x^2} + 2D_3 \frac{\partial^4 f}{\partial x^2 \partial z^2} + D_2 \frac{\partial^4 f}{\partial z^2}\right) = 0 \qquad (3.103)$$

where D_{ij} is the stiffness of the lattice cylinder from the bending moment.

$$D_1 = Q_{11}\frac{h^3}{12}; \quad D_2 = Q_{22}\frac{h^3}{12}; \quad D_3 = \frac{h^3}{12}(Q_{12} + 2Q_{66}) \qquad (3.104)$$

Stiffness constants Q_{ij} are determined from the following equation:

$$Q_{11} = \frac{E_1}{1 - \mu_{12}\mu_{21}}; \quad Q_{22} = \frac{E_2}{1 - \mu_{21}\mu_{12}}; \quad Q_{12} = \frac{\mu_{21}E_1}{1 - \mu_{12}\mu_{21}} = \frac{\mu_{12}E_2}{1 - \mu_{21}\mu_{12}}$$

$$Q_{66} = G_{12}$$

where
E_1 = Modulus of elasticity in warp-x direction
E_2 = Modulus of elasticity in fill-y direction
$\mu_{12}\mu_{21}$ = Poisson's ratio (parameters E and μ are presented in Table 3.18)
h = Height of the lattice cylinder
g = Density of the fiber
η = Acceleration due to gravity

Vibration characteristics of fiberglass are given in Table 3.19.
In the case of free vibration lattice cylinders we use the boundary conditions:

- If $x = 0$; $x = R$; $f = 0$;

$$\frac{\partial^2 f}{\partial x^2} + \mu_{21}\frac{\partial^2 f}{\partial z^2} = 0 \qquad (3.105)$$

- If $z = 0$; $z = h$; $f = 0$;

$$\frac{\partial^2 f}{\partial x^2} + \mu_{12}\frac{\partial^2 f}{\partial z^2} = 0 \qquad (3.106)$$

where R is the outside radius of the lattice cylinder and h is the height of the cylinder.

These boundary conditions are known by the function of deflections [61]:

$$f_{mn} = \sin\frac{m\pi x}{R}\sin\frac{n\pi y}{h} \tag{3.107}$$

Here, m and n are whole digits that are determined as a number of semi-waves in the x, z directions.

By inputting Equations (3.106) and (3.110) in Equation (3.105) to designate $k = R/h$; k is the present geometrical parameter (relationship of radius cylinder to height).

We can determine natural frequencies f_{mn} as:

$$f_{mn} = \frac{\pi^2}{h^2}\left(\frac{g}{h\eta}\right)^{1/2}\left[D_1\left(\frac{m}{k}\right)^4 + 2D_3 n^2\left(\frac{m}{k}\right)^2 + D_2 n^4\right]^{1/2} \tag{3.108}$$

The frequency of the basic tone ($m = 1$, $n = 1$) will be:

$$f_{11} = \frac{\pi^2}{R^2}\left(\frac{g}{h\eta}\right)^{1/2}(D_1 + 2D_3 k^2 + D_2 k^4)^{1/2} \tag{3.109}$$

The frequency of the second tone ($m = 2$, $n = 2$) will be:

$$f_{22} = \frac{4\pi^2}{R^2}\left(\frac{g}{h\eta}\right)^{1/2}(D_1 + 2D_3 k^2 + D_2 k^4)^{1/2} \tag{3.110}$$

The frequency of the third tone ($m = 3$, $n = 3$) will be:

$$f_{33} = \frac{9\pi^2}{R^2}\left(\frac{g}{h\eta}\right)^{1/2}(D_1 + 2D_3 k^2 + D_2 k^4)^{1/2} \tag{3.111}$$

3.7.6 Experimental Investigation

The main failure mode of an actual lattice attached fitting includes both the global buckling of the total structure and the failure due to shear stress.

Hou and Gramoll [54] showed that the results of testing of conical lattice structures were not stable. The low failure was due to microbuckling and is commonly referred to as fiber kinking. Fiber kinking generally occurs because of a weak matrix, which is due in part to the epoxy not curing completely or a deficiency of the hardening agent during the manufacturing process.

The fiber density that was selected was $g = 1770$ kg/m³.

The acceleration due to gravity was $\eta = 9.81$ m/s².

D_{ij} was a stiffness of lattice cylinder from the bending moment.

$D_1 = 145.2$ kg/m², $D_2 = 50.57$ kg/m², and $D_3 = 120.9$ kg/m².

All parameters we were used for natural frequencies determination (Equations 3.109–3.111). The value of the natural frequencies for lattice cylinders depends on the variation of the geometrical parameters $k = 1, 1/2, 1/3$, which are shown in Figure 3.23.

$k = R/H$	f_{11}
1	435.6
2	300
3	273.8
	f_{22}
1	1742
2	1200
3	1095.2
	f_{33}
1	3920.4
2	2700
3	2464.2

Correlated natural frequencies of lattice cylinders are f_{11}, f_{22}, f_{33}. Geometrical parameters change from $k = 1; k = 1/2 ; k = 1/3$.

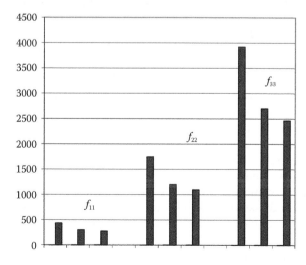

FIGURE 3.23
Correlated natural frequencies of lattice cylinders.

3.7.7 Conclusions

1. This work tried to predict a failure load in lattice cylinders fabricated from carbon–epoxy composites.

2. This work also attempted to develop a methodology to reduce vibration that was capable of activating an electrical circuit that transferred mechanical energy to electrical energy.

3. Finally, this work attempted to determine the natural frequencies of free vibration of lattice cylinders and the knowledge of force frequencies to help designers avoid resonance phenomena for lattice carbon–epoxy structures.

3.8 Dynamic Stability of the Lattice Structures Manufacturing of Carbon Fiber–Epoxy Composites, Including the Influence of Damping Properties

3.8.1 Introduction

Lattice structures that are made of carbon fiber–reinforced plastics have been used in aerospace and interstage structures in launch vehicles. These lattice structures have been found under different combinations of applied compressive loads and bending moments, changeable temperatures, and moisture. Flexibility of cylindrical and conical carbon fiber reinforced models increase fatigue strength of shells.

The purpose of this research was to investigate the dynamic conditions behavior lattice cylinders manufacturing from carbon fiber–epoxy composites. Also, our task was to evaluate the force and free vibration, including non-elastic resistance and damping properties, variable parameters, and strength characteristics of carbon fiber. Additionally, the mechanical vibration was calculated and created a system that registered the force frequencies and reduced the free vibration due to the maintenance of the appropriate electrical circuit and transformed electrical signals to ultrasound signals, which was beneficial in avoiding parametric resonance.

Carbon–epoxy composite materials have been used extensively in upper-stage structures of satellite vehicles to improve payload performance and reduce costs. The payload is improved by the light weight but also the high compressive strength of the carbon–epoxy composite structures. For example, the interstage structure connecting the third-stage rocket to the payload of the Japanese H-Z launch vehicle is a triangular-lattice cylinder that is made of CFRPs.

The lattice structure is referred to as the payload attachment fitting. The success of lattice cylinders for this type of application is essentially the result of their relatively high strength/weight ratio as compared to that of semi-monocular cylinders, and the capability to accommodate the mounting of equipment such as electric and pyrotechnic devices. The efficiency of a CFRP lattice cylinder increases as the number of fibers in each of its layers is aligned with the longitudinal layer axis. In the work of Hou and Gramoll [54], a new filament winding fabrication method for lattice shells was disclosed and had significant potential for improving the performance of lattice cylinders that are used for spacecraft applications.

The lattice helical structure is shown in Figure 3.20, where the fibers lay-up in 1, 2, and 3 directions. Experimental tests were done to investigate the failure behavior of both cylindrical and conical composite lattice structures in static conditions [54].

The dynamic aspects of the lattice structures behavior in the manufacturing of carbon–epoxy composites were investigated by Golfman [62].

3.8.2 Critical Damping Coefficients

Nonelastic resistances are different in different directions. The energy that absorbs and disperses heat is also different in different directions. This phenomenon in mechanics was described by the logarithmic coefficient of attenuations or critical damping coefficients.

In Table 3.20 [58], vibration characteristics of fiberglass materials with different fiber relations and the critical damping coefficients are shown.

The attenuations of vibration of the lattice structures can be characterized by the tensor of force order, where a number of independent critical damping coefficients was equal to 4 [61].

TABLE 3.20

Vibration Characteristics of Fiberglass Materials

Fiber Relations, x, y Axis	Angle (°)	Modulus of Elasticity, $E \times 10^6$ (kg/mm²)	Durability, $\sigma_1 \times 10^2$, (kg/cm²)	Logarithmic Decrement of Attenuation, δ (%)
1:10	0	4.31	22.0	1.02
	45	1.77	3.31	4.5
	90	2.01	2.83	3.7
1:01	0	3.18	13.6	1.4
	45	1.88		4.2
Fabric material 1:10	0	4.2	29.2	1.5
	45	2.1	9.7	9.75
	90	2.06	6.6	5.2
1:01	0	3.12	9.9	2.4
	45	1.89	6.88	7.1

$$\delta_{1111}; \delta_{1122}; \delta_{1212}; \delta_{2222} \qquad (3.112)$$

If the system coordinates rotate on angle φ, we can create six independent, not equal damping coefficients that can be determined through the principal decrements [58]

$$\delta'_{1knm} = \sum_p \sum_q \sum_m \sum_s \alpha_{ip}\alpha_{kq}\alpha_{nm}\alpha_{ms}\delta_{pqms} \qquad (3.113)$$

where $p, q, m, s = 1,2$;

$$\alpha_{11} = \cos\varphi; \ \alpha_{12} = \sin\varphi; \ \alpha_{21} = -\sin\varphi; \ \alpha_{22} = \cos\varphi.$$

The damping coefficient δ_{1122} has lower value than the other three and is therefore reduced to zero. So including this assumption, Equation (3.113) will be as shown below:

$$\delta'_{1111} = \delta_{1111}\cos^4\varphi + \delta_{2222}\sin^4\varphi + \delta_{1212}\sin^2 2\varphi$$

$$\delta'_{2222} = \delta_{1111}\cos^4\varphi + \delta_{2222}\sin^4\varphi + \delta_{1212}\sin^2 2\varphi$$

$$\delta'_{1122} = (\delta_{1111} + \delta_{2222} - 4\delta_{1212})\sin^2\varphi\cos^2\varphi$$

$$\delta'_{1212} = 1/4(\delta_{1111} + \delta_{2222})\sin^2 2\varphi + \delta_{1212}\cos^2 2\varphi \qquad (3.114)$$

$$\delta'_{1222} = 1/2(\delta_{2222}\cos^2\varphi - \delta_{1111}\sin^2\varphi)\sin 2\varphi - \delta_{1212}\sin 4\varphi$$

$$\delta'_{1222} = 1/2[(\delta_{2222}\cos^2\varphi - \delta_{1111}\sin^2\varphi)\sin 2\varphi - \delta_{1212}\sin 4\varphi]$$

$$\delta'_{2111} = 1/2[(\delta_{2222}\cos^2\varphi - \delta_{1111}\sin^2\varphi)\sin 2\varphi - \delta_{1212}\sin 4\varphi]$$

In this case, we need to know the logarithmic decrements of samples under angles $\varphi = 0°$, $45°$, and $90°$.

The first equation of the system pace (Equation 3.114) results in:

$$\delta'_{1111}(\varphi = 0°) = \delta_{1111}$$

$$\delta'_{1111}(\varphi = 90°) = \delta_{2222}$$

$$\delta'_{1111}(\varphi = 45°) = \delta_{45} = \frac{1}{4}(\delta_{1111} + \delta_{2222}) + \delta_{1212} \qquad (3.115)$$

Thus,

$$\delta_{1212} = \delta_{45} - \frac{1}{4}(\delta_{1111} + \delta_{1212})$$

Therefore, if we know damping coefficients along the warp, fill, and diagonal directions, δ_{1212} can be characterized as the shear deformation attenuation.

3.8.3 Theoretical Investigation

We now consider the case of lattice cylinders manufacturing from carbon fiber–epoxy composites, where the linear relation on the principal stresses is aligned with basic warp and fill directions.

The equation for dynamic stability, which includes the effect of attenuation as shown in [61]:

$$\nabla_1^4 \nabla_2^4 \Phi + \frac{c}{K}\frac{d^4\Phi}{d\alpha^4} - \frac{P(t)}{KQ_{1111}h} \times \frac{d^2}{d\alpha^2}\left(\nabla_1^4\Phi\right) + \frac{\rho R^2}{KQ_{1111}} \times \frac{d^2}{dt^2}\left(\nabla_1^4\Phi\right) + R(\Phi)\frac{R^2}{KQ_{1111}}$$

(3.116)

where ρ is the density of carbon fiber material and ∇_1^4 and ∇_2^4 are differential operators:

$$\nabla_1^4 = \frac{d^4}{d\alpha^4} + b\frac{d^4}{d\alpha^2 d\theta^2} + a\frac{d^4}{d\theta^4}$$

$$\nabla_2^4 = \frac{d^4}{d\alpha^4} + d\frac{d^4}{d\alpha^2 d\theta^2} + a\frac{d^4}{d\theta^4}$$

(3.117)

Coefficients a, b, c, and d were determined as:

$$a = \frac{Q_{2222}}{Q_{1111}}; \quad b = \frac{Q_{2222}}{Q_{1111}} - 2\frac{Q_{1122}}{Q_{1111}} - \frac{Q_{1122}^2}{Q_{1111}\,Q_{1212}}$$

$$c = \frac{Q_{2222}}{Q_{1111}} - \frac{Q_{1122}^2}{Q_{1111}^2}; \quad d = 2\frac{Q_{1122} + Q_{1212}}{Q_{1111}}$$

(3.118)

Here, Q_{iknm} are the stiffness constants for components are satisfied by the symmetrical conditions:

$$Q_{iknm} = Q_{kinm} = Q_{ikmn} = Q_{nmik}$$

(3.119)

Ten primary stiffness constants were formulated in Sections 1.2 and 3.4. Stress function Φ in Equation (3.112) is shown as:

$$\Phi(\alpha,\theta,t) = \sum_{\lambda=1}^{\alpha}\sum_{n=1}^{\alpha} P_{\lambda n}(t)(\sin \lambda\alpha \sin \theta n)$$

(3.120)

where
  $P(t)$ = Pulse load acting perpendicular to the top surface of the lattice cyl-
     inder
  t = Time of loading force
  θ = Angular coordinate in tangential direction
  α = Coordinate in radial direction
  λ = Digital number of semiwaves in hoop direction
  n = Digital number of semiwaves in helical direction

Following the assumption of viscosity resistance $R(\Phi)$, which is equal to:

$$R(\Phi) = \frac{\rho f}{\pi} \frac{\partial}{\partial t}\left[\delta'_{iknm} \frac{\partial^4}{\partial \chi i \partial \chi k \partial \chi n \partial \chi m}\right] \tag{3.121}$$

where
  f = Natural frequency
  ρ = Density of carbon fiber material
  δ'_{iknm} = Damping coefficients (see Equation 3.113)
  K = Relationship of chord length L to height h, and $K = L/h$

3.8.4 Free Vibration of the Lattice Structures

The natural frequency f for a lattice cylinder can be found by solving the differential equation for an orthotropic structure [61]:

$$\frac{\partial^2 \omega}{\partial t^2} + \frac{g}{h\eta}\left(D_1 \frac{\partial^4 \omega}{\partial x^2} + D_3 \frac{\partial^4 \omega}{\partial x^2 \partial z^2} + D_2 \frac{\partial^4 \omega}{\partial z^2}\right) = 0 \tag{3.122}$$

where D_{ij} is the stiffness of the buckling lattice cylinder from the bending
moments.

$$D_1 = Q_{11}\frac{h^3}{12}; \quad D_2 = Q_{22}\frac{h^3}{12}; \quad D_3 = \frac{h^3}{12}(Q_{12} + 2Q_{66}) \tag{3.123}$$

Stiffness constants Q_{ij} are determined from Equation (2.2), where
  h = Height of the panel
  g = Density of the fiber
  η = Acceleration due to gravity
  L = Length of the chord

In the case of the free vibration of the lattice cylinder, we can use the same
boundary conditions that we used as the function of deflections (Section 3.7).
We can determine natural frequencies f_{mn} for the basic tone from Equation

(3.109), for the second tone from Equation (3.110), and for the third tone from Equation (3.111). The natural frequencies change when the ultimate load increases from 4000 to 10,000 lb (see Figure 3.24).

The phenomena of losses due to dynamic stability and the appearance of parametric resonance in the lattice structures includes by coincidence the free as well as forced frequencies. In the panels are the compressed force in the vertical direction by force p_{mn} shown as:

$$p_{mn} = p_0 \sin\theta t - p_0 \cos\theta t \qquad (3.124)$$

It is therefore required to solve the Matue equation relative to deflection:

$$\frac{\angle^2 \omega}{\angle t^2} + \omega_{min}^2 \left(1 - \frac{P_0}{P_{min}} \cos Yt \right) \omega(t) = 0 \qquad (3.125)$$

Here, f_{min} are the frequencies of free vibration determined by Equations (3.109) through (3.111). P_{min} is a critical value of the compressed force determined by Hou and Gramoll [54].

The mechanical lattice structure consists of point masses and structural spring elements, which are shown in Figure 3.25.

The critical value of the compressed force can be determined using Equation (3.126):

$$p_{mn} = \frac{\pi^2}{h^2} \left(\frac{g}{h\eta} \right)^{1/2} \left[D_1 \left(\frac{m_4}{k} \right) + 2D_3 n^2 \left(\frac{m_2}{k} \right) + D_2 n^4 \right] (1 + \delta) \qquad (3.126)$$

Ultimate load (lb)

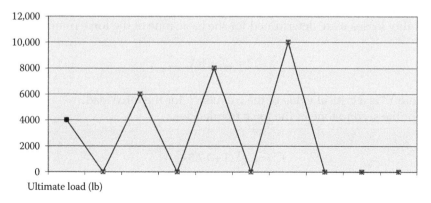

FIGURE 3.24
The natural frequencies change force vibration of the lattice structures.

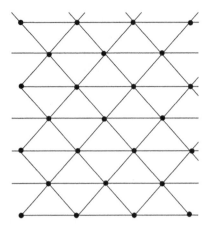

FIGURE 3.25
Mechanical lattice structure consisting of point masses and spring elements.

where $k = L/h$ and δ is the damping coefficient. Now, we input a coefficient:

$$\lambda_{mn} = \frac{p_0}{2p_{\min}} \tag{3.127}$$

Equation (3.127) can be shown as:

$$\frac{\angle^2 \omega}{\angle t^2} + \omega_{\min}^2 (1 - 2\lambda \cos Yt)\omega(t) = 0 \tag{3.128}$$

In the work of Goldenblat and Kopnov [61], the boundary conditions for stability losses were determined for the basic tone of the force frequencies:

$$Y^* = 2f_{mn}(16\lambda_{mn}) \tag{3.129}$$

where Y^* is a critical value of the frequency for the force load.
For the second tone of the force frequencies:

$$Y^* = f_{mn}\left[(1 + 1/3\lambda_{mn}^2)\right]^{1/2}$$

or $\tag{3.130}$

$$Y^* = f_{mn}\left[(1 - 2\lambda_{mn}^2)\right]^{1/2}$$

For the third tone of the force frequencies:

$$Y^* = \frac{2}{3} f_{mn} \left[\left(1 - \frac{9\lambda_{mn}^2}{869\lambda_{mn}} \right) \right]^{1/2} \tag{3.131}$$

The damping coefficient:

$$\delta = \frac{f}{2\pi} x \frac{a^0 n^4 + b^0 n^2 \lambda^2 + c^0 \lambda^4}{an^4 + bn^2 \lambda^2 + \lambda^4} \tag{3.132}$$

Here, the coefficients are:

$$a^0 = a\delta_{2222}; \ b^0 = 2b\delta_{1212}; \ c^0 = \delta_{1111} \tag{3.133}$$

For the basic tone frequencies with attenuation, if $n = 1$ and $\lambda = 1$.

$$\text{Coefficient } \delta = \frac{f}{2\pi} x \frac{a^0 + b^0 + c^0}{a + b + 1} \tag{3.134}$$

Solving Equation (3.128) as:

$$T(t) = X(t) \ Y(t) \tag{3.135}$$

where $X(t)$ and $Y(t)$ are the time functions, which are not known. Equation (3.135) is represented as:

$$X''Y + XY'' + 2\delta XY' + f^2(1 - 2j\cos\Omega t)XY'(Y' + \delta Y) = 0 \tag{3.136}$$

It is required that in Equation (3.136), the coefficient attached to X' be zero, and the result is two differential equations:

$$X''Y + f^2(1 - 2j\cos\Omega t)XY \ XY'' + 2\delta XY' = 0 \tag{3.137}$$

$$Y' + \delta Y = 0 \tag{3.138}$$

We seek Y from Equation (3.139):

$$Y = Ce^{-\delta t} \tag{3.139}$$

If we replace Y in Equation (3.136), we get:

$$X'' + f^2 \left(1 - \frac{\delta}{f^2} - 2j\cos\Omega t \right) X = 0 \tag{3.140}$$

Here, f and Ω are natural and force frequencies. Equation (3.140) is a Matue equation that differs from the basic vibration equation only by an additional member for the attenuation effect.

$$f_\delta = (f^2 - \delta^2)^{1/2} \tag{3.141}$$

3.8.5 Mechanical Vibrations with Damping Effect

The reader may need to refer to the earlier discussion of lattice structures and Newton's second law of motion in Section 3.7.4 and Ref. [60].

In Figure 3.26, the model accelerometer (pos. 2) is installed on the lattice structure (pos. 1). The sensor uses a shear-mode, piezoceramic element that generates an ultrasound signal that goes throughout the linear actuator (pos. 3) and transforms to the electrical circuit board. The critical damping coefficient δ for the carbon–epoxy composite of the lattice cylinder can be determined as the relationship between the potential energy, W, and the energy lost during one deformation cycle, dW.

$$\delta = \frac{dW}{W} = m\omega\lambda = m2\pi f^{1-v}\lambda \tag{3.142}$$

where
  m = Mass of the lattice cylinder
  ω = Natural circular frequency; $\omega = 2\pi f^{1-v}$
  λ = Coefficient of internal friction
  f = Frequency of the cycle of variation of the deformation
  v = Exponent that is dependent on frequency f

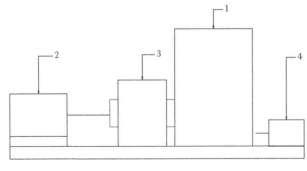

1. Lattice structure
2. Model accelerometer
3. Linear actuator
4. Electrical circuit board
   (PCB INTEL)

FIGURE 3.26
Scheme of transform ultrasound signal to electrical signal.

According to Bok, $v = 0$, whereas according to Fokht, $v = 1$ [57]. Fokht's hypothesis concerning the proportionality of the novelistic stress to the frequency is not confirmed by experiment, whereas the Bok hypothesis is in better agreement with experimental results at least in a rather wide range of frequencies. The modulus of elasticity for fiberglass in the warp-x and fill-y directions is shown in Table 3.20 [62].

Also shown is the critical damping coefficient δ for fiberglass for different angles relative to a warp/fill directions. The critical damping coefficient was also determined in the process of determining the free vibration of the patterns. The coefficient of internal friction λ was found in the process of testing the fiberglass for durability.

The ability of ultrasonic waves to travel in a web direction over a minimum time was also established by Golfman [59]. The velocity of ultrasonic wave propagation was determined as:

$$V_0 = \frac{L}{t} 10^3 = Lf 10^3$$

(3.143)

where
L = Length between the two acoustic heads
t = Time it takes for ultrasonic oscillations to reach from one head to the other
f = Frequency of ultrasonic wave propagation

Under the real dynamic conditions, the internal friction resist ultrasound wave propagation and we have delay time propagation.

$$V_d = Lf 10^3 \lambda$$

(3.144)

The velocity of ultrasonic waves in lattice structures will be:

$$V = V_0 - V_d = Lf 10^3 (1 - \lambda)$$

(3.145)

We determined the modulus of elasticity under angle α as:

$$E\alpha = V_\alpha^2 \rho (1 - \mu_{1\alpha}\mu_{2\alpha})$$

(3.146)

Substituting Equation (3.145) into Equation (3.146), we found the dynamic modulus of elasticity [41]:

$$E\alpha^d = V_\alpha^2 \rho (1 - \mu_{1\alpha}\mu_{2\alpha}) = L^2 f^2 \, 10^6 (1 - \lambda)^2 \rho (1 - \mu_{1\alpha}\mu_{2\alpha})$$

(3.147)

3.8.6 Electrical Analogies of Mechanical Vibration

The vibrations in the mechanical system can also be reduced if we maintain an appropriate electrical circuit. The source of alternating voltage $E = E_0\sin\Omega t$ generates a current $I = dq/dt$ in the circuit, where q denotes the electric charge. The drop in electric potential (voltage) around the circuit is $L(dI/dt)$ across the inductor, resistor R, and q/C across the capacitor. The sum of the voltage drops around the circuit and equals the applied voltage according to Kirchhoff's second law:

$$\sum_{L=1}^{L=n} L\frac{\partial^2 q}{\partial t^2} + \sum_{R=1}^{R=n} R\frac{\partial q}{\partial t\, t} + \sum_{C=1}^{C=n} C\frac{1}{q} = E_0\sin\Omega t \tag{3.148}$$

Equation (3.149) is identical in mechanical form since L, R, C are constants. Here,

$$\sum_{L=1}^{L=n} L \text{ is the sum of inductors, } \sum_{R=1}^{R=n} R \text{ is the sum of resistors,}$$

$$\text{and } \sum_{C=1}^{C=n} C \text{ is the sum of capacitors}$$

The mechanical–electrical analogs can be established by comparing Equations (3.87) and (3.149).

The particular solution that applies to the steady-state vibration of the system should be a harmonic function of time, such as:

$$q = A\sin(\Omega t - \phi) \tag{3.149}$$

We find the first and second derivative:

$$\partial q/\partial t \text{ and } \partial^2 q/\partial t$$

$$\partial q/\partial t = A\Omega t\cos(\Omega t - \phi) \tag{3.150}$$

$$\partial^2 q/\partial t^2 = -A\Omega^2 t^2\sin(\Omega t - \phi)$$

By substituting Equation (3.150) into Equation (3.148), we get the amplitude from the electrical circuit.

$$A_e = \frac{E_0\sin\Omega t}{\displaystyle\sum_{L=1}^{L=n} L\Omega^2 t^2\sin(\Omega t - \phi) - \sum_{R=1}^{R=n} R\Omega t\cos(\Omega t - \phi) + 1\Big/\sum_{C=1}^{C=n} C\sin(\Omega t - \phi)} \tag{3.151}$$

Substituting two boundary conditions:

$$\Omega t - \phi = 0 \text{ or } \Omega t/2$$

into Equation (3.150), we get:

$$A_{e1} = \frac{E_0 \sin \Omega t}{-\sum R \Omega t}; \quad A_{e2} \frac{E_0 \sin \Omega t}{\sum L \Omega^2 t^2 + 1/\sum C} \qquad (3.152)$$

In the case of the transfer of mechanical energy into electrical energy, $A_{m1} = A_{e1}$ and $A_{m2} = A_{e2}$. The source of alternative voltage can be found as:

$$E_{01} = -\frac{pR}{\delta}; \quad E_{02} = \frac{p(\mathring{a}L\Omega^2 t^2 + 1/\mathring{a}C)}{-m\Omega^2 t^2 + Q_{11}} = \frac{p(\mathring{a}L\Omega^2/f^2 + 1/\mathring{a}C}{-m\Omega^2/f^2 + Q_{11}} \qquad (3.153)$$

In Equation (3.148), we can substitute time t for natural frequency f and as a result we find a material properties structure: mass of lattice cylinder (m), critical damping coefficient (δ), stiffness of lattice cylinder (Q_{11}) and vibration characteristics: forcing frequency (Ω), and natural frequency (f).

Now again following Equation (3.153), a value for the hydrostatic pressure (p) correlates with the alternative voltage (E_{01}, E_{02}) and the vibration of lattice cylinders can be reduced by managing the sum of the inductances (L), resistances (R), and reciprocal of capacitances ($1/C$).

The electrical circuit for electronic countermeasures compensation vibration is shown in Figure 3.27.

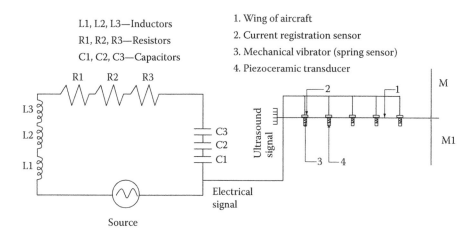

FIGURE 3.27
Active vibration cancellation system.

TABLE 3.21

h (cm)	R (cm)	L (cm)	γ (grad/rad)	D_1 (kg cm²)	D_2 (kg cm²)	D_3 (kg cm²)
0.25	40	10	30/52	0.06×10^6	0.029×10^6	1.62×10^6
0.50	50	20	40/69	0.518×10^6	0.237×10^6	0.255×10^6
0.75	60	30	50/87	1.74×10^6	0.79×10^6	0.852×10^6
1.0	70	40	60/1.04	4.14×10^6	1.9×10^6	2.0×10^6
1.25	80	50	70/1.22	8.1×10^6	3.7×10^6	4.0×10^6
1.50	90	60	80/1.39	14×10^6	6.4×10^6	6.93×10^6
1.75	100	70	90/1.57	22.19×10^6	10.17×10^6	10.97×10^6

The electrical signal is transformed to the ultrasound signal by piezo-ceramic transducer (pos. 1). The current registration sensor (pos. 2) passes the ultrasound signal to the current graphite provider element (pos. 3) filament, winding simultaneously in process fabrication the lattice structure (pos. 4). Therefore we can reduce the free vibration frequencies acting directly on lattice structure.

Optical fibers are transmitted via electrical signals to a computer, and vibrations are reduced by managing the circuit board parameters.

3.8.7 Experimental Investigation

The main failure mode is determined in a method similar to that described in Section 3.7.6.

The basic stiffness parameters for carbon epoxy shells are shown in Table 3.21.

All the parameters that were used were determined for natural frequencies. According to Golfman et al. [58], the damping coefficient δ_{1111} in warp direction was 0.01825, δ_{2222} in fill direction was 0.01933, and δ_{1212} in 450° diagonal direction was 0.02. The vibration characteristics of carbon fiber epoxy are shown in Figure 3.28.

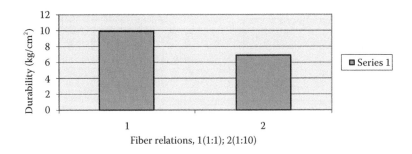

FIGURE 3.28
Vibration characteristics of carbon–fiber epoxy.

The active vibration cancellation system is connected to cancel vibration to create an opposite vibration moment and transform electrical circuit to a mechanical vibration system for wing aircraft and is shown in Figure 3.26. The moment of mechanical vibration of wing M (pos. 1) has been transformed to electrical signals by sensors (pos. 2). Electrical signals with the same parameters (frequencies, amplitudes, and time domain) will reach mechanical vibrators (pos. 3) and the equivalent opposite moment M_1 will be created.

3.8.8 Conclusions

1. The natural frequencies changed when the ultimate load increased from 4000 to 10,000 lb.

2. The natural frequencies of lattice cylinders changed when geometrical parameters changed from $k = 1; f_{11} = 500$ 1/s; $k = 1/2; f_{22} = 1000$ to 1500 1/s; $k = 1/3; f_{33} = 2500$ to 4000 1/s.

3. Damping effect significantly reduces the force frequency. However, frequencies of force vibration with a changing relation to the length of chord to the height of the cylinder from 1/2.5, 1/5, 1/10 specifically reduced 2.2 times and coincided with free vibration frequencies. Specifically, it appeared that when the relation k was 1/10, the height of the skin cylinder increased by four times.

4. The natural frequencies of free vibration of the lattice cylinders and the force frequencies was determined to help designers avoid resonance phenomena for lattice carbon–epoxy structures.

5. A methodology to reduce vibration which was capable of activating an electrical circuit that transferred mechanical energy to electrical energy was developed and helped designers reduce vibration by 50%.

6. A system that registers the force frequencies and reduces the free vibration due to transform electrical signals to ultrasound signals and cancellation vibration was created.

References

1. Shopper, G. A., and D. B. Curliss. 2002. Model-based design for composite materials life management. Paper number AIAA-2002-5516. Proceedings of the 9th AIAA/ISSMO Symposium on Multidisciplinary Analysis and Optimization Conference, September 4–6, 2002, Atlanta, GA.

2. Schoeppner, G. A., G. P. Tandon, and K. V. Pochiraju. 2008. Predicting thermo-oxidative degradation and performance of high temperature polymer matrix composites. In *Multiscale Modeling and Simulation of Composite Materials and Structures*, ed. Y. Kwon, D. H. Allen, and R. Talreja. New York: Springer-Verlag.

3. Schoeppner, G. A., G. P. Tandon, and E. R. Ripberger. 2007. Anisotropic oxidation and weight loss in PMR-15 composites. *Composites Part A: Applied Science and Manufacturing* 38 (3): 890–904.

4. Bowles, K. J., and G. Nowak. 1988. Thermo-oxidative stability studies of Celion 6000/PMR-15 unidirectional composites, PMR-15, and Celion 6000 fiber. *Journal of Composite Materials* 22 (10): 966–985.

5. Parvatareddy, H., J. Z. Wang, D. A. Dillard, T. C. Ward, and M. E. Rogalski. 1995. Environmental aging of high high-performance polymeric composites: effects on durability. *Composites Science and Technology* 53 (4): 399–409.

6. Parvatareddy, H., P. H. Wilson Tsang, and D. A. Dillard. 1996. Impact damage resistance and tolerance of high-performance polymeric composites subjected to environmental aging. *Composites Science and Technology* 56 (10): 1129–1140.

7. Bowles, K. J. 1998. Comparison of graphite fabric reinforced PMR-15 and avi-mid N composites after long term isothermal aging at various temperatures. NASA/TM-1998-107529.

8. Miller, S., D. Papadopoulos, P. Heimann, L. Inghram, and L. McCorkle. 2007. Graphite sheet coating for improved thermal oxidative stability of carbon fiber reinforced/PMR-15 composites, *Composites Science and Technology* 67 (10): 2183–2190.

9. Bortman, J., and B. A. Szabo. 1992. Nonlinear models of fastened structural connections. *Computers and Structures* 43: 909–923.

10. Lambert, J. C., and B. J. Merritt. 1995. Automated stress analysis—reducing stress analysis time by an order of magnitude. *MSC 1995 World Users' Conf. Proc.* Paper No. 41, May 1995.

11. Hyer, M. W. 1987. Effects of pin elasticity, clearance, and friction on the stresses in a pin-loaded orthotropic plate. *Journal of Composite Materials* 21 (3): 190–206.

12. AD-TR-61-153. Load Deflection Characteristics of Joints. Appendix B, pp. 158–170.

13. Lekhnitskii, C. G. 1968. *Anisotropic Plates*, 2nd ed. (Translated from Russian by S. W. Tsai and T. Cheron). New York: Gordon and Breach Science Publications.

14. Golfman, Y. 1991. Strength criteria for anisotropic materials. *Journal of Reinforced Plastics and Composites* 10 (6): 542–556.

15. Icardi, U., and L. Fererro. 2008. A comparison among several recent criteria for the failure analysis of composites. *Journal of Advanced Materials* 40 (4): 73–111.

16. Mises, R. 1928. Mechanik Der Plastischen Formanderunk von Kristallen Zeitschrift Fur Angew. *Math & Mech* B8: H3. Berlin.

17. Hill, R. 1950. *The Mathematical Theory of Plasticity*. Oxford University Press.

18. Hill, R. 1963. Elastic properties of reinforced solids; some theoretical principles. *Journal of the Mechanics and Physics of Solids* 11: 357–372.

19. Tsai, S. W., and E. M. Wu. 1971. A general theory of strength for anisotropic materials. *Journal of Composite Materials* 5: 58–80.

20. Azzi, V. D., and S. W. Tsai. 1965. Anisotropic strength of composites. *Experimental Mechanics* 5: 283–288.

21. Z. Hashin. 1987. Analysis of orthogonally cracked laminates under tension. *Journal of Applied Mechanics* 54: 872–879.

22. Pagano, N. J., and S. R. Soni. 1980. Strength analysis of composite turbine blades. *Journal of Reinforced Plastics Composites* 7: 558–581.
23. Ashkenazi, E. K. 1965. Problems of the anisotropy of strength. *Mechanics of Polymers* 1 (2): 79–82.
24. Ashkenazi, E. K. 1968. *Strength of Wood in Complicated Stress Analysis.* Leningrad Forest Academy.
25. Ashkenazi, E. K., and F. P. Pekker. 1968. Leningrad: Forest Academy.
26. Goldenblat, I. I., and V. A. Kopnov. 1965. Strength criteria for anisotropic materials. *Mechanica IMZGA, Izvestia of Science Academy USSR* 6: 77–83.
27. Goldenblat, I. I., and V. A. Kopnov. 1968. *Criteria of Strength and Plasticity of Construction Materials.* Moscow: Mashinostroenie.
28. Malmeister, A. 1966. Geometry of theories of strength. *Mechanics of Polymers, MKPLA* 2 (4): 324–326, 519–534.
29. Golfman, Y., E. K. Ashkenazi, L. P. Roshcov, and N. P. Sedorov. 1974. Machine components of fiberglass for ship building. *Shipbuilding Issue,* Leningrad.
30. Timoshenko, S., and J. N. Goodier. 1957. *Theory of Elasticity.* New York, NY: McGraw-Hill Book Company.
31. Rabinovich, A. L. 1946. On elastic constants and strength of anisotropic materials. Central Aerodynamic Institute, No. 582.
32. Kerchtein, I. M., P. D. Stepinov, and P. M. Ogebilov. 1969. In question of evaluation anisotropic the momentary and the availability strength fiberglass. *Mechanics of Polymers, MKPLA* 2: 243–247.
33. Pappo, A., and I. Ivenson. 1972. Strength anisotropic materials in triaxial stress conditions. *Rocket Technic and Cosmonautic World* 4: 128–137.
34. Parton, V. Z., and Perlin, P. L. 1984. Mathematical methods of the theory of elasticity. *Magazine World.*
35. Golfman, Y. 1991. Strength criteria for anisotropic materials. *Journal of Reinforced Plastics and Composites* 10 (6): 542–556.
36. Soebroto, H., and F. K. Ko. 1989. Composite preform fabrication by 2-D braiding. In *Proceedings of the Fifth Annual ASM/ESD Advanced Composite Conference,* 307–316. Dearborn, MI, September 25–28.
37. Redman, C. J., and C. D. Douglas. Theoretical prediction of the tensile elastic properties of braided composites. 38th International SAMPE Symposium, Anaheim, CA, May 10–13, 1993.
38. Soebroto, H., D. S. Hager, C. Pastore, and F. K. Ko. 1990. Engineering design of braided structural fiberglass composites. In *Proceedings of the 35th International SAMPE Symposium,* 687–696. April 2–5, 1990.
39. Rozen, B. W., S. N. Chatterjee, and J. J. Kibler. 1977. An analysis model for spatially oriented fiber composites. *Composite Materials: Testing and Design,* ASTM STP 617.
40. Golfman, Y. 1993. Ultrasonic non-destructive method to determine modulus of elasticity of turbine blades. *SAMPE Journal* 29 (4): 31–35.
41. Golfman, Y. 2001. Non-destructive evaluation of aerospace components using ultrasound and thermography technologies. *Journal of Advanced Materials* 33 (4): 21–26.
42. Golfman, Y. 2002. Non-destructive evaluation of parts for hovercraft and ekranoplans. *Journal of Advanced Materials* 34: 3–7.
43. Murioz, R. S., and R. A. Lopez Anido. 2009. Monitoring of marine grade composite double plate joints using embedded fiber optic strain sensors. *Journal of Advanced Materials* 89 (2): 224–234.

44. Beck, M. 2006. Thermoplastic composite materials, low cost manufacturing methods for thermoplastics composites. Boston Chapter SAMPE Meeting.
45. Ticona performance driven solution replaces aluminum in Airbus Wing Fortron polyphenelene sulfide in composites. SAMPE Meeting, 2009. http://www.ticona.com.
46. Premium Aerotec in Augsburg receives autoclave for fuselage structure of new Airbus A350XWB. *JEC Composites Weekly E Letter N.315 Germany Report*. March 3, 2010.
47. Favaloro, M. 2000. Fibers made from Fortran® PPS. SAMPE Meeting, Boston, MA.
48. Buck, M. E. 2004. Thermo-Lite™ Thermoplastic Composites, Phoenix TPC. Boston SAMPE Meeting.
49. Hinrichsen, J., and C. Bautista. 2001. The challenge of reducing both airframe weight and manufacturing cost. *Air and Space Europe* 3 (3–4): 119–121.
50. Nohara, L. B., E. L. Nohara, A. Moura, M. R. P. Goncalves, M. L. Costa, and M. C. Resende. 2006. Study of crystallization behavior of poly (phenylene sulfide). *Polimeros* 16 (2): 104–142.
51. Golfman, Y. 2010. Dynamic stability of jetliners structures manufacturing of fiber/epoxy composites. *Journal of Advanced Materials* 42 (1): 28–39.
52. Imbirchumun, C. A., and A. A. Chichiturum. 1960. About stability and vibration of reinforced composite plates. *A.N. Ezvestia USSR, Mechanics & Mashinostroenia* 1: 113–122.
53. Aluminum Oxide (Al$_2$O$_3$) Properties. http:/www.accuratus.com/alumox.html.
54. Hou, A., and K. Gramoll. 2000. Fabrication and compressive strength of the composite attachment fitting for launch vehicle. *Journal of Advanced Materials* 32 (1): 39–45.
55. Golfman, Y. 1991. Strength criteria for anisotropic materials. *Journal of Reinforced Plastics & Composites* 10 (6): 542–556.
56. Golfman, Y. 2001. Fiber draw automation control. *Journal of Advanced Materials* 34 (2): 35–40.
57. Kushul, M. Y. 1964. *The Self-Induced Oscillations of Rotors*. Consultants Bureau, New Jersey.
58. Golfman, Y., E. Ashkenazi, P. Roshcov, and N. Sedorov. 1974. *Machine Components of Fiberglass for Shipbuilding*. Leningrad: Shipbuilder.
59. Golfman, Y. 1993. Ultrasonic non-destructive method to determine modulus of elasticity of turbine blades. *SAMPE Journal* 29 (4): 31–35.
60. Sandor, B. I., and K. J. Richter. 2000. Engineering Mechanics Statics & Dynamics, 2nd ed. Englewood Cliffs, NJ: Prentice Hall.
61. Goldenblat, I. I., and V. A. Kopnov 1968. *Resistance of Fibreglass*. Moscow: Masshinostroenie.
62. Golfman, Y. 2003. Dynamic aspects of the lattice structures behavior in the manufacturing of carbon–epoxy composites. *Journal of Advanced Materials* 35 (2): 3–8.

4

Interlaminar Shear Stress Analysis

4.1 Interlaminar Shear Stress Analysis of Composites: Carbon Fiber–Epoxy Sandwich Structures

4.1.1 Introduction

Developing effective low-cost methods for improving interlaminar shear strength of two-dimensional (2-D) laminate composites by reinforcement matrix polymers using carbon or fiberglass is the purpose of this research.

The 2001 American Airlines Flight 587 Airbus A300 carbon fiber–epoxy composite tail failure has not yet been fully evaluated in terms of material, manufacturing, and structural load aspect.

In the last 10 years, the composite tail sandwich structures used have been carbon–epoxy prepreg systems (BMS 8256 or BMS 8212) for the skin panels and standard Nomex® honeycomb (BMS 8124) for the core [1].

It was shown by Zeng et al. [2] that in the process of curing, the acting shear forces appeared and created crush conditions. The honeycomb core crush number (HCCN) is maximized when HCCN approaches 1.0, while at the low level of core crush, the HCCN approaches 0.0.

The scheme of forces acting in the core honeycomb is shown in (Figure 4.1).

Early attempts to manufacture sandwich structures with honeycomb inside were performed in 1969. The microstructure was improved [3] when the author did preliminary work on the polymerization of glass–fiber honeycomb. The first curing of glass–fiber honeycomb that was produced using a short regime that had a low pressure of 50 kg/cm^2, with a time of polymerization of 2 min on 1 mm of thickness on the section of the structure, and a temperature of 140°C. The second curing of skin layers with preliminary polymerization of the honeycomb structure had an improved microstructure and increased strength capability on the order of 20% to 30%.

The contour of a profile panel is shown in Figure 4.2, where glass–fiber prepreg was wound on removable mandrels.

The physical and mechanical properties of the carbon fiber–epoxy material Quantum Composites Lytex® 4149 55% carbon fiber–epoxy are shown in Table 4.1 [4]. The honeycomb core material properties and quartz fiber

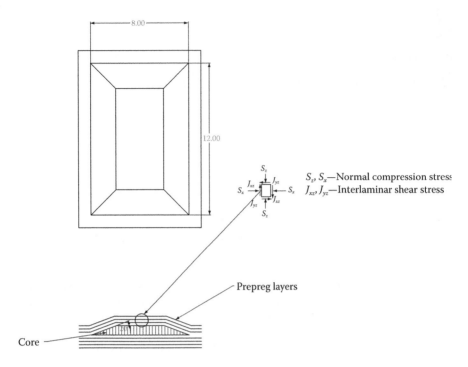

FIGURE 4.1
Boeing core panel design.

composite material properties, Astroquartz II, are shown in Tables 4.1, 4.2, and 4.3 [4].

4.1.2 Theoretical Investigation

The differential equation that determined stress function F for the leading and trailing panels is shown as [5]:

$$\frac{1}{G_{yz}^s + G_{yz}^h}\frac{\partial^2 F}{\partial x^2} + \frac{1}{G_{xz}^s + G_{xz}^h}\frac{\partial^2 F}{\partial y^2} = \frac{2Q\left(\mu_{zy}^s + \mu_{zy}^h\right)yQ}{\left(E_z^s + E_z^h\right)I} - \frac{\varphi'(Y)+C}{2\left(g_{xz}^s + G_{xz}^h\right)I} \quad (4.1)$$

where

E_z^s, E_z^h = Modulus of normal elasticity for skin layers (s) and honeycomb (h)

I = Moment of inertia for skin and honeycomb layers

F = Stress function is acting in flexural bending and twisting of the leading and trailing panels

μ_{zy}^s and μ_{zy}^h = Poisson's ratio for skin layers and honeycomb

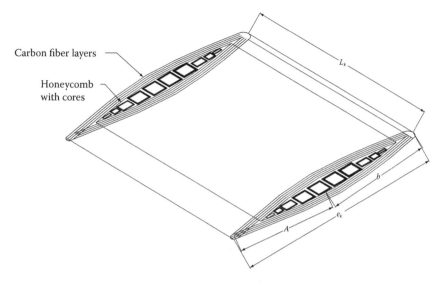

FIGURE 4.2

The contour of a profile panel, with core manufactured using a removable mandrel. Removable mandrels create an empty space which allows for cooling during flight.

TABLE 4.1

Carbon Fiber Epoxy Material Properties

	Metric	**English**	**Comments**
Physical Properties			
Density	1.45 g/cm³	0.0524 lb/in³	ASTM D792
Mechanical Properties			
Tensile strength, ultimate	289 MPa	41,916 psi	ASTM D638
Elongation at break	0.487%	0.487%	ASTM D638
Modulus of elasticity	55.1 GPa	7992 ksi	In tension ASTM D638
Flexural modulus	34.5 GPa	5004 ksi	ASTM D690
Flexural yield strength	613 MPa	88,908 psi	ASTM D790
Compressive yield strength	275 MPa	39,885 psi	ASTM D695
Shear modulus	2.9 GPa	421 ksi	Interlaminar, ASTM D5379
Shear modulus	11 GPa	1595 ksi	In plane, ASTM D5379
Shear strength	65.5 MPa	9500 psi	Interlaminar, ASTM D5379
Shear strength	206 MPa	29,878 psi	In plane ASTM D5379
Impact strength, Izod notched	9.6 J/cm	18 ft lb/in	Notched, ASTM D5379
Compressive modulus	31.7 GPa	4598 ksi	ASTM D695

TABLE 4.2

Honeycomb Core Material Properties

	Metric	English	Comments
Physical Properties			
Density	0.032 g/cm³	0.32 g/cm³	Core density
Mechanical Properties			
Shear modulus	0.331 GPa	48 ksi	Ribbon direction
Shear strength	1.14 MPa	165 psi	Ribbon direction
Compressive yield strength	1.41 MPa	204 psi	

Q = External dynamic load that has been acted upon in the center of the point of hydrodynamic stress function

φ' = Derivative of profile function

C = Constant of a twist determination

If the lateral surfaces are free from all external forces, the stress function F for the contour will be equal to zero and all cross sections will satisfy the boundary conditions.

From Equation (4.1), the values of the stress function F along the boundary can be calculated for all cross sections. The shear stresses for the skin plates can be found using the following equation:

$$\tau_{xz}^{s} = \frac{\partial F}{\partial y} - \frac{Q}{2I_{s}}\left[x^{2} - \varphi(y)\right]$$

$$\tau_{yz}^{s} = \frac{\partial F}{\partial x}$$

(4.2)

TABLE 4.3

Quartz Fiber Composite Core Material Properties (Astroquartz II)

	Metric	English	Comments
Physical Properties			
Density	0.048 g/cm³	0.048 g/cm³	Core density
Mechanical Properties			
Modulus of elasticity	0.172 GPa	25 ksi	In compression
Shear modulus	0.117 GPa	17 ksi	Ribbon direction
Shear strength	1.28 MPa	186 psi	Ribbon direction
Compression yield strength	1.55 MPa	225 psi	

The shear stresses for the honeycomb can be found using Equation (4.2), where F is a stress function and $\varphi(y)$ is a function of profile; I_s and I_h are moments of inertia for skin layers and honeycomb layers.

Now it follows that:

$$\lambda = \frac{G_{YZ}^s + G_{YZ}^h}{G_{XZ}^s + G_{XZ}^h}; \quad q = \frac{2\left(G_{YZ}^s + G_{YZ}^h\right)\left(\mu_{zy}^s + \mu_{zy}^h\right)}{E_z^s + E_z^h}; \quad B = \frac{Q}{2I} \tag{4.3}$$

From here on, λ and q will be called the coefficients of anisotropy for a sandwich carbon fiber structure. Therefore, Equation (4.1) can be given as:

$$\frac{\partial^2 F}{\partial x^2} + \lambda \frac{\partial^2 F}{\partial y^2} = q \frac{Q}{I} y - \lambda B \varphi'(y) + C \tag{4.4}$$

Thus, the contour of profile for the leading and trailing panels is represented as:

$$f(y) = \lambda B \varphi(y), \text{ to } f'(y) = \lambda B \varphi'(y) \tag{4.5}$$

Finally, Equation (4.4) can be shown as:

$$\frac{\partial^2 F}{\partial x^2} + \lambda \frac{\partial^2 F}{\partial y^2} = q \frac{Q}{I} y - f'(y) + C \tag{4.6}$$

Now, we introduce a new stress function, $\phi(u,v)$ in which

$$u = \lambda/x, \ v = y, \text{ and } F(x,y) = \phi(x/\lambda,y) \tag{4.7}$$

After Equation (4.6) is differentiated with respect to x and y, the result is:

$$\frac{\partial^2 F}{\partial x^2} = \lambda \frac{\partial^2 \phi}{\partial u^2}; \quad \frac{\partial^2 F}{\partial x^2} = \frac{\partial^2 \phi}{\partial u^2} \tag{4.8}$$

In Equation (4.6), stress function F is replaced with a new stress function ϕ using the differential expression in Equation (4.8)

$$\frac{\partial^2 \phi}{\partial u^2} + \lambda \frac{\partial^2 \phi}{\partial v^2} = q \frac{Q}{I} v - f'(v) + C \tag{4.9}$$

We can designate:

$$\lambda \phi(u,v) = \phi_1(u,v) \tag{4.10}$$

As a result, we use a new stress function $\varphi_1(u,v)$ and input it into Equation (4.9):

$$\frac{\partial^2 \phi_1}{\partial u^2} + \lambda \frac{\partial^2 \phi_1}{\partial v^2} = q \frac{Q}{I} v - f'(v) + C \tag{4.11}$$

The differential Equation (4.11) was solved in 1935 by Prof. D. Pinov [6] for a symmetrical aviation isotropic profile.

$$\phi_1(u,v) = \frac{Q}{8I} \left[u^2 + av^3 + bv^2 + \frac{1}{3a} \right] * \left[\left(b - 1 + 4q \right) \left(b + 1 \right) - 12C_1 a \right] y + C \left\{ v + \frac{1}{3a} \left(b - 1 + 4q \right) \right] \tag{4.12}$$

In Equation (4.7), a new index stress function u was given by λ/x and y was given by y while in a search for function F, where it was acting in flexure in leading and trailing panels, was found to be:

$$F(x,y) = \frac{Q}{8I} \left[\lambda x^2 + av^3 + bv^2 + \frac{1}{3a} \left\{ \left[\left(b - 1 + 4q \right) \left(b + 1 \right) - 12C_{1a} \right] y + C \right\} \left[v + \frac{1}{3a} \left(b - 1 + 4q \right) \right] \tag{4.13}$$

The first multiplayer is a contour of profile

$$\left[\lambda x^2 + av^3 + bv^2 + \frac{1}{3a} \left[\left(b - 1 + 4q \right) \left(b + 1 \right) - 12C_{1a} \right] y + C \right. \tag{4.14}$$

Here, a, b, C_1, and C are arbitrary coefficients. It is assumed that $a = -k$, and the result is (Equation 4.14)

$$[\lambda x^2 - k(y^3 + \lambda_2 y^2 + \lambda_1 y + \gamma_0)] = 0 \tag{4.15}$$

where $\gamma_0, \gamma_1, \gamma_2$ are designated coefficients of y, and given as:

$$\gamma_2 = \frac{b}{a}; \quad \gamma_1 = \frac{1}{3a} \left[\left(b - 1 + 4q \right) \left(b + 1 \right) - 12C_{1a} \right]; \quad \lambda_0 = \frac{C}{a} \tag{4.16}$$

In Equation (4.15), we replace:

$$y^3 + \gamma_2 y^2 + \gamma_1 y + \gamma_0 = S(y) \tag{4.17}$$

Pinov [6] showed that $S(y)$ can be taken as:

$$S(y) = (y - a)(y - \beta) \quad (\beta < a) \tag{4.18}$$

where a and β are the distance to the edge points of the length profile.

Comparing the coefficients in Equations (4.17) and (4.18) and solving for γ results in:

$$\lambda_2 = -(2a + \beta)$$
$$\lambda_1 = 2a\beta + a^2 \tag{4.19}$$
$$\lambda_0 = -\beta a^2$$

Therefore, if coefficients a and β are known, we can determine coefficients γ_2, γ_1, γ_0 using Equation (4.19), and coefficients a, b, C, and C_1 also can be determined. The contour of profile represented in Equation (4.20) is derived from Equation (4.15) as:

$$\lambda x^2 - k(y - a)^2(y - \beta) = 0 \tag{4.20}$$

The profile function can be described by using Equation (4.20), which was determined for asymmetrical profiles.

$$f_{1,2}(x) = \frac{1}{4}(y_2 - y_1)^2$$

where

$$f_1(x) = a^2 \frac{x}{L_k}\left(1 + \frac{x}{L_k}\right) \quad x < 0$$
$$\tag{4.21}$$
$$f_1(x) = a^2 \frac{x}{L_k}\left(1 - \frac{x}{L_k}\right) \quad x > 0$$

Here, $a = y_2$, y_1 is a thickness of points, and L_k is a width for every cross section (Figure 4.2). In Equation (4.20), the coefficient of profile k can be shown as:

$$k = \frac{27}{16}\frac{e_k^2 \lambda}{(a - \beta)^2} \tag{4.22}$$

where
e_k = Maximum thickness of profile
a = Distance from maximum thickness to exit edge
β = Distance from maximum thickness to entry edge
λ = Coefficient of anisotropy

All designations are given in Figure 4.2.

The minimum set was determined from Equation (4.20):

$$S'(y) = (y - \alpha)(3y - 2\beta - \alpha) = 0 \tag{4.23}$$

The abscissa of maximum is a root of Equation (4.23).

$$y_m = \frac{2\beta + \alpha}{3} \tag{4.24}$$

Obviously the thickness of the contour would be on the 1/3 length from the entrance edge (see Figure 4.2).

$$e_k = \frac{4}{3 * 3^{1/2}} (\alpha - \beta) \left[k / \lambda (\alpha - \beta) \right]^{1/2} \tag{4.25}$$

Therefore, the stress function F includes the equation of aviation symmetrical profiles as expressed in Equation (4.14).

After simplification, the stress function F from Equation (4.14) can be given as:

$$F(x, y) = \frac{Q}{8I\lambda} \left[\lambda x^2 - k(y - \alpha)^2 (y - \beta) \right] \left[y + \frac{1}{3a} (b - 1 + 4q) \right] \tag{4.26}$$

We have determined the derivations of the stress function F:

$$\frac{\partial F}{\partial x} = \frac{Qx}{4I} \left[y + \frac{1}{3a} (b - 1 + 4q) \right]$$

$$\frac{\partial F}{\partial x} = \frac{Q}{8I\lambda} \left\{ -2k(y - \alpha)(y - \beta) \left[y + \frac{1}{3a} (b - 1 + 4q) \right] - k(y - \alpha)^2 \left(y + \frac{1}{3a} (b - 1 + 4q) \right) \right.$$

$$\left. - k(y - \alpha)^2 (y - \beta) + \lambda x^2 \right) \tag{4.27}$$

The interlaminar shear stresses for the skin plate can be found using Equations (4.2) and (4.28).

$$\tau'_{xz} = \frac{Q}{8I\lambda} \left\{ -2k(y - \alpha)(y - \beta) \left[y + \frac{1}{3a} (b - 1 + 4q) \right] - k(y - \alpha)^2 \left(y + \frac{1}{3a} (b - 1 + 4q) \right) \right\}$$

$$- k(y - \alpha)^2 (y - \beta) + \lambda x^2 - \frac{Q}{2I} \left(\lambda x^2 - k(y - \alpha)^2 (y - \beta) \right)$$

$$\tau'_{yz} = -\frac{Qx}{4I} \left(y + \frac{1}{3a} (b - 1 + 4q) \right) \tag{4.28}$$

The interlaminar shear stresses for the honeycomb can be found using Equations (4.3) and (4.28).

$$\tau_{xz}^h = \frac{Q}{8I\lambda}\left\{-2k(y-\alpha)(y-\beta)\left[y + \frac{1}{3a}(b-1+4q)\right] - k(y-\alpha)^2\left(y + \frac{1}{3a}(b-1+4q)\right)\right.$$

$$\left. -\left(k(y-\alpha)^2(y-\beta) + \lambda x^2\right) - \frac{Q}{2I}\left(\lambda x^2 - k(y-\alpha)^2(y-\beta)\right)\right.$$

$$\tau_{yz}^h = \frac{Qx}{4I}\left(y + \frac{1}{3a}(b-1+4q)\right)$$ (4.29)

The relationship between the moment $M(C)$ and the transverse force Q are given as Equation (4.30) and described by Golfman [7,8].

$$\iint (x\tau_{yz} - y\tau_{xz})\partial x\partial y <> 0$$ (4.30)

4.1.3 Experimental Results

We can calculate the maximum and minimum shear stresses in critical points on the boundary between skin and honeycomb layers. We have also determined the minimum shear stresses in a boundary between skin and honeycomb layers considering that the derivation of shear stresses is equal to zero.

$$\frac{\partial \tau_{xz}}{\partial x} = 0; \quad \frac{\partial \tau_{yz}}{\partial x} = 0;$$

$$\frac{\partial \tau_{xz}}{\partial y} = 0; \quad \frac{\partial \tau_{yz}}{\partial y} = 0;$$ (4.31)

The conditions would be satisfied if the boundary between skin and honeycomb layers is satisfied:

$$\frac{\partial^2 \tau_{xz}}{\partial x^2}\frac{\partial^2 \tau_{xz}}{\partial y^2} - \frac{(\partial^2 \tau_{xz})^2}{\partial x \partial y} > 0$$

$$\frac{\partial^2 \tau_{yz}}{\partial x^2}\frac{\partial^2 \tau_{yz}}{\partial y^2} - \frac{(\partial^2 \tau_{yz})^2}{\partial x \partial y} > 0$$ (4.32)

The analysis of Equation (4.32) has shown that both conditions were not satisfied and that inequalities were less than zero. Therefore, the extreme conditions were not determined, and must be found by iteration using a counter or interpolation function. The extreme conditions on the boundary layers were determined using the method of intertwined multiplayer proposed by Lagrange [9].

The following three equations were used.

- The counter equation:

$$\Theta(x,y) = \lambda x^2 - k(y - \alpha)^2(y - \beta) = 0 \qquad (4.33)$$

- The two equations of liaison:

$$\frac{\partial \Phi(x,y)}{\partial x} = \frac{\partial \tau_{xz}}{\partial x} + L \frac{\partial \Theta}{\partial x} = 0$$

$$\frac{\partial \Phi(x,y)}{\partial y} = \frac{\partial \tau_{xz}}{\partial y} + L \frac{\partial \Theta}{\partial y} = 0 \qquad (4.34)$$

where
Φ = Auxiliary function
L = Lagrange multiplayer

From Equation (4.29):
If $x = 0$; $y = \alpha$; τ_{yz}^s, τ_{yz}^h are equal to zero, τ_{xz}^h and τ_{xz}^s are minimum.
If $x = 0$, $y = \beta$; τ_{yz}^s, τ_{yz}^h are equal to zero, τ_{xz}^h and τ_{xz}^s are maximum.

$$\tau_{xz}^s = \frac{Q}{8I\lambda}\left\{ k(\beta - \alpha)^2 \left(\beta + \frac{1}{3a}(b - 1 - 4q) \right) \right\}$$

$$\tau_{xz}^h = \frac{Q}{8I\tau}\left\{ k(\beta - \alpha)^2 \left(\beta + \frac{1}{3a}(b - 1 - 4q) \right) \right\} \qquad (4.35)$$

4.1.4 Conclusions

1. The interlaminar shear stresses that were determined in the leading/trailing panels are within the shape of the symmetrical aviation profiles.

2. The traditional method of distribution of shear stresses with the maximum on the neutral axis and minimum on the contour profile does not work according to the formula by Djuravsky.

3. The interlaminar shear stresses inside the field on the neutral axis were equal to zero and were matched to the contour of the shape of the profile.

4. Stress function was found satisfactory at the boundary conditions. The stress shear was distributed between the prepreg layers and the honeycomb in the case of a simultaneously acting bending moment and a twisting transverse force.

5. Mechanical properties of honeycomb core materials and quartz fiber composites are significant but less than the same of shear strength between carbon fibers layers and honeycomb. So failure first applied in the core, second in boundary between core and carbon fiber layers. Third failure will come in skin carbon fiber layers.

4.2 Interlaminar Shear Strength between Thermoplastics: Rapid Prototyping of Pultruded Profiles and Skin Carbon Fiber–Epoxy Layers

4.2.1 Introduction

The rapid prototyping or pultrusion of thermoplastic profiles opens up new horizons for honeycomb used as mandrels for aviation parts. For example, laser scanning systems digitize the sections requiring a larger volume of data with more than 13,000 points per second. The rapid prototyping and pultrusion of profiles can be used for mandrel manufacturing for wind turbine blades and wind propellers for helicopters and aircraft tails. The purpose of this chapter is to investigate the adhesion forces between rapid prototyping or pultrusion of thermoplastic honeycomb and carbon fiber–epoxy skin layers and estimate the interlaminar shear strength in the border between the skin and honeycomb layers.

A new concept for the design and manufacturing of wing structures that was used for knitting, weaving, braiding, and through-the-thickness stitching for reinforcement textiles was developed in the NASA Advanced Composites Technology program [10]. The braiding process for winding carbon–epoxy layers for the automation and manufacturing of composite wings can use fiber compounds in the mandrel. The compounds are based on milled carbon fiber with as much as 60% carbon fiber in Nylon 6 & 66 and PPS, and at least 40% in others, including PP, ABS, PC, POM, and PSul [11]. The addition of carbon fibers not only increases the flexural modulus (rigidity) of the compound, but also increases the tensile strength.

The resin film infusion process has been identified as a cost-effective fabrication technique for producing damage-tolerant textile composites. This process has been used in the fabrication of a three-dimensional simulated model for complex composites for aircraft stiffened wing structures [12]. Thermoplastic mandrels made from honeycomb for these structures have been shown to reduce costs in this project, which is discussed in this chapter.

4.2.2 Advanced Technology Development

One rapid prototyping process has been developed and is based on shape deposition manufacturing (SDM) techniques [13]. A variety of castable materials can be used to make parts from the molds. Molded SDM was originally developed to make ceramic or polymer parts.

Figure 4.3 shows an example of the mold SDM process sequence for the construction of simple parts that can be made in three layers from CAD systems. Each step in the figure includes one material deposition and one material shaping operation. The process steps are as follows:

1. The first layer of the part contains no undercut features, so the temporary part material is deposited first. It is then shaped to define features on its top surface.
2. The temporary part material is covered with mold material.
3. The second layer is undercut on both sides so that the mold material must be deposited first.
4. The temporary part material is deposited into the cavity left in the mold material. Temporary part material geometry is replicated from the mold material surfaces.
5. The third layer contains an undercut feature on the left and a non-undercut feature on the right. The first step is therefore to deposit the left-hand-side mold material.
6. ... followed by the temporary part material. ...
7. ... and finally the right hand side material.
8. A final layer of mold material is deposited to close the top of the mold cavity. A casting reservoir and sprue are machined into this layer.
9. The temporary mold material is removed by etching to leave the empty mold cavity.
10. The part material is cast into the mold cavity and cured or allowed to set.
11. Remove the mold (11a) before performing finishing operations such as sprue removal (12a) or use the mold materials as a fixture while performing finishing operations (11b). The mold material is removed afterwards (12b). See Figure 4.3.

When traditional fiberglass wing structures are laid up in open molds, each layer of fiberglass is coated with a layer of epoxy resin or vinyl ester resin applied with a spray gun or roller and worked into the laminate with a squeegee. The fiberglass is a combination of stitched directional E-glass and Kevlar (see Figure 4.4).

The best known RTM process is the vacuum resin transfer molding process (VARTM), which controls the distribution of resins through the molds and is

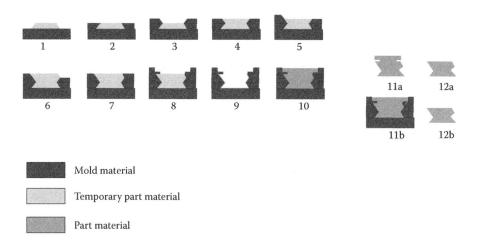

Mold material

Temporary part material

Part material

FIGURE 4.3
Mold material deposition process.

very important. The infusion molding process (SCRIMP), a patented system by Seeman Composites, is performed under high vacuum, whereby all of the air is removed from constructed porous (fibrous usually) dry materials [14]. During and after this process, and after this material is compacted by atmospheric pressure, a resin matrix is introduced to completely encapsulate all the materials within the evacuated area.

The main difference between SCRIMP and vacuum bagged prepreg is that with the SCRIMP method, the fabrics, preforms, and cores are placed in the

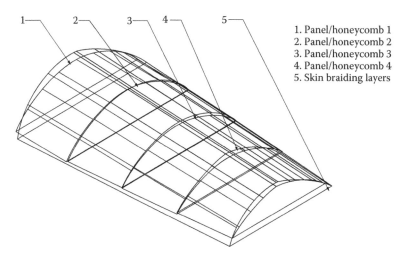

1. Panel/honeycomb 1
2. Panel/honeycomb 2
3. Panel/honeycomb 3
4. Panel/honeycomb 4
5. Skin braiding layers

FIGURE 4.4
Panel assembly wing structure.

mold to dry before the application of any resin, and a high vacuum is used to both compact the laminate and also to draw (infuse) the resin into the composite.

Advanced technology and development has been used to develop wing structures using the above method by a combination of thick skin layers and middle core layers.

The skins together can be thinner than the total thickness of its single-skin counterpart. Cores, however, must be quite thick, and so the total thickness of the cored-skin laminate is more than a single-skin laminate. Most of the core materials provide very good flotation and the thicker core has an attractive side aspect. This makes the core-skin laminates stiffer, and because cores are lightweight, the core-skin laminate weighs less than the single-skin laminates while providing flotation to the aircraft. The most common core materials utilized in building boat hulls and decks include balsa wood, polyvinylchloride (PVC) foam, and styreneacrylonitrile (SAN). Pultruded wing profiles have been created by the combination of carbon–glass composites and thermoplastic resins. Honeycomb matrix materials are presented in Table 4.4, while thermoplastic resin systems are presented in Table 4.5 [16].

Following Tables 4.4 and 4.5, carbon–nylon appears to be a more acceptable material for honeycomb wing construction. Thermoplastic resin system is polyamide (Nylon 6), semicrystalline polymer absorbs moisture and has an average chemical/solvent resistance, low density and high thermomechanical properties. This material has been pultruded for wing profiles.

4.2.3 Stress Analysis

Internal members such as longitudinal and transverse bulkheads support an aircraft wing structural laminate. Bulkhead stiffeners subdivide the wing laminate into panels. Each panel works under air pressure, where air pressure lifts the wing on wave impacts from the air or temporary ground contact if the wing height is just low enough. Under these loads, the panels bend

TABLE 4.4

Honeycomb Composite Materials

Carbon–Nylon	Glass–Nylon	Aramid–Nylon
Carbon–PPS	Glass–PPS	Aramid–PPS
Carbon–PEI	Glass–PEI	Aramid–PEI
Carbon–PEEK	Glass–PEEK	
Carbon–PP	Glass–PP	
Carbon–PMMA	Glass–PMMA	
Carbon–HDPE	Glass–HDPE	

TABLE 4.5

Thermoplastic Resin Systems

Polyamide (PA) (Nylon 6)	Polyethermide (PEI)	Polyphenylene Sulfide (PPS)
Average chemical/solvent resistance	Poor chemical/solvent resistance	Average chemical/solvent resistance
Semicrystalline polymer	Amorphous polymer	Semicrystalline polymer
Density = 1.5 g/cm³	Density = 1.27 g/cm³	Density = 1.35 g/cm³
Absorbs moisture	Average moisture absorption	Very low moisture absorption
Process temperature = 525°F (275°C)	Process temperature = 600°F (315°C)	Process temperature = 625°F (330°C)

Polymethylmethacrylate	Polyetheretherketone (PEEK)	Polypropylene (PP)
Poor chemical/solvent resistance	Excellent chemical/solvent resistance	Fair chemical/solvent resistance
Amorphous polymer	Semicrystalline polymer	Semicrystalline polymer
Density = 1.19 g/cm³	Density = 1.29 g/cm³	Density = 0.91 g/cm³
Very low moisture absorption	Very low moisture absorption	Low moisture absorption
Process temperature = 400°F (205°C)	Process temperature = 725°F (385°C)	Process temperature = 350°F (175°C)

in x, y directions and acting stresses appear with the laminate as a result of lift forces. The panel assemblies are shown in Figure 4.4.

The maximum normal stresses act on the ends of the wings as a result of cutting forces Q_x, Q_y. However, the maximum flexure appears in the end of the wing.

In a single-skin laminate, the supercharge surface is in tension, the suction surface is in compression, and the shear stress appears between skin layers and honeycomb. Tension and compression are generally easy to visualize. Shear is the tendency of the inside surfaces of the skin laminate to slide against each other in opposite directions. Shear is highest in the border between the skin and honeycomb layers.

To resist all these stresses without fracturing, a single-skin laminate must be relatively thick and heavy. In a cored-skin laminate, the outside and inside skins experience the tension and compression stresses, and the core experiences the shear stress. In one study [17], it was shown that maximum shear stress appears between skin laminates and honeycomb.

Therefore, the bending moments and cutting forces can be determined from Equation (4.36), if we postulate for the distribution stresses following the Kirchhoff–Love assumption [18].

It is assumed that the norm in the border of an aircraft structure remains normal and unstretched after deformation.

$$M_x = -\left(D_1 \frac{\partial^2 w}{\partial x^2} + D_{12} \frac{\partial^2 w}{\partial y^2} + 2D_{16} \frac{\partial^2 w}{\partial x \partial y}\right)$$

$$M_y = -\left(D_{12} \frac{\partial^2 w}{\partial y^2} + D_{22} \frac{\partial^2 w}{\partial x^2} + 2D_{26} \frac{\partial^2 w}{\partial x \partial y}\right)$$

$$M_{xy} = -\left(D_{16} \frac{\partial^2 w}{\partial x^2} + D_{26} \frac{\partial^2 w}{\partial y^2} + 2D_{66} \frac{\partial^2 w}{\partial x \partial y}\right) \tag{4.36}$$

$$Q_x = -\left|D_{11} \frac{\partial^3 w}{\partial x^3} + 3D_{16} \frac{\partial^3 w}{\partial x^2 \partial y} + (D_{12} + 2D_{66}) \frac{\partial^3 w}{\partial y x \partial y^2} + D_{26} \frac{\partial^3 w}{\partial y^3}\right|$$

$$Q_y = -\left|D_{16} \frac{\partial^3 w}{\partial x^3} - 3D_{26} \frac{\partial^3 w}{\partial x \partial y^2} + (D_{12} + 2D_{66}) \frac{\partial^3 w}{\partial x^2 \partial y} + D_{22} \frac{\partial^3 w}{\partial y^3}\right|$$

where

M_x = Bending moment acting along the longitudinal axis x

M_y = Bending moment acting along the transverse axis y

M_{xy} = Bending moment acting along the 45° angle relative to axis x, y

Q_x, Q_y = Cutting forces acting in x, y directions

w = Flexure function

The forces acting on the aircraft wing are shown in Figure 4.5.

If directions of the acting force fit with the axis of symmetric elasticity material, the stiffness parameters are:

$$D_{11} = D_1; \; D_{22} = D_2; \; D_{12} + 2D_{66} = D_3; \; D_{16} = D_{26} = 0$$

$$D' = \frac{1}{8}\left(3D_1 + 2D_3 C^2 + 3D_2 C^4\right) \tag{4.37}$$

The stiffness parameters of the wing structure can be determined as:

$$D_1 = \frac{E_{11}^s h_1^3 + E_{11}^h h_2^3}{12\left[1 - \left(\mu_{12}^s \mu_{21}^s + \mu_{12}^h \mu_{21}^h\right)\right]}; \quad D_2 = \frac{E_{22}^s h_1^3 + E_{22}^h h_2^3}{12\left[1 - \left(\mu_{12}^s \mu_{21}^s + \mu_{12}^h \mu_{21}^h\right)\right]};$$

$$D_{12} = \frac{G_{12}^s h_1^3 + G_{12}^h h_2^3}{12\left[1 - \left(\mu_{12}^s \mu_{21}^s + \mu_{12}^h \mu_{21}^h\right)\right]}; \quad D_3 = D_1\left(\mu_{21}^s + \mu_{21}^h\right) + 2D_{12} \tag{4.38}$$

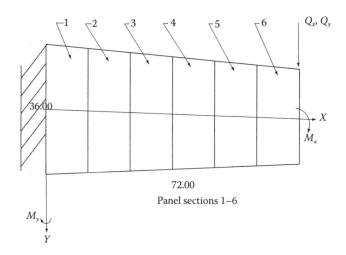

FIGURE 4.5
Forces acting on the aircraft wing.

where

$E^s_{11}, E^s_{22}, G^s_{12}, \mu^s_{12}, \mu^s_{21}$ = Modulus of normal and shear elasticity and Poisson's ratio for skin layers in axial and transverse directions

$E^h_{11}, E^h_{22}, G^h_{12}, \mu^h_{12}, \mu^h_{21}$ = Modulus of normal and shear elasticity and Poisson's ratio for honeycomb in axial and transverse directions

h_1, h_2 = Thickness of skin and honeycomb layers

The precise solution of a flexure aircraft wing if we imagine it as an elliptic plate is [18]:

$$w = \frac{p_x a^4}{64 D'} \left(\frac{x^2}{a^2} - \frac{y^2}{b^2} \right) \qquad (4.39)$$

where

$$D' = \frac{1}{8} \left[3D_{11} + 2(D_{12} + 2D_{66})C^2 + 3D_{22}C^4 \right]; \qquad C - \frac{a}{b} \qquad (4.40)$$

where a is a width section and b is a height section (see Figure 4.4).

We input Equation (4.39) into Equation (4.36) and we obtain bending moments and cutting forces as:

$$M_x = -\frac{p_x a^2 D_1}{16D'}\left[\left(3+(\mu_{12}^s+\mu_{12}^h)C^2\right)\frac{x^2}{a^2}+\left(1+3(\mu_{21}^s+\mu_{21}^h)C^2\right)\frac{y^2}{b^2}1-(\mu_{21}^s+\mu_{21}^h)C^2\right]$$

$$M_y = -\frac{p_y a^2 D_2}{16D'}\left[\left(3(\mu_{12}^s+\mu_{12}^h)C^2\right)\frac{x^2}{a^2}+\left((\mu_{21}^s+\mu_{21}^h)+3C^2\right)\frac{y^2}{b^2}-1-C^2\right]$$

$$M_{xy} = \frac{P_{xy}D_3}{16D'}*\frac{4C}{ab}xy \tag{4.41}$$

$$Q_x = \frac{p_x^a}{8D'}\left\{D_1\left[3-(\mu_{12}^s+\mu_{12}^h)C^2\right]+2D_3C^2\right\}\frac{x}{a}$$

$$Q_y = \frac{p_y^a}{8D'}\left\{D_2(\mu_{12}^s+\mu_{12}^h)+3C^2\right]+2D_3C^2\right\}\frac{y}{b}$$

Now, we add the significant cutting forces and moments M_x, M_y, M_{xy}, Q_x, and Q_y from Equation (4.36) to Equations (4.37) and (4.38) and we get an equation of compatibility.

$$D_1\frac{\partial^4 w}{\partial x^4}+2D_3\frac{\partial^4 w}{\partial x^2\partial y^2}+D_2\frac{\partial^4 w}{\partial y^2}=$$
$$p_z+h\frac{\partial^2\varphi}{\partial y^2}\frac{\partial^2 w}{\partial x^2}-2h\frac{\partial^2\varphi}{\partial x\partial y}\frac{\partial^2 w}{\partial x\partial y}+h\frac{\partial^2\varphi}{\partial x^2}\frac{\partial^2 w}{\partial y^2}+\frac{h}{b}\frac{\partial^2\varphi}{\partial x^2}+\frac{h}{L}\frac{\partial^2 w}{\partial y^2} \tag{4.42}$$

For the composite wing structure $L > b$ and the equation of compatibility deformation, Equation (4.42) can be converted to:

$$a_2\frac{\partial^4\varphi}{\partial x^4}+a_{12}\frac{\partial^4\varphi}{\partial x^2\partial y^2}+a_1\frac{\partial^4\varphi}{\partial y^4}=\left(\frac{1}{L}\frac{\partial^2 w}{\partial x^2}\right) \tag{4.43}$$

and the equation of compatibility for the external forces will also be converted to:

$$a_2\frac{\partial^4\varphi}{\partial x^4}+a_{12}\frac{\partial^4\varphi}{\partial x^2\partial y^2}+a_1\frac{\partial^4\varphi}{\partial y^4}=\left(\frac{1}{L}\frac{\partial^2 w}{\partial x^2}\right) \tag{4.44}$$

Here, L is the length of the wing aircraft and b is the width of the wing aircraft. There are also lateral p_x and p_z equal distribution forces and shear stresses τ_{xy} and τ_{yx}.

The axial flexure function w is:

$$w(x,y) = w_0 w_1(x,y); \text{ and} \tag{4.45}$$

the stress function is the sum of the tensile (compression) stress φ_0 and the function of bending stress φ_b, which is explained by the loss of wing structure stability.

$$\varphi(x,y) = \varphi_0(x,y) + \varphi_b(x,y) \tag{4.46}$$

$$\text{and } \varphi_b(x,y) = \varphi_b^c(x,y)\varphi_b^0 \tag{4.47}$$

where

    $\varphi_b(x,y) =$ Function of bending stresses dependent only on the coordinate system

    φ_b $=$ Function of the bending stresses including only the amplitude vibration

The function of the tensile (compression) stress is represented by the equation:

$$\varphi_0(x,y) = -\frac{1}{2}p_x y^2 - \frac{L}{2h}p_z x^2 + xy\tau_{xy} \tag{4.48}$$

Assuming that $\tau_{xy} = \tau_{yx}$, the tensile (compression) stress depends on longitudinal and transverse deformations; therefore, in Equation (4.48), in the case of compression stress we propose a minus sign.

We input Equations (4.45) through (4.47) into Equation (4.48) and integrate this equation. The result is a system from two equations connecting stress function φ_b^0 and displacement w_0:

$$k_1 \varphi_b^0 = -\frac{w_0}{L}k_2$$

$$k_3 w_0 = -w_0 hPx k_6 - w_0 LPz k_4 - 2hw_0\tau_{xy}k_5 + \frac{h}{L}\varphi_b^0 k_7 \tag{4.49}$$

If we input the first of Equation (4.49) into the second, we get:

$$hk_6 Px + Lk_4 p_z + 2hk_3\tau_{xy} = -k_3\frac{hk_2 k_7}{Lh_1} \tag{4.50}$$

This is the general equation for the investigation of loss stability in wing structure.

Coefficients $k_1, k_2, k_3, k_4, k_5, k_6,$ and k_7 can be found by using the following equation:

$$k_1 = \int_0^L \int_0^{2\pi L} \left(D_2 \frac{\partial^4 \varphi_b^c}{\partial x^4} + D_{12} \frac{\partial^4 \varphi_b^c}{\partial x^2 \partial y^2} + D_1 \frac{\partial^4 \varphi_b^c}{\partial y^4} \right) w_1 \partial x \partial y$$

$$k_2 = \int_0^L \int_0^{2\pi L} \left(\frac{\partial^2 w_1}{\partial x^2} \right) w_1 \partial x \partial y$$

$$k_3 = \int_0^L \int_0^{2\pi L} \left(D_1 \frac{\partial^4 w_1}{\partial x^4} + 2D_3 \frac{\partial^4 w_1}{\partial x^2 \partial y^2} + D_2 \frac{\partial^4 w_1}{\partial y^4} \right) \varphi_b^c \partial x \partial y$$

(4.51)

$$k_4 = \int_0^L \int_0^{2\pi L} \left(\frac{\partial^2 w_1}{\partial y^2} \right) \varphi_b^c \partial x \partial y$$

$$k_5 = \int_0^L \int_0^{2\pi L} \left(\frac{\partial^2 w_1}{\partial x^2 \partial y^2} \right) \varphi_b^c \partial x \partial y$$

$$k_6 = \int_0^L \int_0^{2\pi L} \left(\frac{\partial^2 w_1}{\partial x^2} \right) \varphi_b^c \partial x \partial y$$

4.2.4 Influence of Acting Forces on the Stability of an Aircraft Wing

In the case of the symmetrical form of loss stability, we can show below a simple proposal for the function of flexure and bending stress function. If a wing structure could be bent only in the longitudinal direction and the lateral compression force and shear stresses are absent, we get

$$w(x) = w_0 \sin \frac{m\pi x}{L}; \quad \varphi_b(x) = \varphi_b^c \sin \frac{m\pi x}{L} \tag{4.52}$$

where
  m = Digital number of semiwaves in the longitudinal direction
  n = Digital number of semiwaves in the lateral direction
  $p_x = 0$, and $\tau_{xy} = \tau_{yx} = 0$

Therefore, Equation (4.52) can be replaced by

$$p_x = -\frac{k_3}{hk_6} - \frac{k_2 k_7}{Lk_1 k_6} \tag{4.53}$$

Coefficients $k_1, k_2, k_6,$ and k_7 can be determined from

$$k_1 = D_2 \pi L^2 \left(\frac{m\pi}{L}\right)^4 \; ; \; k_{2,6,7} = \pi L^2 \left(\frac{m\pi}{L}\right)^2 \; ; \; k_3 = D_1 \pi L^2 \left(\frac{m\pi}{L}\right)^4 \quad (4.54)$$

We can replace the above coefficients in Equation (4.53):

$$p_x = \frac{D_1}{h}\left(\frac{m\pi}{L}\right)^2 + \frac{1}{D_2 b}\left(\frac{m\pi}{L}\right)^2 \quad (4.55)$$

Now we can find the critical significance of the longitudinal compression pressure for an orthotropic wing structure by minimizing the parameter $(m\pi/L)$ [18]:

$$p_x = \frac{2}{LD_2}\left(\frac{D_1 D_2}{h}\right)^{1/2} = \frac{h}{L}\frac{(E_{11}E_{22})^{1/2}}{\left\{3\left(1 - \mu_{12}\mu_{21}\right)\right\}^{1/2}} \quad (4.56)$$

In a second example, we consider that the wing structure could be compressed only in the lateral direction and where a longitudinal compression force and shear stresses are absent. We propose that the function of flexure and bending stress function are distributed in nonsymmetrical form.

$$w(x) = w_0 \sin\frac{\pi x}{L} \sin\frac{ny}{b} \; ; \; \varphi_b(x) = \varphi_b^c \sin\frac{\pi x}{L} \sin\frac{ny}{b} \quad (4.57)$$

The critical value of the lateral compression pressure has been done by Golfman [20].

In the case of the acting longitudinal and lateral forces equation where $\tau_{xy} = \tau_{yx}$ equals zero, Equation (4.58) can be assigned as:

$$h\,k_6 Px + Lk_4 p_z = -k_3 - \frac{h}{L^2}\frac{k_2 k_7}{k_1} \quad (4.58)$$

If a wing structure was found under the acting torsion moments $M_k = 2\pi R^2 h\tau_{xy}$, Equation (4.58) is:

$$\tau_{xy} = -\frac{k_3}{2hk_5} - \frac{k_2 k_7}{2L^2 k_1 k_5} \quad (4.59)$$

In the linear performance stability of a wing structure we can approximate the analogy to an isotropic shell [18].

$$w(x,y) = w_0 \sin\frac{\pi x}{L}\sin\frac{n(y-\gamma_x)}{b} \tag{4.60}$$

where γ_x is a tangent of angle fibers that are normal to the x coordinate. For a convenient analysis, we get:

$$w(x,y) = \varphi_b(x,y) = \sin\frac{\pi x}{L} * \sin\frac{n(y-\gamma x)}{b} \tag{4.61}$$

Now, we can calculate all the constants from Equation (4.58) using Equations (4.59) and (4.61).

$$k_1 = \frac{\pi Lb}{2}\left\{ D_2\left[\left(\frac{\pi}{b}\right)^4 + 6\left(\frac{\pi}{b}\right)^2\left(\frac{n\gamma}{L}\right)^2 + \left(\frac{n\gamma}{L}\right)^4\right] + D_{12}\left(\frac{n}{L}\right)^2\left[\left(\frac{\pi}{b}\right)^2 + \left(\frac{n\gamma}{L}\right)^2\right] + D_1\left(\frac{n}{L}\right)^4 \right\} \tag{4.62}$$

$$k_3 = \frac{\pi Lb}{2}\left\{ D_1\left[\left(\frac{\pi}{b}\right)^4 + 6\left(\frac{\pi}{b}\right)^2\left(\frac{n\gamma}{L}\right)^2 + \left(\frac{n\gamma}{L}\right)^4\right] + 2D_3\left(\frac{n}{L}\right)^2\left[\left(\frac{\pi}{b}\right)^2 + \left(\frac{n\gamma}{L}\right)^2\right] + D_2\left(\frac{n}{L}\right)^4 \right\}$$

$$k_2 = k_7 = -\frac{\pi Lb}{2}\left[\left(\frac{\pi}{b}\right)^2 + \left(\frac{n\gamma}{L}\right)^2\right]; \quad k_5 = -\frac{\pi b}{2}\frac{n^2\gamma}{L} \tag{4.63}$$

If we minimize the parameter $\pi/b = 1$, and avoid the value of the fourth-order $(n\gamma/R)^4$, the coefficients k_1 through k_7 will be:

$$k_1 = \frac{\pi L}{2}\left[\frac{1}{E_2} + 6\left(\frac{n\gamma}{L}\right)^2\right]$$

$$k_3 = \frac{\pi Lb}{2}\left\{ D_1\left[1 + 6\left(\frac{n\gamma}{L}\right)^2\right] + 2D_3\frac{n^4\gamma^2}{L^2} + D_2 \right\}$$

$$k_2 = k_7 = -\frac{\pi L}{2}\frac{n^2\gamma^2}{L}$$

$$k_5 = -\frac{\pi ln^2\gamma}{L} \tag{4.64}$$

We have minimized shear stress by parameters n and γ, and the critical significance of the shear stress will be [18]:

$$(\tau_{xy})_c = 0.74 \left(\frac{E_1}{E_2} \right)^{3/8} \frac{E_2}{(1 - \mu_{12}\mu_{21})} \frac{h}{4} \left(\frac{Lh}{b^2} \right)^{1/4} \tag{4.65}$$

Equation (4.64) can be implemented if the geometrical dimensions of a shell and the mechanical characteristics of the material can follow the conditions of Equation (4.65), where:

$$c = \frac{1}{12(1 - \mu_{12}\mu_{21})} \left(\frac{h}{L} \right)^2$$

$$\left(\frac{E_2}{E_1} \right)^{1/2} c^{1/2} < \left(\frac{\pi L}{b} \right)^2 < \left(\frac{E_2}{E_1} \right)^{1/2} c^{-1/2}$$

$$\left(\frac{E_1}{E_2} \right)^{1/2} c^{1/2} < \left(\frac{\pi L}{b} \right)^2 < \left(\frac{E_1}{E_2} \right)^{1/2} c^{-1/2} \tag{4.66}$$

4.2.5 Experimental Investigation

The virtual engineering analysis will be developed and divided into the following steps:

1. The stiffness parameters will be determined using Equation (4.58).
2. The critical values of longitudinal and lateral pressure found in one study [20] will be used.
3. The bending moments M_x, M_y, M_{xy} and cutting forces Q_x, Q_y will be determined by Equation (4.41).
4. The coefficients k_1 through k_7 responsible for the lost stability wing structure will be found.
5. The critical significance of the shear stress will be determined by Equation (4.66).

Tables 4.6 and 4.7 show the physical and mechanical properties of carbon fiber–epoxy skin layers and honeycomb core materials. Acting bending and torsion moments facilitate finding interlaminar shear strength.

The modulus of elasticity is 55.1 GPa (7992 ksi). The interlaminar shear strength of carbon fiber–epoxy has a value of 65.5 MPa, which is significantly higher than honeycomb shear strength, where E_1^0, E_1^{90}, E_2^0, E_2^{90} are the moduli

TABLE 4.6

Carbon Fiber Epoxy Material Properties

	Metric	English	Comments
Physical Properties			
Density	1.45 g/cm³	0.0524 lb/in³	ASTM D792
Mechanical Properties			
Tensile strength, ultimate	289 MPa	41,916 psi	ASTM D638
Elongation at break	0.487%	0.487%	ASTM D638
Modulus of elasticity	55.1 GPa	7992 ksi	In tension, ASTM D638
Flexural modulus	34.5 GPa	5004 ksi	ASTM D690
Flexural yield strength	613 MPa	88,908 psi	ASTM D790
Compressive yield strength	275 MPa	39,885	ASTM D695
Shear modulus	2.9 GPa	421 ksi	Interlaminar ASTM D5379
Shear modulus	11 GPa	1595 ksi	In-plane, ASTM D5379
Shear strength	65.5 MPa	9500 psi	Interlaminar ASTM D5379
Shear strength	206 MPa	29,878 psi	In-plane, ASTM D5379
Impact strength, Izod notched	9.6 J/cm	18 ft lb/in	Notched, ASTM D256
Compressive modulus	31.7 GPa	4598 ksi	ASTM D695

of normal elasticity in the x, y directions for the skin layers and honeycomb, and ρ_1, ρ_2 are the density of materials for skin layers and honeycomb.

Following Golfman [19], the moduli of normal elasticity for skin layer wing manufacturing by carbon–epoxy composites in x and y directions are $E_1^0 = 8.48 \times 10^6$ MPa (12.3×10^6) psi, and $E_1^{90} = 6.55 \times 10^6$ MPa (9.5×10^6) psi. The stiffness of the wing by the bending moments was determined as: $D_1 = 145.2$ kg/m², $D_2 = 50.57$ kg/m², $D_3 = 120.9$ kg/m², and $D' = 1843.5$ kg/m².

Densities of the skin layers and honeycomb are $\rho_1 = 1.998$ g/cm³ and $\rho_2 = 0.95$ g/cm³. The length of the wing is 48 ft (14.4 m), width b is equal to 3 ft (36 in), and high section h is equal to 1 ft (6 in).

All parameters will be calculated with the C language using data in Tables 4.6 and 4.7.

TABLE 4.7

Honeycomb Core Material Properties

	Metric	English	Comments
Physical Properties			
Density	0.032 g/cm³	0.032 g/cm³	Core density
Mechanical Properties			
Shear modulus	0.331 MPa	48 ksi	Ribbon direction
Shear strength	1.14 MPa	165 psi	Ribbon direction
Compressive yield strength	1.41 MPa	204 psi	

4.2.6 Conclusions

1. Carbon fiber–epoxy sandwich wings were developed from rapid prototyping to pultrusions.
2. Carbon–nylon profiles for honeycomb absorbing moisture has average chemical–solvent resistance, low density, and high thermomechanical properties.
3. A method for the determination of bending moments and cutting forces was developed by estimation of minimal compression pressure for lateral and longitudinal directions.
4. Acting bending moments and torsion moments in Equation (4.41) allowed us to find the interlaminar shear stress.

4.3 Developing a Low-Cost Method to Reduce Delamination Resistance in Multilayer Protection Systems

4.3.1 Introduction

Composite materials are playing a key role in the development of lightweight integral armor for military vehicles. One common class of advanced composites are composed of layers of resin-impregnated fibers bonded together under heat and pressure. The structural properties of these composites are fiber-dominated, which provides excellent strength in the x and y planes but minimal strength in the z direction. Due to the inherently poor through-thickness properties of composites, their use to date has been limited.

Z-Fiber® reinforcement of composite laminates has been developed and is a way to reduce delaminating of fiber-reinforced polymeric composites. Z-Fiber preforms have been developed and the process used to manufacture them has been patented. The reason for looking to optimal technology for manufacturing 3-D multifunctional composites as armor for military vehicles is low cost.

Z-Fiber preforms are composed of a number of a small-diameter Z-Fiber rods inserted into an elastic medium. The rods are subsequently transferred from the elastic medium into an uncured composite using an ultrasonic horn in z direction. Z-Fibers transfer a 2-D structure into a 3-D structure. This process appears for small thickness plates (0.04–1.75 in) as a bridge between prepreg's 2-D technology and braiding and winding 3-D technologies.

The army has a critical need for new multifunctional material solutions for armored combat systems such as armored personnel carriers, self-propelled howitzers, up-armored wheeled vehicles, and tanks. Three types of impact damage that can destroy these vehicles and equipment are described by Evans and Boyce [21]. The effects from the low-speed impact of a stone being thrown up from a runway and the surface damage are often not significant enough to be detected visually. The ballistic fragments cause visible damage in the surface and within the components as a result of high-speed impact. The ballistic fragments, cracks and internal cracks during curing and cooling processes are all significant. The third type of damage applies to aircraft fuel tanks such as "wet wings." A ballistic fragment penetrates one side of the fuel tank and decelerates rapidly, imparting most of its energy into the fluid.

A shock wave in the fluid, known as hydraulic or hydrodynamic ram, can severely damage the fuel tank [22]. The hydrodynamic ram causes severe deformation of the skins, and internal stiffeners can become detached. Composite structures are thought to be particularly vulnerable to this type of damage. Z-Fiber rods are typically composite but can also be metal for specific applications. Some of the more common materials used for the rods include SiC/BMI, T650/BMI, T300/epoxy, P100/epoxy, S-glass-epoxy, titanium, boron fibers, stainless steel, and aluminum [23]. Z-Fiber preform shapes include block, strip, and grid. The difference is a thickness insertion. Z-Fiber preforms can be designed for use in the reinforcement of 0.04- to 1.75-in-thick composites. Standard area densities of reinforcement range from 0.75% (damage tolerance) to 4% (fastener replacement/stiffener attachment). Greater reinforcement densities of up to 10% are used for high-end application only.

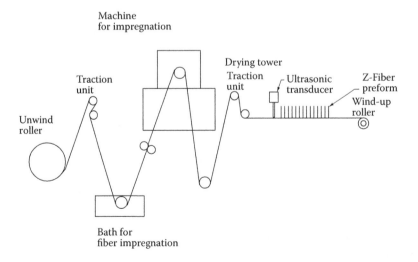

FIGURE 4.6
Schematic of a solution-dip prepreg operation.

Figure 4.6 shows a schematic of a solution-dip 3-D prepreg operation using Z-Fiber. An ultrasound transducer inserts a fiber in the z-direction.

In Figure 4.7, an automatic conveyer for Z-Fiber insertion in dry textile is shown.

One study [24] describes the impregnation process for prepregs and braided composites. The reinforcing fibers can be either unidirectionally aligned or woven into a fabric while the matrix can be a thermosetting or a thermoplastic polymer.

4.3.2 New Technology Concept

Multilayers for the ballistic protection systems are divided into tough and plasticity layers. Every layer has its own eigenfrequencies and stiffness. The full energy from the impact loads is distributed between tough and plasticity layers. Glass and polyester is a typical plastic layer, and ceramic alumina epoxy layers are a typical tough type of layer. Four thermoset resins were used as the matrix: orthophthalic polyester, isophthalic polyester, vinyl ester, and reinforced epoxy [25].

Coinjection resin transfer molding (CIRTM) and diffusion-enhanced adhesion (DEA) are two processes that were invented and developed to address the cost and performance barriers that hinder the introduction of composite materials for combat ground vehicle application [26]. When applied in tandem, these two composite processing technologies enable the manufacture of lightweight composite/ceramic integral armor, offering significant cost-reduction and performance enhancement over existing defense industry practices. CIRTM was developed for single-step manufacturing of integral armor through simultaneous injection of multiple resins into a multilayer preform.

The process achieves excellent bonding between the layers—an important aspect of the CIRTM process. It also develops an understanding of the resin flow and cure kinetics to aid in process optimization. Furthermore, the work

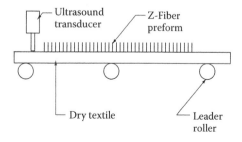

FIGURE 4.7
Automation conveyer Z-Fiber insertion in dry textile.

has enabled the production of new composite structures, including stitched structures with improved ballistic response.

In Figure 4.8, we see a sample of multilayer protection systems. The unwind rollers (pos. 1, 2, 3) simultaneously draw polyethylene (pos. 9) and prepreg (textile + epoxy resin) (pos. 8). The three layers are pulled by tensile rollers (pos. 4) and a dancer roller (pos. 5). Curing this multilayer system occurs when the layers are pulled through the oven (pos. 6). For the cooling process a camera was used (pos. 7). The prepreg layers (pos. 8) have a preliminary insertion Z-Fiber.

4.3.3 Virtual Stress Approach in Multifunctional Layers

The linear extension of the first layer can be calculated as [27]:

$$\Delta l_1 = \frac{V_{impact}}{2}\left(\frac{m_1}{E_1 L_1}\right)^{1/2} \tag{4.67}$$

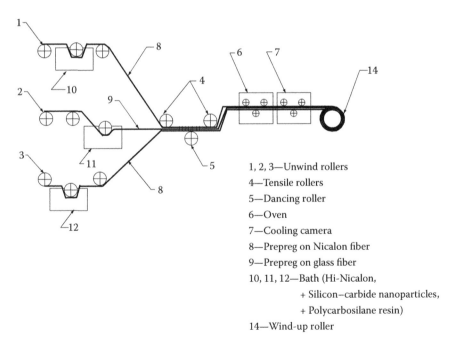

1, 2, 3—Unwind rollers
4—Tensile rollers
5—Dancing roller
6—Oven
7—Cooling camera
8—Prepreg on Nicalon fiber
9—Prepreg on glass fiber
10, 11, 12—Bath (Hi-Nicalon,
                          + Silicon–carbide nanoparticles,
                          + Polycarbosilane resin)
14—Wind-up roller

FIGURE 4.8
Automation process manufacturing hybrid textile-based multilayered flexible system with fiber inserted perpendicular prepreg layers.

where
  V_{impact} = Velocity of the impact load
  E_1 = Modulus of elasticity for the first layer
  m_1 = Mass of the first layer
  L_1 = Length of the first layer

We developed this approach for multifunctional layers. The linear extension of the second layer is calculated as:

$$\Delta l_2 = \frac{V_{impact}}{2} \left(\frac{m_2}{E_2 L_2} \right)^{1/2}$$ (4.68)

where
  V_{impact} = Velocity of the impact load
  E_2 = Modulus of elasticity for the second layer
  m_2 = Mass of the second layer
  L_2 = Length of the second layer

The linear extension of the third layer can found as:

$$\Delta L_3 = \frac{V_{impact}}{2} \left(\frac{m_3}{E_3 L_3} \right)^{1/2}$$ (4.69)

where
  V_{impact} = Velocity of the impact load
  E_3 = Modulus of elasticity for the third layer
  M_3 = Mass of the third layer
  L_3 = Length of the third layer

The deformations of the first, second, and third layers can be designated as:

$$\varepsilon_{z_1} = \frac{\Delta l_1}{L_1}; \quad \varepsilon_{z_2} \frac{\Delta l_2}{L_2}; \quad \varepsilon_{z_3} \frac{\Delta l_3}{L_3}$$ (4.70)

The stress components of the first and third layers are:

$$\sigma_{z_1} = \frac{\varepsilon_{z_1} E_1 + \mu_{zx} E_2 + \mu_{zy} E_3}{1 - \mu_{zx} \mu_{zy} \mu_{xy}}$$ (4.71)

The volume of the prepreg layers consists of the volume of resin and volume of fiber:

$$V_p = V_R + V_f \tag{4.72}$$

where P_R is the mass of resin, P_f is the mass of fiber, ρ_R is the density of resin, and ρ_f is the density of fiber.

$$V_R = \frac{P_R}{\rho_R} \; ; \quad V_f = \frac{P_f}{\rho_f} \tag{4.73}$$

The stress components of the second (polyethylene) layer are:

$$\sigma_{z_2} = \frac{\varepsilon_{z_2} E_2 \mu_{zx} + \mu_{zy} E_3}{1 - \mu_{zx}\mu_{zy}\mu_{xy}} \tag{4.74}$$

Lattice structures have been worked under mechanical vibrations. The appropriate equations of motion following Newton's second law in x, y, z directions becomes [28]:

$$m\frac{\partial^2 x}{\partial z^2} + \delta_1 \frac{\partial x}{\partial z} + Q_{11}x = P\sin\Omega t \tag{4.75}$$

$$m\frac{\partial^2 y}{\partial z^2} + \delta_2 \frac{\partial y}{\partial z} + Q_{22}y = P\sin\Omega t \tag{4.76}$$

$$m\frac{\partial^2 z}{\partial x^2} + \delta_3 \frac{\partial z}{\partial x} + Q_{33}z = P\sin\Omega t \tag{4.77}$$

where

m	= Mass of the lattice cylinder
$\delta_1, \delta_2, \delta_3$	= Critical damping coefficients, which have different values in directions x, y, z
Q_{11}, Q_{22}, Q_{33}	= Stiffness of the lattice cylinder in diagonal directions x, y, z
P	= Dynamic impact load
Ω	= Forcing frequency
t	= Time of wave propagation
x, y, z	= Representative active displacements of the lattice structure cells
Q_{11}, Q_{22}, Q_{33}	= Constants for components and are satisfied by the diagonal directions x, y, z

The dynamic impact loads changed constantly as accelerations $a_1 = \partial^2 x/\partial z^2$, $a_2 = \partial^2 y/\partial z^2$, $a_3 = \partial^2 z/\partial z^2$, and velocities $v_1 = \partial x/\partial z$, $v_2 = \partial y/\partial z$, and $v_3 = \partial z/\partial x$.

The particular solution that applies to the nonsteady–state vibration of the system should be the harmonic function of the time, such as [29]:

$$x = A_x \sin^2(\Omega t - \phi_1)$$
$$y = A_y \cos^2(\Omega t - \phi_2) \qquad (4.78)$$
$$z = A_z \sin^2 \cos^2(\Omega t - \phi_3)$$

Here, ϕ is a phase angle reflecting of a different phase between the applied impact force and the resulting vibration, and is determined as [25]:

$$A_x = \frac{P \sin \Omega t}{2\delta_1 \sin(\Omega t - \phi_1) + Q_{11} \sin^2(\Omega t - \phi_1) - 2m \cos(\Omega t - \phi_1)} \qquad (4.79)$$

We input Equation (4.79) into Equations (4.75), (4.76), and (4.77) and get the amplitude of vibration in directions x, y, z.

$$A_x = \frac{P \sin \Omega t}{2\delta_1 \sin(\Omega t - \phi_1) + Q_{11} \sin^2(\Omega t - \phi_1) - 2m \cos(\Omega t - \phi_1)} \qquad (4.80)$$

$$A_y = \frac{P \sin \Omega t}{2\delta_2 \cos(\Omega t - \phi_2) + Q_{22} \cos^2(\Omega t - \phi_2) - 2m \sin(\Omega t - \phi_2)} \qquad (4.81)$$

$$A_z = \frac{P \sin \Omega t}{4\delta_3 \sin(\Omega t - \phi_3) \cos(\Omega t - \phi_3) - Q_{33} \sin^2(\Omega t - \phi_3) \cos^2(\Omega t - \phi_3) + 4m \cos 2(\Omega t - \phi_3)} \qquad (4.82)$$

The critical damping coefficients δ_1, δ_2, δ_3 have an anisotropic behavior and are precisely described by Golfman [30]. From the basic equations of the mechanical motion properties (Equations 4.75, 4.76, and 4.77), we can find a maximum displacement of the lattice structure cells in the x, y, z directions.

$$\Delta l = \frac{P \sin \Omega t - ma - \delta_1 V_{impact}}{Q_{11}} \qquad (4.83)$$

$$\Delta w = \frac{P \sin \Omega t - ma - \phi_2 V_{impact}}{Q_{22}} \qquad (4.84)$$

$$\Delta h = \frac{P \sin \Omega t - ma - \delta_3 V_{impact}}{Q_{33}} \qquad (4.85)$$

where V_{impact} is the velocity of the dynamic impact loads, which is applied when vehicles enter from the start, usually simulated by virtual engineering.

4.3.4 Feature of Design of the Lattice Structures

In the work of Martinsson and Movchan [31], a simple intuitive method was developed for designing lattice materials with such intervals, called bandgaps, around certain prescribed frequencies. The process starts with a mechanical lattice consisting of point masses and rubber bands organized in the geometry illustrated in Figure 4.9, and we then want to introduce a band gap near the frequency *f*.

We constructed a mechanical oscillator consisting of a mass suspended by three rubber bands in the following constellation (see Figure 4.10).

Using *m* to denote the mass and *k* to denote the spring constant of the rubber bands of this oscillator, one finds that its eigenfrequency is $\sqrt{(3k/2m)}$. If we choose *k* and *m* so that $\sqrt{(3k/2m)}$ and incorporate the basic oscillator into the repeating cell of the mechanical structure, the result is 3-D helical cylinders that have eigenfrequencies that can be found as:

$$f_1 = \sqrt{(3Q_{11}/2m)}; \ f_2 = \sqrt{(3Q_{22}/2m)}; \ f_3 = \sqrt{(3Q_{33}/2m)} \tag{4.86}$$

4.3.5 Force Vibration of the Lattice Structures

The phenomena of losses due to dynamic stability and the appearance of a parametric resonance in the lattice structures includes by coincidence free as

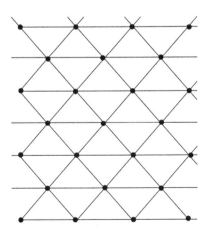

FIGURE 4.9
Mechanical lattice structure consisting of point masses and rubber bands.

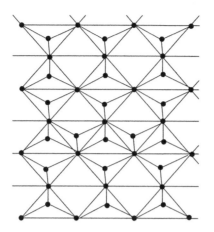

FIGURE 4.10
Mechanical lattice structure constructed of a mechanical oscillator consisting of a mass suspended by three rubber bands.

well as forced frequencies. In the lattice helical structures is the compressed force in the vertical direction 3 by force p_{mn}, shown as:

$$p_{mn} = p_0 \sin\theta t - p_0 \cos\theta t \qquad (4.87)$$

In order to solve the Matue equation relative to deflection:

$$\frac{\partial^2 f}{\partial t^2} + f^2 \min\left(1 - \frac{P_0}{P_{min}} \cos\Omega t\right) f(t) = 0 \qquad (4.88)$$

where
  f_{mn} = Frequencies of free vibration determined by equation in lieu of Equation (4.88)
  P_0 = Value of hydrostatic pressure as determined by Golfman [24]
  P_{mn} = Critical value of the compressed force as determined by Golfman [24]
  The critical value of the compressed force can be determined using Equation (4.58) [32].

$$p_{mn} = \frac{\pi^2}{h^2}\left(\frac{g}{ha}\right)^{1/2}\left[D_1\left(\frac{m^4}{k}\right) + 2D_3 n^2\left(\frac{m^2}{k}\right) + D_2 n^4\right]^{1/2}(1+d) \qquad (4.89)$$

where $k = L/h$, and d is the correction coefficient.
  D_{ij} is the stiffness of the helical lattice structures from the bending moment.

$$D_1 = Q_{11}\frac{h^3}{12}; \quad D_2 = Q_{22}\frac{h^3}{12}; \quad D_3 = \frac{h^3}{12}(Q_{12} + 2Q_{66}) \qquad (4.90)$$

where stiffness constants Q_{ij} are determined from Equation (1.2.3), and
$\quad h$ = Height of the cylinder
$\quad g$ = Density of the fiber
$\quad a$ = Acceleration due to gravity
$\quad L$ = Length of the chord; $L = 2\pi(R - r)$
$\quad R$ = Outside radius of the cylinder
$\quad r$ = Inside radius of the cylinder

Now, we input a coefficient:

$$\lambda_{mn} = \frac{p_0}{2p_{min}} \tag{4.91}$$

Because λ_{mn} is a coefficient of a compression force relationship, so Equation (4.88) can be shown as:

$$\frac{\partial^2 f}{\partial t^2} + f_{min}^2(1 - 2\lambda \cos \Omega t) f(t) = 0 \tag{4.92}$$

In the study of Goldenblat and Kopnov [32], the boundary conditions for stability losses were determined for the basic tone of the force frequencies:

$$\Omega^* = 2f_{mn}(16\lambda_{mn}) \tag{4.93}$$

Here, Ω^* is a critical value of the frequency for the force load.
$\quad$ For the second tone of the force frequencies:

$$\Omega^* = f_{mn}\left[(1 + 1/3\lambda_{mn}^2)\right]^{1/2} \tag{4.94}$$

or

$$\Omega^* = f_{mn}\left[(1 - 2\lambda_{mn}^2)\right]^{1/2}$$

Equation (4.95) indicates two boundaries for stability losses.
$\quad$ For the third tone of the force frequencies:

$$\Omega^* = f_{mn}\left[(1 - 2\lambda_{mn}^2)\right]^{1/2}$$

$$\Omega^* = \frac{2}{3}f_{mn}\left[\left(1 - \frac{9\lambda_{mn}^2}{8 \pm 9\lambda_{mn}}\right)\right]^{1/2} \tag{4.95}$$

4.3.6 Experimental Investigation

The main failure mode of a helical lattice cylinder with attached fittings includes both the bending of the structure and the failure due to general buckling. Hou and Gramoll [33] showed that the results in testing of conical lattice structures were not stable. The low failure was due to microbuckling and is commonly referred to as fiber kinking. Fiber kinking generally occurs because of a weak matrix, which is due in part to the epoxy not curing completely or a deficiency of the hardening agent during the manufacturing process.

The fiber density that was selected in Gramoll's work [21] was $g = 1770$ kg/m³, while the acceleration due to gravity was $a = 9.81$ m/s². D_{ij} was the stiffness of the lattice structures from the bending moment and $D_1 = 145.2$ kg/m², $D_2 = 50.57$ kg/m², and $D_3 = 120.9$ kg/m².

We analyzed Equation (4.90) for basic tone frequencies and saw that force frequencies are two times higher than eigenfrequencies, so parametric resonance is not possible. For the second tone of the force frequencies, the opportunities to fit with eigenfrequencies are real, because the frequencies for the force and free vibrations are closed. A correlation exists between the force and free frequencies.

We simulated the impact particulate debris of the lattice structure. We also selected a small cell from the helical lattice cylinder. The cell has an equal length, width, and height of 1.0 in and a mass m of 0.000253 lb. The event that was simulated is actually the dropping of the cell from a height of 100 in, which resulted in an impact velocity of 278 in/s when under the influence of a gravity field of strength, 386.4 in/s². The Algor software predicted a maximum deformation (ΔL) of 0.000694 in, which compares very favorably with the value of 0.000699 given in Equation (4.96) [27].

$$ma + \delta v + Q_{1,2,3}d = 0 \tag{4.96}$$

where

m	= Mass of the cell of the helical lattice cylinder
δ	= Critical damping coefficient of the small cell element
$Q_{1,2,3}$	= Stiffness of the small cell element
a	= Acceleration of the gravity of a small cell element
v	= Impact velocity of a small cell element
d	= Displacement in the 1, 2, 3 directions of a small cell element

4.3.7 Conclusions

1. We developed a concept for the automation process for manufacturing of multilayer armor materials.

2. We investigated the theory based on the construction of a mechanical oscillator consisting of a mass suspended by three rubber bands.

3. We analyzed the eigenfrequencies by incorporating the basic oscillator into the repeating cell of the mechanical structure.

4. We found that the basic tone force frequencies were two times higher than the eigenfrequencies, and therefore parametric resonance was not possible and the eigenfrequencies with force frequencies were compared.

5. In the case of the second tone of the force frequencies, the opportunities to fit with eigenfrequencies were real, because the frequencies (the force and free vibrations) are very close so the possibility of parametric resonance exists.

6. The calculations of the third tone of the force frequencies shows that failure conditions do not exist because free and force frequencies do not fit.

7. The impact particulate debris of the lattice structures can damage local areas but do not influence microbuckling structures. The maximum deformation of the small cell elements can be predicted.

Appendix

This research was written as an investigation of a low-cost method seeking the optimal technology for armor lattice structures. Dynamic aspects of the lattice structures' behavior in the manufacturing of carbon epoxy composites was published in the *Journal of Advanced Materials* in 2003 and this is a continuation of this work to develop strength criteria for anisotropic materials.

References

1. Buehler, F. U., and J. C. Seferis. 2001. Consistency evaluation of a qualified glass fiber prepreg system. *Journal of Advanced Materials* 34 (2): 41–50.
2. Zeng, S., J. C. Seferis, K. J. Ahn, and C. L. Pederson. 1994. Model test panel for processing and characterization studies of honeycomb composite structures. *Journal of Advanced Materials* 25 (2): 9–21.
3. Golfman, Y. 1969. Investigation influence the features of molding technology for fiber glass propeller blades on their strength. PhD dissertation, Shipbuilding Institute, Leningrad, USSR.
4. Quantum Composites Lytex Properties, Honeycomb Core Material Properties, MatWeb.com Database Internet, 2002, http://www.placecore.com/product-honeycomb-cores.htm.

5. Lekhnitskii, S. G. 1968. *Anisotropic Plates*. Translated from the second Russian edition by S. W. Tsai and T. Cheron. New York: Gordon & Breach.

6. Pinov, D. 1935. *Solving Problem of Bending Polynomial Stress Function*. Moscow: Central Institute of Aerodynamics, 209.

7. Golfman, Y. 1996. The interlaminar shear stress analysis of composites in marine front. SAMPE Technical Conference, Arizona.

8. Golfman, Y. 2004. The interlaminar shear stress analysis of composites sandwich/carbon fiber epoxy structures. *Journal of Advanced Materials* 36 (2): 16–21.

9. Fichtengolts, G. M. 1966. Course of Differential and Integration Calculation. Moscow Leningrad University.

10. Dow, M. 1997. The Advanced Stitching Machine: Making Composite Wing Structures of the Future. FS-1997-08-31-LaRC, August. http://oea.larc.nasa.gov/PAIS/ASM.html.

11. Premix Thermoplastics. 2004. Premix Technologies Introduces Full Line of Carbon Fiber Reinforced Compounds, 13/10/2004, Milton, WI.

12. MacRae, D. Resin film infusion (RFI) process simulation of complex wing structures. Virginia Tech College of Engineering, http://www.sv.vt.edu/comp_sim/macrae/macrae.html.

13. Park, B.-H. Rapid Prototyping Laboratory Stanford University, http://www-rpl.stanford.edu/research_detail.asp?

14. Lazarus, P. 1994. Infusion, *Professional Boat Builder*. N31/October/November.

15. Hovercraft Hull-core Materials, http:/4wings.com.phtemp.com/tip/core.html.

16. Buck, M. E. 2004. Low cost manufacturing methods for thermoplastic composites (continuous fiber). SAMPE Meeting, Boston Section, 10/14/04.

17. Golfman, Y. 2004. The interlaminar shear stress analysis of composites sandwich/carbon fiber/epoxy structures. *Journal of Advanced Materials* 36 (2): 16–21.

18. Goldenblat, I. I., and V.A. Kopnov. 1968. *Resistance of Fiberglass*. Moscow: Mashinostroenie.

19. Golfman, Y. 2001. Nondestructive evaluation of aerospace components using ultrasound and thermography technologies. *Journal of Advanced Materials* 33 (4): 21–25.

20. Golfman, Y. 2007. The interlaminar shear strength between thermoplastics rapid prototyping or pultruded profiles and skin carbon/fiber/epoxy layers. *Journal of Advanced Materials, Special Edition* 3: 21–28.

21. Evans, D. A., and J. S. Boyce. 1989. Transverse reinforcement methods for improved delamination resistance. 34th International SAMPE Symposium, vol. 34, pp. 271–282.

22. Jacobson, M. J., R. M. Heitz, and J. R. Yamane. 1986. Survivable composite integral fuel tanks, v1 Testing and Analysis, AFWAL-TR-85-3085. Northrop Corporation. www.ftp.rta.nato.int/public/PubFulltext/.

23. Aztex, Inc. Z-Fiber Preform. http:/www.aztex-z-fiber.com.

24. Golfman, Y. 2004. Impregnation process for prepregs and braided composites. *Journal of Advanced Materials Special Edition* 3: 34–37.

25. Abrate, S. 1998. *Impact on Composite Structures*. Cambridge, UK: Cambridge University Press, pp. 135–160.

26. University of Delaware. Diffusion-Enhanced Adhesion and Co-Injection Transfer Molding. www.stormingmedia.us.

27. Algor Inc. (technical staff). 1998. *The Theoretical Basis of Event Simulation*. Pittsburgh, PA: Algor Inc.

28. Golfman, Y. 2003. Dynamic aspects of the lattice structures' behavior in the manufacturing of carbon–epoxy composites. *Journal of Advanced Materials* 35 (2): 3–8.
29. Golfman, Y. 2001. Fiber draw automation control. *Journal of Advanced Materials* 34 (2): 35–40.
30. Golfman, Y. 2003. Dynamic stability of the lattice structures in the manufacturing of carbon fiber/epoxy composites including the influence of damping properties. Submitted to the *Journal of Advanced Materials*, Special Edition 3: 11–20.
31. Martinsson, P. G., and A. B. Movchan. 2002. Vibration of lattice structures and photonic bandgaps. Submitted to *Quarterly Journal of Mechanics and Applied Mathematics*. http://acnath.colorado.edu/faculty/martin.
32. Goldenblat, Y. Y., and V. A. Kopnov. 1968. *Resistance of Fiberglass*. Moscow: Mashinostroenie.
33. Hou, A., and K. Gramoll. 2000. Fabrication and compressive strength of the composite attachment fitting for launch vehicles. *Journal of Advanced Materials* 32 (1): 39–45.

5

Fatigue Strength, Stress, and Vibration Analysis

5.1 Fatigue Strength Prediction for Aerospace Components Using Reinforced Fiberglass or Graphite–Epoxy

5.1.1 Introduction

The application of anisotropic composites, such as reinforced fiberglass or graphite–epoxy materials in a large aviation and marine novel construction will be critical in the twenty-first century.

The purpose of this research is the prediction of fatigue strength of composite structures under long-term service, which is dependent on several technological factors.

The Advanced General Aviation Technology Experiments (AGATE) was founded in 1994 and is a cost-sharing industry–university–government partnership whose mission is to develop technology that will stimulate the U.S. general aviation industry. Initiated by the National Aeronautics and Space Administration (NASA), the AGATE consortium has more than 70 members from industry, universities, the Federal Aviation Administration (FAA), and other government agencies.

The AGATE Advanced Materials Program has directed the creation of composite material allowable that has been approved and witnessed by the FAA for use in the next-generation single-pilot, four-seater, and near-all-weather light airplanes; the first two being Cirrus SR20 and Lancair's Columbia 300. The two companies that manufactured these aircraft have been contributing industrial members to the Advanced Materials Program. A major goal of the AGATE program has been to produce a "standard" FAA approved composite material qualification methodology within the general aviation community. A part of this methodology has concentrated on dynamic and fatigue strength prediction.

5.1.2 Fatigue Strength Prediction

In the large aviation and marine novel components fabricated from composites (i.e., fiberglass or graphite–epoxy material), there is a significantly lower strength than on samples. Some authors attribute this to "scaling effects" [1] and this strength construction is shown as technological defects (see Equation 1.1).

Strength reduction can be explained as a result of the influence of techno-logical factors in resin formulation and curing, and fiber distortion. We consider these technological factors as the defects of structural laminates and the technological behavior as shrinkage and warpage, which results from thermal stress during the molding process.

Bailey et al. [2] has shown, using a simple equilibrium model, that the thermal residual stress transverse to the fibers in a constrained 90° ply can be expressed (see Equation 1.2). This stress is introduced upon cooldown from the curing temperature due to the mismatch in the coefficient of thermal expansion of the adjacent piles in a laminate.

From the prediction strength of every layer we can approximate the average strength of all construction and answer the question of how long this construction will be serviceable [8].

We consider that every layer of construction has strong orthotropic properties and the construction has a homogeneous structure and is equally impregnated by epoxy or other isotropic resin. An optimal structure has oriented fibers whose direction of reinforcement coincides with the direction of acting normal stresses along axes x, y, z (see Figure 5.1).

Fatigue strength can be predicted as a linear correlation between compression strength and the function $\Phi(\sigma)$ (see Equation 1.9).

$$\sigma_{-1} = \sigma_s \Phi(\sigma)$$

Here, σ_{-1} is the fatigue strength in x, y, z directions and σ_s is the compression strength in x, y, z directions.

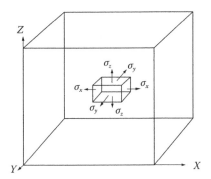

FIGURE 5.1
Fiber orientation in orthotropic composites; stress components coincide with fiber orientation x, y, z.

Function $\Phi(\sigma)$ can be shown as the Weibull distribution function (see Equation 1.10).

$$\Phi(\sigma) = 1 - P(t)$$

Here, $P(t)$ is the probability of collapse in the local part of the construction from compression strength. We assume that the general strength equals unity and $P(t)$ has been subordinated to the normal distribution law (see Equation 1.11).

$$P(t) = \frac{1}{(2\pi)^{1/2}} e^{-t^2/2}$$

where parameter t is equal to $t = \dfrac{\sigma_{bi} - \sigma_{bm}}{S_j}$ (see Equation 1.12), and

σ_{bi} = Current strength in x, y, z directions
σ_{bm} = Middle strength in x, y, z directions
S_j = Sample standard deviation for each environment via (see Equation 1.13)

$$S_j^2 = \frac{1}{n_j - 1} \sum_{l=1}^{n_j} (\sigma_{bi} - \sigma_{bm})^2$$

Here, n_j is the number of testing samples, and calculating the sample mean σ_{bm} (see Equation 1.14)

$$\sigma_{bm} = \frac{1}{n_j} \sum_{n=1}^{nj} \sigma_{bi}$$

For a single test condition (such as 0° compression strength), the data was collected for each environment being tested. The number of observations in each environmental condition was n_j, where j represents the total number of environments being pooled. If the assumption of normality was significantly violated, the other statistical model should be investigated to fit the data. In general, the Weibull distribution provides the most conservative basic value.

In the work of Talreja [3], the Weibull distribution function is given in Equation (1.15).

$$\Phi(X, A, B, C) = 1 - \exp\left\{-\left(\frac{X-A}{B}\right)\right\}$$

Here, the parameters are: $X > 0$; $B > 0$; $C > 0$; X, A, B, C each are equal to a discrete symbol.

For strength distribution, we designate:

$$X = \sigma_{bi}; A = \sigma_{bm}; B = S_j; C = N$$

where N is the base of testing. Therefore, Equation (1.15) for fatigue strength is represented by Equation (1.16).

$$\Phi(\sigma_{bi}, \sigma_{bm}, S_j, N) = 1 - \exp\left\{-\left(\frac{\sigma_{bi} - \sigma_{bm}}{S_j}\right)^N\right\}$$

If we consider that

$$\Phi(\sigma_{bi}, \sigma_{bm}, S_j, N) = 1 - P(t) = \exp\left\{-\left(\frac{\sigma_{bi} - \sigma_{bm}}{S_j}\right)^N\right\} \tag{5.1}$$

we get the logarithmic equation (see Equation 1.18).

$$\ln[1 - P(t)] = N\ln(\sigma_{bm} - \sigma_{bi}) - N\ln S_j$$

It is in Equation (1.18) that shows a straight line in logarithmic coordinates. Base of testing N can determine the inclination of this straight line. Therefore, the period of testing N will be determined as Equation (1.19).

$$N = \frac{\ln[1 - P(t)]}{\ln(\sigma_{bm} - \sigma_{bi} - \ln S_j)}$$

If we propose that the shear stress is responsible for the delaminating of the composite, the period of testing N will be determined as:

$$N_1 = \frac{\ln[1 - P(t_1)]}{\ln(\tau_{bm} - \tau_{bi}) - \ln S_j} \tag{5.2}$$

where

$P(t_1)$ = Probability of structure collapse responsible from interlaminar shear stress

τ_{bi} = Current shear stress acting in interlaminar layers

τ_{bm} = Middle significant shear stress

S_{τ} = Middle square deviation

The middle square deviation can be found using the following equation:

$$S_{\tau} = \frac{1}{n-1}(\tau_{bi} - \tau_{bm})^2 \tag{5.3}$$

Shear strength samples molded by fiberglass, graphite–epoxy, or silicon–graphite ceramic matrix reinforced fiber Sylramic, CG Nicalon, and Nextel N720 were tested using two methods: the quad lap shear test (peel test) and the tear test. Shear strength silicon ceramic reinforced by Hi-Nicalon fiber compatibility test is shown in Figure 5.2.

If the interlaminar shear strength is responsible for collapse construction, fatigue strength can be predicted as:

$$\tau_{-1} = \tau_s \Phi(t) \tag{5.4}$$

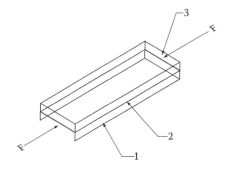

Quad lap shear test

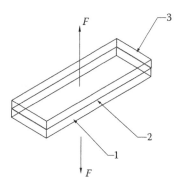

Tear test

1—3-D woven Nicalon ceramic layer (thickness 1 mm)

2—Mitigator suppressor layer

3—3-D woven carbon–glass layer (thickness 1 mm)

FIGURE 5.2
Shear strength Nicalon fiber compatibility.

Function $\Phi(\tau)$ is shown as:

$$\Phi(t) = 1 - P(t_1) \tag{5.5}$$

Here, $P(t_1)$ is the probability of collapse of the local part of construction from the interlaminar shear strength.

$$\text{Parameter } t_1 = \frac{\tau_{bi} - \tau_{bm}}{S_\tau} \tag{5.6}$$

The Weibull distribution function is shown as the example of the Laplace function. The Laplacian transformation is a powerful method for solving linear differential equations arising in engineering mathematics [4].

Thus, we designate:

$$P(t) = L(t) = \int_0^\infty e^{-bt} f(t)\,dt \tag{5.7}$$

The function $L(t)$ is called the Laplacian transform of the original function $f(t)$. Furthermore, the original function $f(t)$ is called the inverse transform function and will be denoted by $L^{-1}(F)$, that is,

$$f(t) = L^{-1}(F) \tag{5.8}$$

$$L(e) = \frac{1}{(2\pi)^{1/2}} \int_0^t e^{-t_2/2} \tag{5.9}$$

Function $L(e)$ is a two parameter function that represents the law of normal distribution function and the prediction probability with variation strength parameters σ_{bi} and t (Equation 1.12).

5.1.3 Static and Dynamic Fatigue Strength

Reinforced fiberglass or graphite–epoxy is very sensitive to "static fatigue," which is the equal static load repeatedly applied to the material over a long period of time.

The limit of fatigue for graphite–epoxy on the base of 1000 cycles equals 0.6 to 0.7 from the limit of the static strength. Bending does not increase due to beneficial elastic properties. Loops of hysteresis from loading and unloading are practically identical. Residue of deformation accumulated after testing samples for 1000 cycles has no significant value. Dynamic testing has the support significance of internal heat.

Internal heat depends on harmonic frequencies and the amplitude of vibration, and the heat of the composites is connected with hysteresis losses. The quantity of heat is increased when there is an increase in frequency of vibration. If the frequency is 1000 cycles per minute, the internal heat of the sample will be 50 to 70°C.

If the frequency decreases to 300 cycle/min the internal heat simultaneously is reduced to 25 to 30°C. The very important characteristic of dynamic properties is the internal dispersion of energy (critical damping coefficient [5]). During the harmonic vibration, which can reach 1×10^6 cycles, there is no visible cracking. It does not mean that there is no internal cracking.

It is very important to use nondestructive ultrasonic evaluation to determine the dynamic modulus of elasticity [6], and not visible cracking propagation.

The criteria of quality for dynamic fatigue after 1×10^6 to 5×10^6 cycles have shown crack visibility. However, the samples with visibility cracks after testing have very high strength characteristics. Therefore, the value of predicting fatigue strength is somewhat lower than test results. Correlation between dynamic fatigue and the number of cycles in the case of harmonic bending in the graphite–epoxy composites exists.

5.1.4 Experimental Investigation

Table 5.1 represents the parameters of the compression strength for graphite–epoxy composites [7]. We tested six samples and determined the sample standard deviations (Equation 1.12).

Also, we determined three parameters: t (Equation 1.12), probability $P(t)$ using tabulating function (Table 5.2) and function $\Phi(\sigma)$ following Equation (5.1).

Fatigue strength prediction has a good correlation with compression strength (see Figure 5.3).

TABLE 5.1

Fatigue Compression Strength Prediction for Graphite–Epoxy Composites

$\sigma_s* 10^{-3}$ (MPa/psi)	$\sigma_m* 10^{-3}$ (MPa/psi)	$S_j* 10^{-3}$ (MPa/psi)	Parameter t (Equation 1.12)	$P(t)$	$\Phi(t)$	Fatigue Strength, $\sigma_{-1}*$ 10^{-3} (MPa/psi)
0.37/54			0.412	0.3668	0.6332	0.24/34.2
0.36/52			0.24	0.3885	0.6115	0.22/31.8
0.34/50	0.34/49.16	0.024/3.46	0.07	0.3980	0.6000	0.20/30.0
0.33/48			0.009	0.3989	0.6010	0.198/28.8
0.32/46			0.025	0.3876	0.6124	0.19/28.17
0.31/45			0.035	0.3765	0.6235	0.19/28.05

TABLE 5.2

Density of Probability Function (First Approach)

t	0	1	2	3	4	5	6	7	8	9
0.0	3989^{-4}	3989	3989	3988	3986	3985	39,982	3980	3977	3973
0.1	3970^{-4}	3965	3961	3956	3951	3945	3939	3932	3925	3918
0.2	3910^{-4}	3902	3894	3885	3876	3867	3857	3847	3836	3825
0.3	3814^{-4}	3802	3790	3778	3765	3752	3739	3726	3711	3697
0.4	3683^{-4}	3668	3653	3637	3621	3605	3589	3572	3555	3538
0.5	3521^{-4}	3503	3485	3467	3448	3429	3410	3391	3372	3352
0.6	3332^{-4}	3312	3292	3271	3251	3230	3209	3187	3166	3144

Table 5.3 presents the parameters of interlaminar shear strength for graphite–epoxy composites. We tested six samples and determined the sample standard deviation (Equation 1.13).

We also determined three parameters: t_1 (Equation 5.6), probability $P(t_1)$ using tabulating function (Table 5.2), and function $\Phi(\tau)$ following Equation (5.1).

Fatigue strength prediction using the Laplacian function Equation (5.9) is presented in Table 5.4.

Tabulating Laplacian functions is given in Table 5.5. The number of samples, N, the sample mean σ_{bm}, τ_{bm}, and sample standard deviation S_j and S_τ were selected from the pooled data. The normal distribution B basis and A basis allowable was calculated using the pooled mean, standard deviation, and tolerance factors for each environment.

5.1.5 Concluding Remarks

1. A methodology for the prediction of the fatigue strength of anisotropic materials such as fiberglass, graphite–epoxy, and ceramic matrix composites was developed.

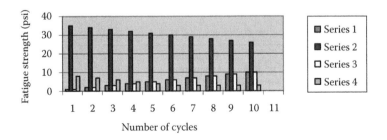

FIGURE 5.3
Correlation between fatigue strength and the number of cycles in harmonic bending of graphite–epoxy composites.

TABLE 5.3

Fatigue Interlaminar Shear Strength Prediction for Graphite–Epoxy Composites (First Approach)

$\tau_s * 10^{-3}$ (MPa/psi)	$\tau_s * 10^{-3}$ (MPa/psi)	$S_\tau * 10^{-3}$ (MPa/psi)	t_1	$P(t_1)$	$\Phi(t_1)$	$\tau_{-1} * 10^{-3}$ (MPa/psi)
0.07/10.4			0.692	0.3144	0.685	7.12
0.07/10.2			0.384	0.3711	0.628	6.41
0.068/10	0.068/9.95	0.004/0.65	0.076	0.3980	0.602	6.02
0.067/9.8			0.230	0.3885	0.611	5.99
0.066/9.7			0.384	0.3711	0.628	6.09
0.066/9.6			0.538	0.3448	0.655	6.28

2. The probability of local cracking can be predicted using mathematical functions: the first and second approaches of the law of normal distribution.

3. In spite of the visibility of cracks, the samples have a general strength. However, it is very important to use nondestructive evaluation methods for the determination of cracks that are not visible.

5.2 Effect of Thermoelasticity for Composite Turbine Disk

5.2.1 Introduction

Lightweight and high-strength anisotropic composites such as carbon–carbon, graphite–epoxy, and fiberglass under different combinations of applied stress components (biaxïal and triaxial stress conditions) pose a challenge to designers for establishing reliable failure criterion. Stimulated

TABLE 5.4

Fatigue Compression and Interlaminar Strength Prediction for Graphite–Epoxy Composites (Second Approach)

$\sigma_s * 10^{-3}$ (MPa/psi)	$\sigma_m * 10^{-3}$ (MPa/psi)	$S_j * 10^{-3}$ (MPa/psi)	Parameter t (Equation 1.12)	$L(t)$	$\Phi(t)$ (Equation 5.1)	Fatigue Strength, $\sigma_{-1} * 10^{-3}$ (MPa/psi)
0.34/50	0.34/49.16	0.023/3.46	0.070	0.279	0.721	0.29/36.05
$\tau_s * 10^{-3}$ (MPa/psi)	$\tau_s * 10^{-3}$ (MPa/psi)	$S_\tau * 10^{-3}$ (MPa/psi)	t_1	$L(t_1)$	$\Phi(t_1)$	$\tau_{-1} * 10^{-3}$ (MPa/psi)
0.068/10	0.068/9.95	0.004/0.65	0.076	0.299	0.701	0.048/7.01

TABLE 5.5

Density of Probability of the Laplacian Function (Second Approach, Equation 5.9)

t	0	1	2	3	4	5	6	7	8	9
0.0	000	040	080	120	160	199	239	279	319	359
0.1	398	438	478	517	557	596	636	675	714	753
0.2	793	832	871	910	948	987	0.026	0.064	0.103	0.141
0.3	792	217	255	293	331	368	406	443	480	517
0.4	554	591	628	664	700	736	772	808	844	879
0.5	915	950	985	0.019	0.054	0.088	0.123	0.157	0.190	0.224
0.6	257	291	324	357	389	422	454	486	517	549

stress components in a composite turbine disk under the influence of centrifugal loads and temperature are a typical example. In this section we propose to use Hook's law relations between strain and stress components including the effect of thermoelasticity called the Duhamel–Neumann law. Our concept is based on determining the stress parameters using equations from Hook's Duhamel–Neumann law and finding results from positive and negative values depending on changing temperature and coefficient of thermal expansions for a composite turbine disk.

We used the following assumptions:

1. Mechanical stresses calculated from the Timoshenko–Lechnitskii equations are dependent on geometrical and physical parameters and independent of temperature.

2. Under certain conditions a composite material can be designed with a modulus of elasticity, which is independent of temperature and applied stress [9].

Graphite–epoxy composite is a superhybrid resin matrix composite with a high epoxy content that is reinforced with graphite or carbon particles. Carbon fiber reinforced carbon [also called carbon–carbon (C/C)] is a composite material consisting of carbon fiber reinforcement in a matrix of graphite that keeps a high temperature.

Carbon–carbon is well suited to structural applications at high temperatures, or where thermal shock resistance and/or a low coefficient of thermal expansion is needed. Strength prediction of anisotropic (carbon–carbon) turbine disk requires a failure criterion accounting for all of the stress components including the thermal components under uniaxial, biaxial, and triaxial stress conditions. The strength of the composites can be predicted using fourth-order polynomial criteria. Fourth-order polynomial criteria have better approximation to experimental data than second-order polynomials.

The function of thermal stress components can be given using the parametric method.

5.2.2 Theoretical Investigation

When properties of an orthotropic disk have been changed under acting loads and temperature, the stress–strain relations must be assigned in matrix form [10].

$$\sigma_{ij} = Q_{ij} * (\varepsilon_j - \alpha_j * T) \quad (i,j = 1,2,6) \tag{5.10}$$

where
Q_{ij} = Stiffness constants
α_j = Coefficients of temperature expansion
T = Environmental temperature

For orthotropic materials we have 10 stiffness constants (see Equation 2.12). Prof. Parton [11] called this relation the Duhamel–Neumann law. Here, E_1, E_2 are moduli of normal elasticity in warp and fill directions, and μ_{12}, μ_{21}, μ_{23}, μ_{32} are Poisson's ratio of material. The first symbol designates the direction of force and the second symbol designates the direction of transverse deformation. α_1, α_2 are coefficients of thermal expansion in warp and fill directions.

$$\sigma_1^T = Q_{11}\varepsilon_1 + Q_{12}\varepsilon_2 - Q_{13}T$$

$$\sigma_2^T = Q_{21}\varepsilon_1 + Q_{22}\varepsilon_2 - Q_{23}T$$

$$\tau_{12}^T = \tau_{12} - G_{12}\alpha_{12}T \tag{5.11}$$

$$\tau_{21}^T = \tau_{21} - G_{21}\alpha_{21}T$$

Replacing in Equation (5.11) Q_{ij} appropriate significance we get for orthotropic material:

$$\sigma_1^T = \frac{E_1}{1 - \mu_{12}\mu_{21}}\varepsilon_1 + \frac{\mu_{21}E_2}{1 - \mu_{12}\mu_{21}}\varepsilon_2 - \frac{E_1 T}{1 - \mu_{12}\mu_{21}}(\alpha_1 + \mu_{21}\alpha_2)$$

$$\sigma_2^T = \frac{\mu_{12}E_2}{1 - \mu_{12}\mu_{21}}\varepsilon_2 + \frac{E_2}{1 - \mu_{12}\mu_{21}}\varepsilon_1 - \frac{E_2 T}{1 - \mu_{12}\mu_{21}}(\mu_{12}\alpha_1 + \alpha_2)$$

$$\tau_{12}^T = \frac{\sigma_1^T + \sigma_2^T}{2} - G_{12}\gamma_{12}\alpha_{12}T \tag{5.12}$$

$$\tau_{21}^T = \frac{\sigma_1^T + \sigma_2^T}{2} - G_{21}\gamma_{21}\alpha_{21}T$$

where

σ_1^T, σ_2^T = Tensile (compression) stresses acting in planes 1, 2

τ_{12}^T, τ_{21}^T = Shear stresses acting along ground of plane

G_{12}, G_{21} = Modulus of shear

α_{12}, α_{21} = Coefficients of thermal expansions in arbitrary directions

Obviously, τ_{12}^T is not equal to τ_{21}^T. This effect was shown in the work of Goldenblat and Kopnov [12], in which fiberglass tubes were twisted in opposite directions at normal temperature. However, in those fibers that were twisted to the right side, the fiber carried the load. In those fibers that were twisted to the left, the resin carried the load.

   In another study [13], a strength theory was developed and strength criteria in tensor form were designated. These criteria can be used separately for tensile and compression loads. Strength criteria were also assigned in biaxial stress conditions. The deformations were measured by fiber-optic sensors and installed on the surface testing disk. Coefficient relative strength depends on quality materials and are determined experimentally by test patterns.

5.2.3 Experimental Investigation

The correlation between strain and stress following Hook's law, when acting normal stresses coincide with directions of elasticity symmetry (fiber direction) is:

$$\varepsilon_1 = \frac{\sigma_1}{E_{11}} - \frac{\mu_{21}}{E_{22}}\sigma_2$$

$$\varepsilon_2 = \frac{\sigma_2}{E_{22}} - \frac{\mu_{12}}{E_{22}}\sigma_1$$

(5.13)

From Equation (5.13), we replace σ_1, σ_2 on significance getting Prof. Timoshenko's correlate normal stresses with spread of rotation and geometrical parameters (Equation 5.14) for a rotating disk [14]

$$\sigma_1 = \frac{3+\mu_{21}}{8}\rho\omega^2\left(b_1^2 + a_1^2 - \frac{a_1^2 b_1^2}{r^2} - r^2\right)$$

$$\sigma_2 = \frac{3+\mu_{21}}{8}\rho\omega^2\left(b_1^2 + a_1^2 + \frac{a_1^2 b_1^2}{r^2} - \frac{1+3\mu_{12}}{3+\mu_{21}}r^2\right)$$

(5.14)

where

a, b, r = Geometrical dimensions of the gas turbine disk (Figure 5.4)

ρ = Density of material

ω = Velocity of rotation

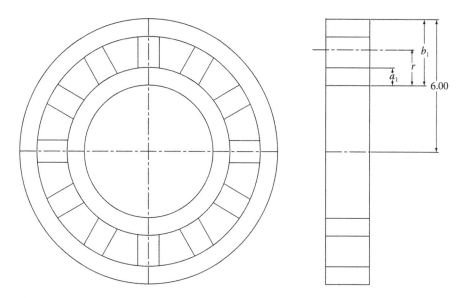

FIGURE 5.4
Gas turbine disk with blades.

Vibration aspects of the turbine disk can be seen if replaced with $\omega = 2\pi f H^2$. f is the frequency of disk vibration and H is the amplitude of vibration.

Deformations (Equation 5.13) after replacing σ_1, σ_2 from Equation (5.14) will be assigned as:

$$\varepsilon_1 = \frac{\rho\omega^2}{8}\left(\frac{3+\mu_{12}}{E_1}A - \frac{\mu_{21}(3+\mu_{21})}{E_2}B\right)$$

$$\varepsilon_1 = \frac{\rho\omega^2}{8}\left(\frac{3+\mu_{21}}{E_2}B - \frac{\mu_{12}(3+\mu_{12})}{E_2}A\right)$$

(5.15)

We designate the geometrical parameters:

$$A = b_1^2 + a_1^2 - \frac{a_1^2 b_1^2}{r^2} - r^2$$

$$B = b_1^2 + a_1^2 + \frac{a_1^2 b_1^2}{r^2} - \frac{1+3\mu_{12}}{1+\mu_{21}}$$

(5.16)

Now, we use Equations (5.15) and (5.16) to determine stress components with vibration aspects

$$\sigma_1^T = \frac{Q_{11}\rho\omega^2}{8}\left(\frac{3+\mu_{12}}{E_1}A - \frac{\mu_{21}(3+\mu_{21})}{E_2}B\right)$$

$$+\frac{Q_{22}\rho\omega^2}{8}\left(\frac{3+\mu_{21}}{E_2}B - \frac{\mu_{12}(3+\mu_{21})}{E_1}A\right) - Q_{13}T$$

$$\sigma_2^T = \frac{Q_{21}\rho\omega^2}{8}\left(\frac{3+\mu_{12}}{E_1}A - \frac{\mu_{21}(3+\mu_{21})}{E_2}B\right)$$

$$+\frac{Q_{22}\rho\omega^2}{\delta}\left(\frac{3+\mu_{21}}{E_2}B - \frac{\mu_{12}(3+\mu_{21})}{E_1}A\right) - Q_{23}T$$

(5.17)

For experimental data, we selected a gas turbine rotation disk with geometrical dimensions $a_1 = 0.4$ in, $b_1 = 4.0$ in, and radius $r = 3.0$ in (see Figure 5.1). The disk was fabricated from graphite–epoxy using a molding process, and had density of material $\rho = 0.1497 \times 10^{-3}$ lb/in^3.

The Poisson ratio was $\mu_{12} = \mu_{21} = 0.036$. The modulus of elasticity in warp direction was $E_1 = 25.1 \times 10^5$ psi, and in fill direction was $E_2 = 4.8 \times 10^5$ psi, which suggested strong anisotropy. Coefficients of thermal expansions were in warp direction $\alpha_1 = 0.34 \times 10^{-3}$ in/in °F, and in fill direction $\alpha_2 = 26.4 \times 10^{-3}$ in/in °F.

Next, the disk was fabricated from carbon–carbon, and had density of material $\rho = 0.195 \times 10^{-3}$ lb/in^3. The modulus of elasticity in warp direction was $E_1 = 12 \times 10^5$ psi, and in fill direction was $E_2 = 10 \times 10^5$ psi, and the Poisson ratio was $\mu_{12} = \mu_{21} = 0.036$.

Coefficients of thermal expansions were in warp direction $\alpha_1 = 3.8 \times 10^{-3}$ in/in °F, and in fill direction $\alpha_2 = 3.71 \times 10^{-6}$ in/in °F. Coefficients of thermal expansions in arbitrary directions were $\alpha_{12} = 2 \sin\theta^*\cos\theta\,(\alpha_1 - \alpha_2)$.

We calculated the stress components for rotation disks fabricated from graphite–epoxy and carbon–carbon when the velocity of rotation changed from 3627 to 4800 and 7000 rad/s. Significances of stress components for graphite–epoxy and carbon–carbon disks are represented in the work of Golfman [15].

Significances of stress components slowly reduced when the temperature increased to 500°F.

5.2.4 Conclusions

1. The mechanical and thermal properties of stress components were calculated for turbine composite disks fabricated from graphite–epoxy and carbon–carbon. All investigations were based on using simultaneously the Hook and Duhamel–Neumann laws.

2. The effect of thermoelasticity reduced stresses when the temperature increased to 500°F. The angles of inclination of the curves depended on physical properties and anisotropic materials.

3. This method of stress determination is very useful for researchers who select structural anisotropic materials to coincide stress components with fiber lay-up. We can predict strength characteristics and predict failure of turbine disks.

5.3 Strength Analysis of Turbine Engine Blades Manufactured from Carbon–Carbon Composites

5.3.1 Introduction

Turbine engine blades manufactured in a laminate form of carbon–carbon composite layers have been oriented in various directions. In the work of Pagano and Soni [16], two approximate analytical models were derived to describe the stress–strain field within each layer of a rotating turbine blade.

At the foundation of the model is the fundamental observation that the very small values of width–thickness ratio preclude the assumption of classical lamination theory. In such cases, the width is insufficient to promote the appropriate stress transfer mechanism necessary to develop the stress distribution given by classical lamination theory. The task of this section was to investigate the influence of the layers' orientation and the airstream's load on the stress distribution. We have assumed that the turbine blade is presented as a curvature plate with variable thickness and forces acting on it (see Figure 5.5).

Acting forces are the airstream's load, rotating forces, and bending and torsion moments. Pagano and Soni [16] assumed that the blade rotates about the X_r axis, which is parallel to X with a constant angular velocity (see Figure 5.6).

We have X, Y, Z Cartesian coordinates and polar coordinates R, θ, X_r. In Figure 5.6, we see a cut equilibrium element from a turbine blade and the normal and shear stress distribution in a top section, where F_z is the axial force, F_y is the tangential force, and $M_z, M_y,$ and M_x are the rotating moments relative to axis X_r. The cross-sectional area is denoted by A and the volume of the region above this plane by V. ρ denotes the mass density, ω is the angular velocity, r is the radius of rotating blade, and the components of the force acting on the cross section A are given by

$$F_z = \rho \omega^2 \int \sin \theta r^2 x \, dr \, d\theta \, dx \qquad (5.18)$$

$$F_y = \rho \omega^2 \int_v^v \cos \theta r^2 x \, dr \, d\theta \, dx \qquad (5.19)$$

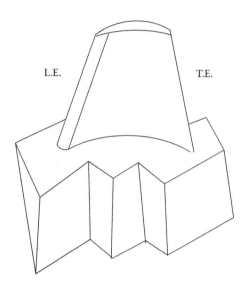

FIGURE 5.5
Rotor blade configuration.

While the moments about the x, y, z axis are:

$$M_x = \rho z_c \omega^2 \int_v r^2 \sin\theta \, dr \, d\theta \, dx \qquad (5.20)$$

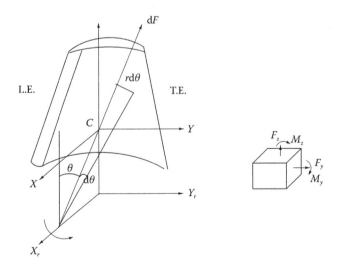

FIGURE 5.6
Turbine blade and coordinate system.

$$M_y = -\rho\omega^2 \int_v xr^2 \cos\theta \, dr \, d\theta \, dx \tag{5.21}$$

$$M_z = -\rho\omega^2 \int_v xr^2 \sin\theta \, dr \, d\theta \, dx \tag{5.22}$$

where z_c is a distance from point C, the center of acting hydrodynamic forces to axis y. θ is the angle of the rotating blade.

Equations (5.18) through (5.22) can be solved as:

$$F_z = -\rho\varpi^2 xr^3/3\cos\theta \tag{5.23}$$

$$F_y = \rho\varpi^2 xr^3/3\sin\theta \tag{5.24}$$

$$M_x^R = -\rho z_c \varpi^2 x \cos\theta r^3 /3 \tag{5.25}$$

$$M_y^R = -\rho\omega^2 x^2 \cos\theta r^3 /6 \tag{5.26}$$

$$M_z^R = -\rho\omega^2 x^2 \cos\theta r^3 /6 \tag{5.27}$$

5.3.2 Theoretical Investigation

In the linear performance relationship between acting forces and linear deformation in matrix form the following was established:

$$
\begin{array}{cccccc}
F_z, F_y & Q_\varepsilon & Q_{ex} & Q_{eR} & Q_{e\theta} & \varepsilon \\
M_z^R & Q_{xe} & Q_x & Q_{xR} & Q_{x\theta} & \Upsilon_x \\
M_y^R & Q_{R\varepsilon} & Q_{Rx} & Q_R & Q_{R\theta} & \Upsilon_R \\
M_x^R & Q_{\theta\varepsilon} & Q_{\theta x} & Q_{\theta y} & Q_\theta & \Upsilon_\theta
\end{array}
\tag{5.28}
$$

where, R, θ, X are the current polar coordinates; ε, Υ_x, Υ_R, Υ_θ are the current deformation existing from the acting forces F_z, F_y, and bending moments M_z, M_y, and torsion moment M_x. Q_{ki} coefficient of stiffness in polar coordinates and $Q_{ki} = Q_{ik}$ (k, $i = \varepsilon$, x, R, θ).

This system follows the theory of Kerhgofa–Klebsha transformed on five independent relations in Equation (1.3)

$$\varepsilon z = \frac{F_z}{E_1 S_1} ; \quad \varepsilon y = \frac{F_y}{E_2 S_2} ; \quad \Upsilon_x = \frac{M_z^R}{E_1 Iz} ; \quad \Upsilon_R = \frac{M_y^R}{E_1 I_R} ; \quad \Upsilon_\theta = \frac{M_x^R}{G_{xz} T_x}$$

where

E_1, E_2 = Modulus of elasticity
S_1, S_2 = Section area
I_z, I_R = Moments of inertia for axes x, R
G_{xz} = Shear modulus
T_x = Geometrical stiffness for torsion

The moment of inertia for axes x, R should be determined as Equation (1.4).

$$I_z = \int_s y^2 \, dS; \quad I_r = \int_s z^2 \, dS$$

The geometrical stiffness for torsion should be determined as:

$$T_x = \int_s (R^2 \theta \, dS)$$

If we assume that stress components in the top section in Figure 5.6 can be found using equations designated in the polar coordinates:

$$\sigma_r = \frac{1}{r} \times \frac{d\Phi}{dr} + \frac{1}{r^2} \times \frac{d^2\Phi}{d\theta^2} + Z(r,\theta) \tag{5.29}$$

$$\sigma_\theta = \frac{d^2\Phi}{dr^2} + Y(r,\theta) \tag{5.30}$$

$$\tau_{r\theta} = \frac{1}{r^2} \times \frac{d\Phi}{d\theta} - \frac{1}{r} \times \frac{d^2\Phi}{drd\theta} = -\frac{d}{dr}\left(\frac{1}{r} \times \frac{d\Phi}{d\theta} \right) \tag{5.31}$$

Here, Φ is the stress function as a function of variables r and θ. We also assume that $Z(r,\theta) = Y(r,\theta) = -\int R \, dr$, where R is a body force, $R_r = F_z$ and $R_\theta = Fy$; therefore, Equation (5.28) will result in the following:

$$\sigma_r = \frac{1}{r} \times \frac{d\Phi}{dr} + \frac{1}{r^2} \times \frac{d^2\Phi}{d\theta^2} - \int F_z(r,\theta) \tag{5.32}$$

$$\sigma_\theta = \frac{d^2\Phi}{dr^2} - \int F_y(r,\theta) \tag{5.33}$$

$$\tau_{r\theta} = \frac{1}{r^2} \times \frac{d\Phi}{d\theta} - \frac{1}{r} \times \frac{d^2\Phi}{drd\theta} = -\frac{d}{dr}\left(\frac{1}{r} \times \frac{d\Phi}{d\theta}\right) \quad (5.34)$$

The stress function Φ following Chou and Pagano [17] can be found as:

$$\Phi = M\psi \quad (5.35)$$

Here, coefficient M has a constant value and ψ is a geometrical profile function. In the work of Golfman [18], the contour of aviation profile ψ is represented by the following equation:

$$\lambda x^2 - k(y - \alpha)(y - \beta) = 0 \quad (5.36)$$

The coefficient of profile K can be found as:

$$K = \frac{27}{16} \frac{e_k^2 \lambda}{\alpha - \beta} \quad (5.37)$$

where
e_k = Maximum length of the profile
α = Distance from exit x to exit edge
β = Distance from exit x to entry edge
λ = Coefficient of anisotropy

$$\lambda = \frac{G_{yz}}{G_{xz}} \quad \text{for skin layers} \quad (5.38)$$

G_{yz}, G_{xz} are the modulus of shear in xz, yz interlaminar directions.
For a composite sandwich carbon fiber epoxy–structure, the modulus of shear can be determined as follows:

$$\lambda = \frac{G_{xz}^s + G_{yz}^h}{G_{xz}^s + G_{yz}^h} \quad (5.39)$$

The shear modulus with the index s is designated for the skin layers; the shear modulus with index h is designated for the sandwich layers.
The functions' relationship between Cartesian and polar coordinates can be shown [17] in Equation (5.40):

$$x = r\cos\theta; \quad y = r\sin\theta; \quad r^2 = x^2 + y^2 \quad (5.40)$$

$$\frac{dr}{dx} = \frac{x}{r} = \cos\theta; \quad \frac{dr}{dy} = \frac{y}{r} = \sin\theta; \quad \frac{d\theta}{dx} = -\frac{y}{r^2} = -\frac{\sin\theta}{r}; \quad \frac{d\theta}{dy} = \frac{x}{r^2} = \frac{\cos\theta}{r}$$

If we replace x and y in Equation (5.40), the contour of aviation profile would be shown as:

$$\psi = \psi r^2 \cos^2\theta - k(r\sin\theta - \alpha)(r\sin\theta - \beta) = 0 \qquad (5.41)$$

Thus the stress function Φ from Equation (5.35) can be shown in polar coordinates as:

$$\Phi = M(\lambda r^2 \cos^2\theta - k(r\sin\theta - \alpha)(r\sin\theta - \beta)) \qquad (5.42)$$

The complementary equation between Cartesian and polar coordinates would follow [19]:

$$\frac{d^2\Phi}{dr^2} + \frac{1}{r}\frac{d\Phi}{dr} + \frac{1}{r^2}\frac{d^2}{d\theta^2} = \frac{d^2\Phi}{dx^2} + \frac{d^2\Phi}{dy^2} \qquad (5.43)$$

The left side of the complementary Equation (5.43) represents the normal radial stresses acting in r direction, while the outside forces represent a combination of tension forces F_z, F_y and bending moments M_z^r and M_y^r.

From the integrating stress function $\Phi = M\psi = M[\lambda x^2 - k(y - \alpha)(y - \beta)]$ relative to x and y we get:

$$\frac{d^2\Phi}{dx^2} = M\left[2\lambda - k(y-\alpha)(y-\beta)\right]; \quad \frac{d^2\Phi}{dy^2} = M\left[\lambda x^2 - 2k + k(\beta+\alpha-\alpha\beta)\right] \quad (5.44)$$

Therefore, by adding this function we can find coefficient M.

$$M[2\lambda - k(y-\alpha)(y-\beta)] + [\lambda x^2 - 2k + k(\beta + \alpha - \alpha\beta)] = F_z + F_y + M_z + M_y$$

$$M = \frac{F_z + F_y + M_z + M_y}{\left[2\lambda - k(y-\alpha)(y-\beta)\right] + \left[\lambda x^2 - 2k + k(\beta+\alpha-\alpha\beta)\right]} \qquad (5.45)$$

We already know the stress function Φ in polar coordinates (Equation 5.41), so we can find all the components necessary to determine the normal and shear stresses (Equation 5.33)

$$\frac{d\Phi}{dr} = 2M\Big[\lambda r \cos^2\theta - k(\sin\theta - \alpha)(\sin\theta - \beta)\Big]$$

$$\frac{d^2\Phi}{dr^2} = 2M\Big[\lambda \cos^2\theta - k(\sin\theta - \alpha)(\sin\theta - \beta)\Big]$$

$$\frac{d\Phi}{dr} = 2M\Big[2\lambda r \cos\theta - k(\sin\theta - \alpha)(\sin\theta - \beta)\Big] \qquad (5.46)$$

$$\frac{d^2\Phi}{d\theta^2} = 2M\Big[-2\lambda r \sin\theta + k(\sin\theta - \alpha)(\sin\theta - \beta)\Big]$$

Now, for the normal and shear stresses we find:

$$\sigma_r = 2M\left(\lambda\cos h^2\theta - \frac{2\lambda}{r\sin\theta} - \frac{\rho\omega^2 r}{3\cos\theta}\right)$$

$$\sigma_\theta = 2M\Big[\lambda\cos^2\theta - k(\sin\theta - \alpha)(\sin\theta - \beta)\Big] - \frac{\rho\omega^2 r}{3\sin\theta} \qquad (5.47)$$

$$\tau_{r\theta} = 2M\left[2\lambda\cos\theta - \frac{k}{r(\cos\theta - \alpha)(\cos\theta - \beta)}\right]$$

In the case of torsion turbine blades, the equation of compatibility can be shown as:

$$\frac{d\tau_{xz}}{dy} - \frac{d\tau_{yz}}{dx} = (-2G_{xz} - 2G_{yz})\Upsilon \qquad (5.48)$$

Here, the shear stresses are represented as:

$$\tau_{xz} = \frac{d\Phi}{dy}; \quad \tau_{yz} = \frac{d\Phi}{dx} \qquad (5.49)$$

The stress function is represented in Equation (5.37). In this case, the necessary components look like

$$\frac{d\Phi}{dy} = M_1\Big[\lambda x^2 - k(2y - \beta - \alpha + \alpha\beta)\Big] \qquad (5.50)$$

$$\frac{d\Phi}{dx} = M_1\Big[2\lambda x - k(y^2 - y\beta - y\alpha + \alpha\beta)\Big]$$

By substituting Equation (5.45) into Equation (5.43), we find coefficient M_1

$$M_1 = \frac{(-2G_{xz} - 2G_{yz})\Upsilon}{\lambda x(x-2) - k\left[y^2 - \alpha(1+y) - \beta(1+y) + 2\alpha\beta\right]} \tag{5.51}$$

5.3.3 Airstream Load

We propose that between airstream load and stress rotating components there exists a linear correlation.

$$\sigma_{r\max} = q_a \sigma r$$

$$\sigma_{\theta\max} = q_a \sigma_\theta \tag{5.52}$$

$$\tau_{r\theta\max} = q_a \tau_{r\theta}$$

Here, q_a = airstream coefficient ($q_a \geq 1$); σ_r, σ_θ, $\tau_{r\theta}$ = normal and shear stresses that were determined using Equation (5.46). Calculations show that if the coefficient q_a increases by two times and normal stresses also increase, shear stresses can be reached with threat to strength. Finally, analog results can be shown if acting only on the bending moments.

This method to determine stress components is very important when we try to predict durability of turbine blades using strength criteria in biaxial stress conditions [20].

$$\frac{\sigma_r^2 + c\sigma_\theta^2 + d\tau_{r\theta}^2 + s\sigma_r\sigma_\theta}{\left(\sigma_r^2 + c\sigma_\theta^2 + d\tau_{r\theta}^2 + \sigma_r\sigma_\theta\right)^{1/2}} \leq [\sigma_{br}] \tag{5.53}$$

Here

$$[\sigma_{br}] = \frac{\sigma_{br}}{K_0}; \quad c = \frac{\sigma_{br}}{\sigma_{b\theta}}; \quad d = \frac{\sigma_{br}}{\tau_{br\theta}}; \quad s = \frac{4\sigma_{br}}{\sigma_{br\theta}^{45}} - c - d - 1$$

and σ_r, σ_θ, $\tau_{r\theta}$ are normal and shear stresses, respectively. Coefficients c, d, s are the relative strengths and are dependent on the quality of materials and are determined experimentally. K_0 is the factor of safety that is the ratio of strength and the resultant durability stress of the material.

5.3.4 Experimental Investigation

We have evaluated and selected two schemes of fiber orientation. The first case was the orientation for layers of perpendicular axis z. The second case was the orientation for layers with a parallel axis z. The second case represents a new technology similar to when fiber orientation is consistent with a force in the F_z direction. Similarly, technology has been used to grow crystals for silicon–carbon turbine blades.

In the second scheme each layer represents an orthotropic component when acting only with normal stresses [21]

$$\sigma_r = \frac{\rho\omega^2 r^2}{9 - k^2}(3 + \mu_{r\theta})\left[\left(\frac{r}{l_1}\right)^{k-3} - 1\right]$$

$$\sigma_\theta = \frac{\rho\omega^2 r^2}{9 - k^2}(3 + \mu_{r\theta})\left[\left(\frac{r}{l_1}\right)^{k-3} k - \frac{k^2 + 3\mu_{r\theta}}{3 + \mu_{r\theta}}\right]$$

(5.54)

These stresses can reach maximum values when radius $r = l_1^{k-3}\left(\dfrac{2}{k-1}\right)^{1/2}$ and maximum radial stress for orthotropic components is:

$$\sigma_{r\max} = \frac{\rho\omega^2 l_1^2}{9 - k^2}(3 + \mu_{r\theta})\left[\left(\frac{2}{k-1}\right)^{k-1/k-3} - \left(\frac{2}{k-1}\right)^{2/k-9}\right]$$

(5.55)

Maximum tangential stress σ_θ can be found when:

$$r = l_1\left[\frac{2(k^2 + 3 + \mu_{r\theta})}{K(3 + \mu_{r\theta})(k-1)}\right]^{1/2}$$

(5.56)

$$\sigma_{\theta\max} = \frac{\rho\omega^2 l_1^2}{9 - K^2}(3 + \mu_{r\theta})\left[\left(\frac{2(k^2 + 3 + \mu_{r\theta})}{k(3 + \mu_{r\theta})(k-1)}\right)^{k-1/k-3} k - \frac{k^2 + 3\mu_{r\theta}}{(3 + \mu_{r\theta})}\left[\frac{2(k^2 + 3 + \mu_{r\theta})}{k(3 + \mu_{r\theta})(k-1)}\right]^{2/k-3}\right]$$

(5.57)

Coefficient $k = (E_\theta/Er)^{1/2}$ and E_r, E_θ are the moduli of elasticity in radial and tangential directions. If $K = 4$ for cord composites, $\mu_{r\theta}$ is the Poisson ratio = 0.23, the length blade = 150 mm, the high blade = 7.5 mm, and the resultant equations are:

$$\sigma_{r\max} = 0.0658\rho\omega^2 l_1^2(r = 0.8171); \quad \sigma_{\theta\max} = 0.596\rho\omega^2 l_1^2(r = 0.8741) \quad (5.58)$$

The significance of the maximum radial and tangential stresses (Equation 5.59) can be influenced by natural frequencies in tangential and radial directions.

The relationship between maximum tangential stress $\sigma_{\theta max}$ and maximum radial stress σ_{rmax} is equal to 9.05 ($\sigma_{\theta max}/\sigma_{rmax} = 9.05$). It means that maximum radial stresses are nine times less than maximum tangential stress.

Following Golfman [22], the relationship between natural frequencies can be determined as:

$$\frac{f_{11}^{\theta}}{f_{11}^{R}} = \left(\frac{E_{\theta}}{E_{R}}\right)^{1/2} \tag{5.59}$$

In the scheme of using cord composites, the coefficient $k = (E_{\theta}/E_{r})^{1/2}$ is equal to 4, so the natural frequencies in the radial direction are four times less than in the tangential direction.

5.3.5 Conclusions

1. Variable methods are offered to calculate the stress components for gas turbine blades in Cartesian and polar coordinates.

2. It is assumed that between airstream load and stress components there exists a linear correlation. The calculations show that if the coefficient q_a increases two times and normal stresses also increase, the shear stresses can reach a threshold of failure. The analog result can only happen if acting only on the bending moments.

3. We recommend that one uses strength criteria for predicting the durability of composite turbine blades.

4. The natural frequencies of the turbine blades can be put in a favorable area outside of the force vibration frequencies to avoid parametric resonance.

5. By changing the scheme ply orientation, we have shown the influence on values of interlaminar shear stress and the increase in the durability service for gas turbine blades.

5.4 Stress and Vibration Analysis of Composite Propeller Blades and Helicopter Rotors

5.4.1 Introduction

Composite propeller blades, helicopter rotors, and fans have been manufactured in a laminate form from carbon fiber reinforced epoxy where the blade skin is made of aramid fiber reinforced epoxy resin. Additional reinforcement in the nose and trailing edge is made for the protection of the laminated layers. The carbon fibers with reinforced epoxy have been oriented in various directions.

Changing the fiber direction of the layers can turn the torsion and bending stiff noses of composite blades. Therefore, the natural frequencies of propeller blades can be favorably placed in an area outside the operational rpm range.

This section relates the evaluation of the force and free vibration frequencies for the purpose of avoiding air and ground resonance.

In the early 1970s, composite propeller blades with reinforced epoxy were utilized throughout the world [23]. Propeller blades and fans were manufactured from fiberglass that was protected by special polymers [24]. For impact protection the leading edge was covered by thin copper inputted in the prepreg package during compression molding. The history of the manufacturing process is very significant because it relates how the configuration and architecture of design was determined.

A new chapter in aviation history opened recently with the maiden flight of the world's first civil tilt rotor, the Bell/Augusta Aerospace BA609 [25].

The rotors were installed in the vertical position and hovered at an altitude of 50 ft, performed left and right pedal turns, both forward and aft flight maneuvers, and four takeoffs and landings.

When the rotors tilt forward in the horizontal position, the aircraft is able to fly as a turboprop fixed-wing airplane.

The transition from helicopter mode to airplane mode takes 2 s, as does the transition from airplane mode to helicopter mode.

For the design and manufacturing of the prototype YUH-60A UTTAS tail rotor the pultrusion process [26] was selected. A fully cured pultruded spar design layout and manufacturing approach was developed. This is important since the pultruded spar consists of the two blades being manufactured simultaneously.

Elastic coupling, which has a significant effect on the dynamic elastic torsion response of the rotor, was considered by Smith [27].

5.4.2 Stress Analysis of Propeller Blades

We have assumed that the propeller blade as presented in Figure 5.7 is a curved plate with variable thickness and forces acting on it. Acting forces include airstream load, rotating forces, and bending and torsion moments. It is assumed that the blade rotates about the Z axis, with a constant angular velocity ω (see Figure 5.7). We have X, Y, and Z Cartesian coordinates and polar coordinates R, θ, and Z.

In Figure 5.7, we see a cut through the equilibrium element for a rotor blade and the normal and shear stress distribution in the top section. Here, F_z is the axial force; F_x, F_y are the tangential forces; and M_z, M_y, and M_x are rotating moments relative to axes Z, Y, X, respectively. The cross-sectional area is

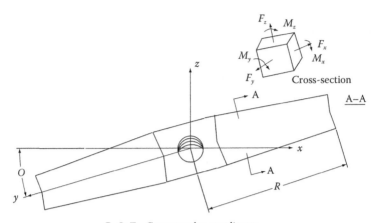

R, O, Z—Current polar coordinates

X, Y, Z—Current Cartesian coordinates

FIGURE 5.7
Helicopter rotor with composite blades.

denoted by A–A and the volume of the region above this plane by V, while ρ is designated the mass density, ω the angular velocity, r is the radius of rotating blade, and the components of the force acting on the cross section A–A are given by

$$F_z = \rho\omega^2 \int_v \sin\theta r^2 x \, dr \, d\theta \, dx \tag{5.60}$$

$$F_x = F_y = \rho\omega^2 \int_v \cos\theta r^2 x \, dr \, d\theta \, dx \tag{5.61}$$

While the moments about the x, y, z axes are:

$$M_x = \rho\omega^2 \int r^2 \sin\theta \, dr \, d\theta \, dx \tag{5.62}$$

$$M_y = \rho\omega^2 \int xr^2 \cos\theta \, dr \, d\theta \, dx \tag{5.63}$$

$$M_z = \rho\omega^2 \int xr^2 \sin\theta \, dr \, d\theta \, dx \tag{5.64}$$

where θ is the angle of the rotating blade.

Equations (5.60) through (5.64) can be solved as:

$$F_z = -\rho\omega^2 x r^3 / 3\cos\theta \tag{5.65}$$

$$F_x = F_y = \rho\omega^2 x r^3 / 3\sin\theta \tag{5.66}$$

$$M_x^R = \frac{-\rho\omega^2 x \cos\theta r^3}{3} \tag{5.67}$$

$$M_y^R = \frac{-\rho\omega^2 x^2 \cos\theta r^3}{6} \tag{5.68}$$

$$M_z^R = \frac{-\rho\omega^2 x^2 \sin\theta r^3}{6} \tag{5.69}$$

5.4.3 Theoretical Investigation

In the linear performance relationship between the acting forces and the linear deformation in matrix form the following was established:

$$
\begin{array}{cccccc}
F_z, F_y, F_x & Q_\varepsilon & Q_{x\varepsilon} & Q_\varepsilon R & Q_{\varepsilon\theta} & \varepsilon \\
M_z^R & Q_{x\varepsilon} & Q_x & Q_{xR} & Q_{x\theta} & \gamma_Z \\
M_y^R & Q_{R\varepsilon} & Q_{Rx} & Q_R & Q_{R\theta} & \gamma_R \\
M_x^R & Q_{\theta\varepsilon} & Q_{\theta x} & Q_{\theta y} & Q_\theta & \gamma_\theta
\end{array}
\tag{5.70}
$$

where R, θ, Z are the current polar coordinates; ε, Y_x, Y_R, Y_θ are the current deformations existing from the acting forces F_z, F_y, F_x and bending moments M_z, M_y, and torsion moment M_x. Q_{ki} is the coefficient of stiffness in the polar coordinates and $Q_{ki} = Q_{ik}$ (k, $I = \varepsilon$, x, R, θ).

This system follows the theory of Kerhgofa–Klebsha and has been transformed on five independent relationships presented in Equation (1.3).

$$\varepsilon_z = \frac{F_z}{E_1 S_1}; \quad \varepsilon_y = \frac{F_y}{E_2 S_2}; \quad \gamma_z = \frac{M_z^R}{E_1 I_z}; \quad \gamma_R = \frac{M_y^R}{E_1 I_R}; \quad \gamma_\theta = \frac{M_x^R}{G_{xz} T_x}$$

Here, E_1, E_2 are the modulus of elasticity; S_1, S_2 are the sections area; I_z, I_R are the moments of inertia for axes x, R; G_{xz} is a shear modulus; and T_x is the geometrical stiffness for torsion. T_x is the moment of inertia for axis x.

These parameters can be determined as Equation (1.4).

$$I_z = \int y^2 \, dS; \quad I_R = \int z^2 \, dS$$

The geometrical stiffness for torsion can be determined as:

$$T_x = \int_S (R^2 \theta \, dS)$$

If we assume that the stress components in the top section can be found using Equations (5.29) through (5.31) designated in the polar coordinates, we get:

$$\sigma_\theta = \frac{d^2\Phi}{dr^2} + Y(r,\theta)$$

$$\tau_{r\theta} = \frac{1}{r^2} \times \frac{d\Phi}{d\theta} - \frac{1}{r} \times \frac{d^2\Phi}{drd\theta} = \frac{d}{dr}\left(\frac{1}{r} \times \frac{d\Phi}{d\theta}\right)$$

where Φ is the stress function and a function of variables r and θ.
   We also assume that:

$$Z(r,\theta) = Y(r,\theta) = \int R \, dr,$$

where R is a body force. $R_r = F_z$ and $R_\theta = Fy$; therefore, Equations (5.32) through (5.34) will result in the following:

$$\sigma_r = \frac{1}{r} \times \frac{d\Phi}{dr} + \frac{1}{r^2} \times \frac{d^2\Phi}{d\theta^2} - \int F_z(r,\theta)$$

$$\sigma_\theta = \frac{d^2\Phi}{dr^2} - \int F_y(r,\theta)$$

$$\tau_{r\theta} = \frac{1}{r^2} \times \frac{d\Phi}{d\theta} - \frac{1}{r} \times \frac{d^2\Phi}{drd\theta} = -\frac{d}{dr}\left(\frac{1}{r} \times \frac{d\Phi}{d\theta}\right)$$

The stress function Φ seen in the work of Chou and Pagano [28] can be found as Equation (5.35)

$$\Phi = M\psi$$

Here, coefficient M has a constant value and ψ is a geometrical profile function.
   In the work of Golfman [29], the aviation profile ψ is represented by Equation (5.36)

$$\lambda x^2 - k(y - \alpha)(y - \beta) = 0$$

The coefficient of profile K can be found as Equation (5.37)

$$K = \frac{27}{16} \frac{e_k^2 \lambda}{\alpha - \beta}$$

where
e_k = Maximum length of profile
α = Distance from exit x to exit edge
β = Distance from exit x to entry edge
λ = Coefficient of anisotropy

Therefore,

$$\lambda = \frac{G_{yz}}{G_{xz}}$$

represents Equation (5.38) for the skin layers, where

G_{yz}, G_{xz} = modulus of shear in xz, yz interlaminar directions

For a composite carbon fiber–epoxy sandwich structure the modulus of shear can be determined as Equation (5.39).

$$\lambda = \frac{G_{xz}^s + G_{yz}^h}{G_{xz}^s + G_{yz}^h}$$

The shear modulus with the index s can be designated for the skin layers; the shear modulus with index h can be designated for the sandwich layers.

The functions' relationship between Cartesian and polar coordinates can be shown in Equation (5.40):

$$x = r\cos\theta; \quad y = r\sin\theta; \quad r^2 = x^2 + y^2$$

$$\frac{dr}{dx} = \frac{x}{r} = \cos\theta; \quad \frac{dr}{dy} = \frac{y}{r} = \sin\theta; \quad \frac{d\theta}{dx} = \frac{y}{r^2} = \frac{\sin\theta}{r}; \quad \frac{d\theta}{r^2} = \frac{x}{r^2} = \frac{\cos\theta}{r}$$

If we replace x and y in Equation (5.40), the contour of the aviation profile can be shown as Equation (5.41)

$$\psi = \lambda r^2 \cos^2\theta - k(r\sin\theta - \alpha)(r\sin\theta - \beta) = 0$$

The stress function Φ from Equation (5.35) can be shown in polar coordinates as:

$$\Phi = M(\lambda r^2\cos^2\theta - k)(r\sin\theta - \alpha)(r\sin\theta - \beta)$$

The complementary equation between Cartesian and polar coordinates, Equation (5.42) would follow [28]

$$\frac{d^2\Phi}{dr^2} + \frac{1}{r}\frac{d\Phi}{dr} + \frac{1}{r^2}\frac{d^2}{d\theta^2} = \frac{d^2\Phi}{dx^2} + \frac{d^2\Phi}{dy^2}$$

The left side of complementary Equation (5.42) represents the normal radial stresses acting in the r direction, while the outside forces represent a combination of tension forces F_z, F_y and bending moments M_z^r and M_y^r.

From the integrating stress function, which is relative to x and y, we get Equation (5.44).

$$\frac{d^2\Phi}{dx^2} = M\left[2\lambda - k(y-\alpha)(y-\beta)\right]; \quad \frac{d^2\Phi}{dy^2} = M\left[\lambda x^2 - 2K + k(\beta+\alpha-\alpha\beta)\right]$$

Therefore, by adding this function we can find coefficient M (see Equation 5.45).

$$M\left[2\lambda - k(y-\alpha)(y-\beta)\right] + \left[\lambda x^2 - 2k + k(\beta+\alpha-\alpha\beta)\right] = F_z + F_y + M_z + M_y$$

$$M = \frac{F_z + F_y + M_z + M_y}{\left[2\lambda - k(y-\alpha)(y-\beta)\right] + \left[\lambda x^2 - 2k + k(\beta+\alpha-\alpha\beta)\right]}$$

We already know the stress function Φ in polar coordinates (Equation 5.35), so we can find all the components necessary to determine the normal and shear stresses (Equation 5.46).

$$\frac{d\Phi}{dr} = 2M\left[\lambda r\cos^2\theta - k(\sin\theta - \alpha)(\sin\theta - \beta)\right]$$

$$\frac{d^2\Phi}{dr^2} = 2M\left[\lambda r\cos^2\theta - k(\sin\theta - \alpha)(\sin\theta - \beta)\right]$$

$$\frac{d\Phi}{d\theta} = 2M\left[\lambda r\cos\theta - k(\sin\theta - \alpha)(\sin\theta - \beta)\right]$$

$$\frac{d^2\Phi}{d\theta^2} = 2M\left[-2\lambda r\sin\theta - k(\sin\theta - \alpha)(\sin\theta - \beta)\right]$$

Now, for the normal and shear stresses we find:

$$\sigma_r = 2M(\lambda \cos^2 \theta - 2\lambda/r \sin \theta - \rho \omega^2 r/3 \cos \theta)$$

$$\sigma_\theta = 2M\left[\lambda \cos^2 \theta - k(\sin \theta - \alpha)(\sin \theta - \beta)\right] - \rho \omega^2 r^3/3 \sin \theta$$

$$\tau_{r\theta} = 2M\left[2\lambda \cos \theta - k/r(\cos \theta - \alpha)(\cos \theta - \beta)\right]$$

In the case of torsion rotor blades, the equation of compatibility can be shown as Equation (5.48)

$$\frac{d\tau_{xz}}{dy} - \frac{d\tau_{yz}}{dx} = (-2G_{xz} - 2G_{yz})\Upsilon$$

Here, the shear stresses are represented as Equation (5.49)

$$\tau_{xz} = \frac{d\Phi}{dy}; \quad \tau_{yz} = \frac{d\Phi}{dx}$$

The stress function is represented in Equation (5.41).
    In this case the necessary components look like:

$$\frac{d\Phi}{dy} = M_1\left[\lambda x^2 - k(2y - \beta - \alpha + \alpha\beta)\right]$$

$$\frac{d\Phi}{dx} = M_1\left[2\lambda x - k(y^2 - y\beta - y\alpha + \alpha\beta)\right]$$

By substituting Equation (5.47) into Equation (5.49), we find coefficient M_1 (see Equation 5.51).

$$M_1 = \frac{(-2G_{xz} - G_{yz})\Upsilon}{\lambda x(x-2) - k\left[y^2 - \alpha(1+y) - \beta(1+y) + 2\alpha\beta\right]}$$

5.4.4 Vibration Analysis

The longitudinal motion of a stable helicopter is found to exhibit two modes, which are damped oscillations. The first mode with light damping and a relatively long period is called the long-period or phugoid mode. The second heavy damped mode is referred to simply as the short-period mode. The equation for forced vibration without damping is:

$$m\frac{\partial^2 \varpi}{\partial x^2} + Q_{11}x = (T - D - W \sin \theta)\sin \Omega t$$

where

   Ω = Forcing frequency acting from turbulence movement
   t = Time of a wave propagation
   m = Mass of the helicopter; all time is constant and does not depend on
        movement and attitude

We assume a periodic force of magnitude: $F = (T - D - W \sin \theta) \sin\Omega t$, where T is the thrust force, D is the drag force, and W is the weight of a helicopter. In the case of free vibration when turbulence movement does not exist $(T - D - W \sin \theta) \sin\Omega t = 0$, which has the following solution as [30]:

$$x = C_1\sin\varpi t + C_2\cos\varpi t \tag{5.71}$$

and where circular frequency $\varpi = (Q_{11}/m)^{1/2}$, where Q_{11} is the stiffness of the rotor blade and can be determined using Equation (5.32), and C_1 and C_2 are arbitrary constants.

We assume $C_1 = A \cos\phi$ and $C_2 = A \sin\phi$, where A is the amplitude of vibration and ϕ is the phase angle of vibration. We input this into Equation 5.72 and the displacement x will be:

$$x = A\cos\phi\sin\varpi t + A\sin\phi\cos\varpi t \tag{5.72}$$

or $x = A\sin(\varpi + \phi)$. We replace circular frequency $\varpi = 2\pi f$, where f is a motion frequency, so $x = \sin(2\pi f + \phi)$, and

$$f = \frac{1}{2\pi}\left(\frac{Q_{11}}{m}\right)^{1/2} \tag{5.73}$$

By differentiating Equation (5.73), we can determine velocity and acceleration of the longitudinal vibration:

$$V = (2\pi f + \phi)\cos(2\pi f + \phi) \tag{5.74}$$

$$a = -(w\pi f + \phi)^2\sin(2\pi f + \phi) \tag{5.75}$$

The equation for force vibration with damping in the longitudinal direction becomes:

$$m\frac{\partial^2\varpi}{\partial x^2} + \delta\frac{\partial\varpi}{x} + Q_{11}x = (T - D - W \sin \theta)\sin \Omega t \tag{5.76}$$

Here, m is a mass of a helicopter and δ is a critical damping coefficient. The particular solution that applies to the steady-state vibration of the helicopter should be a harmonic function of time such as [31]:

$$x_p = A\sin(\Omega t - \phi) \tag{5.77}$$

where A and ϕ are constant.

Substituting x_p in Equation (5.76), we get:

$$-m\Omega^2 A\sin(\Omega t - \phi) + \delta\Omega A\cos(\Omega t - \phi) + Q_{11}A\sin(\Omega t - \phi) = (T - D - W\sin\theta)\sin\Omega t \tag{5.78}$$

Submitting two boundary conditions $\Omega t - \phi = 0$ or $\Omega t - \phi = \pi/2$ results in:

$$A(Q_{11} - m\Omega^2) = (T - D - W\sin\theta)\sin\Omega t \text{ and } \delta\Omega A = (T - D - W\sin\theta)\sin\Omega t \tag{5.79}$$

The phase angle ϕ reflects a different phase between the applied force and the resulting vibration and is determined as:

$$\tan\phi = \frac{\delta\Omega}{Q_{11} - m\Omega^2} \tag{5.80}$$

The sin and cosine function can be eliminated from Equation (5.73):

$$A\delta\Omega + A(Q_{11} - m\Omega^2) = (T - D - W\sin\theta) \tag{5.81}$$

From Equation (5.80) the forcing frequency Ω can be determined if $\tan\phi = 1$ as:

$$\Omega = (Q_{11}A - (T - D - W\sin\theta)/mA)^{1/2} \tag{5.82}$$

The forcing frequency Ω has never been fit with a natural frequency f, which avoids parametric resonance.

Submitting two boundary conditions:

$$\Omega t - \delta = 0; \text{ or } \Omega t - \phi = \pi/2 \text{ results in:}$$

$$\delta\Omega A = (T - D - W\sin\theta)\sin\Omega t \tag{5.83}$$

$$-m\Omega^2 t^2 A + Q_{11}A = (T - D - W\sin\theta)\sin\Omega t$$

Thus the magnitude of amplitude changes from A_{1m} to A_{2m}:

$$A_{1m} = \frac{(T - D - W\sin\theta)\sin\Omega t}{\delta\Omega}; \quad A_{2m} \frac{(T - D - W\sin\theta)\sin\Omega t}{-m\Omega^2 t^2 + Q_{11}} \tag{5.84}$$

The electrical circuit for electronic countermeasures to compensate vibration and the critical damping coefficient δ for the carbon–epoxy composite of the blades can be determined as the relationship between the potential energy W and the energy lost during one deformation cycle dW.

$$\delta = \frac{dW}{W} = m\omega\lambda = m2\pi f^{1-v}\lambda \tag{5.85}$$

where
  m = Mass of a helicopter
  ω = Natural circular frequency; $\omega = 2\pi f^{1-v}$
  λ = Coefficient of internal friction
  f = Frequency of the cycle of variation of the deformation
  v = Exponent and is dependent on the frequency f

According to Bok, $v = 0$, while according to Fokht, $v = 1$ [32]. Fokht's hypothesis concerning the proportionality of the nonelastic stress to the frequency is not confirmed by experiment, while the Bok hypothesis is in better agreement with experimental results, at least in a rather wide range of frequencies.

One study [30] shows the critical damping coefficient δ for fiberglass for different angles relative to a warp/fill direction. The critical damping coefficient has also been determined in the process of determining the free vibration of the patterns. The coefficient of internal friction λ was found in the process of testing the fiberglass for durability. The ability of ultrasonic waves to travel in a web direction over a minimum time was also established by Golfman [33].

The velocity of ultrasonic waves propagation was determined as:

$$V_0 = \frac{L}{t}10^3 = Lf10^3$$

where
  L = Length between the two acoustic heads
  t = Time taken for the ultrasonic oscillations to reach from one head to the other
  f = Frequency of ultrasonic wave propagation

In the real dynamic conditions the internal friction that results from these actions has some delay to ultrasonic wave propagation.

$$V_d = Lf\,10^3\lambda$$

The velocity of ultrasonic waves in lattice structures will be:

$$E\alpha = V_\alpha^2\rho(1-\mu_{1\alpha}\mu_{2\alpha})$$

We determined the modulus of elasticity under angle α as:

$$E\alpha = V_\alpha^2\rho(1-\mu_{1\alpha}\mu_{2\alpha})$$

The transverse displacement S for a rotor blade in polar coordinates can be determined as [31]

$$m\frac{\partial^2 S}{\partial t^2} = F\left(\frac{\partial^2 Sx}{\partial x^2} + \frac{\partial^2 S_R}{\partial R^2} + \frac{\partial^2 S_\theta}{\partial \theta^2}\right) \tag{5.86}$$

where

m = Mass of rotor blade
F = Outside forces, $F = (T - D - W \sin \theta)\sin \Omega t$
S_x, S_R, S_θ = Transverse displacements in the polar coordinates

We can represent the section areas S_x, S_R, S_θ as:

$$S_x = \Upsilon_x X^2; \quad S_R = \Upsilon_R R^2; \quad S_\theta = \Upsilon_\theta \theta^2$$

We can replace the angle deformation following as:

$$\Upsilon_z = \frac{M_z^R}{E_1 I_z}; \quad \Upsilon_R = \frac{M_y^R}{E_1 I_R}; \quad \Upsilon_\theta = \frac{M_x^R}{G_{xz} T_x}$$

We now substitute the angle deformation in Equation (5.46), and after differentiation Equation (5.46) can be shown as:

$$m\frac{\partial^2 S}{\partial t^2} = 2(T - D - W \sin \theta)\sin \Omega t \left(\frac{M_z^R}{E_1 Iz} + \frac{M_y^R}{E_1 I_R} + \frac{M_x^R}{G_{xz} T_x}\right) \tag{5.87}$$

In Equation (5.54), we can substitute time t for natural frequency f and as a result we can find material properties structure as well as mass of rotor blade (m), critical damping coefficient (δ), stiffness of rotor blade (Q_{11}), vibration characteristics, forcing frequency (Ω), and natural frequency (f).

The natural frequency f can be found by solving the differential equation for an orthotropic rotor blade in a polar coordinate:

$$\frac{\partial^2 f}{\partial t^2} + \frac{g}{h\eta}\left(D_1 \frac{\partial^4 f}{\partial x^2} + 2D_3 \frac{\partial^4 f}{\partial x^2 \partial R^2} + D_2 \frac{\partial^4 f}{\partial \theta^2}\right) = 0 \tag{5.88}$$

Here D_{ij} is the stiffness of the rotor blade from the bending moment.

$$D_1 = Q_{11} S_x; \quad D_2 = Q_{22} S_R; \quad D_3 = (Q_{12} + 2Q_{66})S_\theta \tag{5.89}$$

Stiffness constants Q_{ij} are determined from Equation (1.28), only:
Q_{66} = $G_{12}T_x$, where T_x = geometrical stiffness
E_1 = Modulus of elasticity in the warp-x direction
E_2 = Modulus of elasticity in the fill-y direction
$\mu_{12}\mu_{21}$ = Poisson ratio
h = Height of the rotor blade
g = Density of the fiber
η = Acceleration due to gravity

In the case of a free vibration rotor blade, we use the boundary conditions:

$$\text{If } x = 0; \quad x = R; \quad f = 0$$

$$\frac{\partial^2 f}{\partial x^2} + \mu_{21}\frac{\partial^2 f}{\partial R^2} = 0$$

$$\text{If } z = 0; \quad z = h; \quad f = 0$$

$$\frac{\partial^2 f}{\partial x^2} + \mu_{12}\frac{\partial^2 f}{\partial R^2} = 0$$

where h is the height of the section of a propeller blade and R is the radius of a propeller blade. These boundary conditions are known by the function of deflections [28]

$$f_{mn} = \sin\frac{m\pi x}{R}\sin\frac{n\pi y}{h} \tag{5.90}$$

Here, m and n are whole digits and are determined as a number of semi-waves in the x and z directions.

By inputting Equation (5.58) into Equation (5.55), we designate k as the present geometrical parameter (relationship of radius and propeller blade to height).

We can determine natural frequencies f_{mn} as:

$$f_{mn} = \frac{\pi^2}{h^2}\left(\frac{g}{h\eta}\right)^{1/2}\left[D_1\left(\frac{m}{k}\right)^4 + 2D_3 n^2\left(\frac{m}{k}\right)^2 + 2D_2 n^4\right]^{1/2}$$

The frequency of the basic tone ($m = 1$, $n = 1$) will be:

$$f_{11} = \frac{\pi^2}{R^2}\left(\frac{g}{h\eta}\right)^{1/2}(D_1 + 2D_3 k^2 + D_2 k^4)^{1/2}$$

The frequency of the second tone ($m = 2, n = 2$) will be:

$$f_{22} = \frac{4\pi^2}{R^2}\left(\frac{g}{h\eta}\right)^{1/2}\left(D_1 + 2D_3k^2 + D_2k^4\right)^{1/2}$$

The frequency of the third tone ($m = 3, n = 3$) will be:

$$f_{33} = \frac{9\pi^2}{R^2}\left(\frac{g}{h\eta}\right)^{1/2}\left(D_1 + 2D_3k^2 + D_2k^4\right)^{1/2}$$

5.4.5 Experimental Analysis

The main failure mode of actual rotor blades includes bending and torsion moments. In the work of Hou and Gramoll [34], it was shown that the results of testing of conical lattice structures were not stable. The low failure was due to microbuckling and is commonly referred to as fiber kinking. Fiber kinking generally occurs because of a weak matrix, which is due in part to the epoxy not curing completely or a deficiency of hardening agent during the manufacturing process.

In our investigation the fiber density that was selected had $g = 1770$ kg/ m^3, and the acceleration due to gravity was $\eta = 9.81$ m/s^2. D_{ij} is a stiffness of the rotor blades from the bending moments and $D_1 = 145.2$ kg/m^2, $D_2 = 50.57$ kg/m^2, and $D_3 = 120.9$ kg/m^2.

All the parameters that were used for natural frequencies were determined. Small specimen static test results are shown in Table 5.7.

The value of the natural frequencies for the rotor blades depends on the variation of the geometrical parameters k, which are shown in Table 5.8.

The damping coefficient δ_{1111} in the warp direction according to Golfman [33] was 0.01825, δ_{2222} in the fill direction was 0.01933, δ_{1212} in the 45° diagonal direction was 0.02. Finally, the value of the natural frequencies was also dependent on the variation of the geometrical parameters.

In the case of vibration in the longitudinal direction when the mechanical amplitude was changed from $A_{m1} = A_{e1} = 1$ to $A_{m2} = A_{e2} = -1$, the force frequencies are changed to:

$$\Omega_1 = \frac{E_0}{-\sum Rt}; \quad \Omega_2 = -\left(\frac{E_0 - \sum 1/C}{\sum Lt^2}\right)^{1/2} \tag{5.91}$$

TABLE 5.7

Strength and Modulus Correlate with Temperature for Graphite–Epoxy Prepregs

Postcured	Graphite–Epoxy Prepreg Requirements			
	−65°F	RT	160°F	250°F
Flexural	at	200		
Strength, ksi	(0.95 RT)	(0.70 RT)	(0.65 RT)	
Flexural modulus, psi × 10⁶	(0.95–1.05 RT)	16–18 (0.95 RT)		(0.65 RT)
Interlaminar shear strength, psi	(0.95 RT)		12,000 (0.70 RT)	(0.65 RT)
Tensile strength, ksi	–	–	–	–
Tensile modulus, psi × 10⁶	–	–	–	–
Transverse tensile strain, in/in	–	4000	–	–
Postcured	Pultrusion (Average of 3 to 5 Specimens)			
	−65°F	RT	160°F	250°F
Flexural	272	213	130	
Strength, ksi				
Flexural modulus, psi × 10⁶	–	17.1	15.8	13.4
Interlaminar shear strength, psi	–	13,500	11,800	7600
Tensile strength, ksi	192.0	188.3	189.0	–
Tensile modulus, psi × 10⁶	18.0	17.8	17.9	18.1
Transverse tensile strain, in/in	–	–	–	–

We then input the natural frequencies $f = 1/t$ in Equation (5.59), and get:

$$\Omega_1 = \frac{E_0 f}{-\Sigma R} ; \quad \Omega_2 = -f\left(\frac{E_0 - \Sigma 1/C}{\Sigma L}\right)^{1/2} \tag{5.92}$$

If the force frequencies change due to variations, then voltage E_0, the inductance (L), resistance (R), and reciprocal of capacitance ($1/C$) change. The model 352C23ICP accelerometer and measures just 28 × 5.7 mm and weighs only 0.2 g. The sensor employs a shear-mode, piezoceramic element that generates a 5-mV/g output signal with a frequency response from 2 Hz to 10 kHz. The resonance frequency is specified as greater than 70 kHz.

We varied the voltage from 10 to 100 V using a decrement of 10, the natural frequency from 10 to 100 Hz, the resistance from 50 to 100 ohms with a 10-ohm decrement, the inductance from 0.5 to 2.5 millihenrys, and the reciprocal of capacitance ($1/C$) from 1 to 2.5 microfarad. All the terms for the electrical parameters that we used are discussed by Gibilisco [35].

TABLE 5.8

Value of Natural Frequencies for Rotor Blades

Radius (m)	Height (m)	$k = R/h$	$f_{11} \times 10^{-4}$	$f_{22} \times 10^{-4}$	$f_{33} \times 10^{-4}$
2	0.2	10	5.38	21.52	48.42
4	0.2	20	5.28	21.12	47.52
6	0.2	30	5.26	21.04	47.34
8	0.2	40	5.26	21.04	47.34
10	0.2	50	5.23	20.92	47.07

The principal signal transmission to a propeller or helicopter blade by electrical signal transforms to an ultrasound signal is provided by a piezo-ceramic transducer. The current registration sensor passes the ultrasound signal to the current conductor elements, which are molded simultaneously into the blade structure. Therefore, we can reduce the vibration frequency acting directly on a structural blade.

5.4.6 Conclusions

1. For composite helicopter blades we found a solution to determine the normal and shear stresses in the polar coordinates including the blade profile function.

2. The natural frequencies for the second and third tones increase up to 4 to 9 times compared with the natural frequencies for the basic tone and can be put in a favorable area outside the operational rpm range.

3. The circuit device used as a countermeasure is a potential for compensation vibration. This was shown in the use of the reciprocal of capacitance $(1/C)$ and the storage inductance L. Therefore, we can manage force frequencies more flexibly compared to manipulation with only resistance.

4. The resonant frequency was measured by an accelerometer and was specified as greater than 70 kHz.

5. A methodology was developed to reduce vibration that is capable of activating an electrical circuit by transferring mechanical energy into electrical energy and has helped designers reduce vibration by 50%.

References

1. Johnson, D. P. 2001. Thermal cracking in scaled composite laminates. *Journal of Advanced Materials* 33 (1): 3–6.

2. Bailey, J. E., P. T. Curtis, and A. Parvizi. 1979. On the transverse cracking and longitudinal splitting behavior of glass and carbon fiber reinforced epoxy. *Proceedings Royal Society, London* A366: 599–623.
3. Talreja, R. 1987. Fatigue of composite materials. PhD thesis, Technical University of Denmark, Technomic Publishing Co.
4. Kreyszing, E. 1966. *Advanced Engineering Mathematics*. Columbus, OH: John Wiley and Sons.
5. Golfman, Y. 2001. Fiber draw automation control. *Journal of Advanced Materials* 34 (2) 2001.
6. Golfman, Y. 2001. Nondestructive evaluation of aerospace components using ultrasound and thermography technologies. *Journal of Advanced Materials* 33 (4): 21–25.
7. Golfman, Y. 1991. Strength criteria for anisotropic materials. *Journal of Reinforced Plastics & Composites* 10 (6): 542–556.
8. Golfman, Y. 2004. The fatigue strength prediction of aerospace components using reinforced fiber/glass or graphite/epoxy. *Journal of Advanced Materials* 36 (2): 39–43.
9. Wienhold, P. D., R. K. Eby, and J. M. Liu. 1993. Effect of fiber nonlinear elasticity on the temperature and stress dependence of Young's modulus of continuous fiber composites. *Journal of SAMPE Quarterly* 24 (4): 35.
10. Whitney, J. M., I. M. Daniel, and R. B. Pipes. 1982. *Experimental Mechanics of Fiber Reinforced Composite Materials*. Englewood Cliffs, New Jersey: Prentice-Hall.
11. Parton, V. Z., and P. I. Perlin. 1984. *Mathematical Methods of the Theory of Elasticity*. Moscow: MIR.
12. Goldenblat, I. I., and V. A. Kopnov. 1966. *Criteria of Strength and Plasticity of Fiberglass*. Moscow: Building Machinery.
13. Golfman, Y. 1991. Strength criteria for anisotropic materials. *Journal of Reinforced Plastics and Composites* 10 (6): 542–556.
14. Timoshenko, S., and J. N. Goodier. 1991. *Theory of Elasticity*. New York, NY: McGraw-Hill Book Company.
15. Golfman, Y. 1994. Effect of thermoelasticity for composite turbine disk. 26th International SAMPE Technical Conference, Arizona.
16. Pagano, N. J., and S. R. Soni. 1988. Strength analysis of composite turbine blades. *Journal of Reinforced Plastics & Composites* 7 (6): 558–581.
17. Chou, P. C., and N. J. Pagano. 1992. *Elasticity Tensor, Dyadic & Engineering Approaches*. New York, NY: Dover Publications.
18. Golfman, Y. 1996. The interlaminar shear stress analysis of composites in marine front. SAMPE Technical Conference, Arizona.
19. Timoshenko, S. T., and J. N. Goodier. 1957. *Theory of Elasticity*. New York, NY: McGraw-Hill Book Co Inc.
20. Golfman, Y. 1991. Strength criteria for anisotropic materials. *Journal of Reinforced Plastics & Composites* 10 (6): 542–556.
21. Nemeth, Y., C. B. Sarancan, and B. C. Strelyev. 1910. *Strength of Plastics*. Moscow: Machine Builder.
22. Golfman, Y. 2003. Dynamic aspects of he lattice structures behavior in the manufacturing of carbon–epoxy composites. *Journal of Advanced Materials* 35 (2): 3–8.
23. Golfman, Y., L. P. Rochkov, and N. P. Sedorov. 1978. *Polymer Materials Application on the Rotor Blades for Hovercrafts*. Moscow, Russia: Central Scientific Research Institute.

24. Golfman, Y. 1977. Polymers for Fiberglass Protection. Russian patent # 594745, Registered by the Government Committee for Innovation.

25. Leder, B, 2003. World's first civil tiltrotor achieves first flight. Bell Helicopter Press Release. Available at: http:/belhelicopter.com/companyinfo/pressReleases/pr_0307001.html. Accessed March 7, 2003.

26. Bonassar, M. J. 1980. MM&T fiber-reinforced plastic helicopter-tail rotor assembly (pultruded spar). U.S. Army Aviation R&D Command Final Report (for August 1975–October 1978).

27. Smith, E. C. 1994. Vibration and flutter of stiff-inplane elasticity tailored composite rotor blade mathematical & computed modeling. *Special Edition of Rotocraft Modeling* 20 (1–2).

28. Chou, P. C., and N. J. Pagano. 1992. *Elasticity Tensor, Dyadic & Engineering Approaches*. New York, NY: Dover Publications Inc.

29. Golfman, Y. 1966. The interlaminar shear stress analysis of composite in marine front. 32nd International SAMPE Technical Conference, Arizona.

30. Golfman, Y. 2003. Dynamic aspects of the lattice structure behavior in the manufacturing of carbon–epoxy composites. *Journal of Advanced Materials*, 35 (2): 3–8.

31. Golfman, Y. 2001. Fiber draw automation control. *Journal of Advanced Materials* 34 (2): 35–40.

32. Kushul, M. Y. 1964. *The Self-Induced Oscillations of Rotors*. Consultants Bureau, New Jersey.

33. Golfman, Y. 1993. Ultrasonic non-destructive method to determine modulus of elasticity of turbine blades. *SAMPE Journal* 29 (4): 31–35.

34. Hou, A., and K. Gramoll. 2000. Fabrication and compressive strength of the composite attachment fitting for launch vehicles. *Journal of Advanced Materials* 32 (1): 39–45.

35. Gibilisco, S. 2002. *Physics Demystified*. New York: McGraw-Hill.

6

NDE Methods Control Properties

6.1 Ultrasonic Nondestructive Method to Determine the Modulus of Elasticity of Turbine Blades

6.1.1 Introduction

The ability of ultrasonic waves to travel in a web direction over a minimum time for composites was advanced by the author in 1966 [1]. This effect can be used for different applications in composites, ceramics, and metal alloys [2,3]. Nondestructive evaluation of the material properties of a structure makes this method very useful. Research has been conducted to determine the modulus of elasticity of high-speed turbine blades [4].

In this research, a general nondestructive test method for deterring the modulus of elasticity in different directions is discussed.

Predicting the modulus of elasticity of turbine blades using an ultrasonic method probably gives the option of estimating the strength of the blades. Gershberg [5] determined the parameters of elasticity for specimens fabricated from fiberglass. We determined the parameters of elasticity in the blades of screw propellers fabricated from orthotropic fiberglass using a nondestructive ultrasonic method [6]. The advantage of nondestructive evaluation of the properties of new materials used for manufacture of turbine blades is that they can be predicted with little difficulty.

6.1.2 Theory and Application of Ultrasonic Method

Knowledge of the parameters of elasticity is necessary to calculate stress and estimate the strength of turbine blades. If we assume composite turbine blades to be manufactured from orthotropic material having three planes of elasticity symmetry, their technical behavior can be completely characterized by nine classic constants [7], shown in Table 6.1.

For a composite turbine blade, there are nine independent parameters: E_z, E_y, and E_x are the moduli of elasticity in the three principal directions z, y, and x, respectively; G_{zy}, G_{yx}, and G_{xz} are the moduli of shear in the ply orientation zy; and the interlaminar directions are zx and yx. μ_{zy}, μ_{yx}, and μ_{xz} are the Poisson ratios.

TABLE 6.1

Elastic Constants for Orthotropic Materials

$1/E_z$	$-\mu_{yz}/E_y$	$-\mu_{xz}/E_x$	0	0	0
$-\mu_{zy}/E_x$	$1/E_y$	$-\mu_{xy}/E_x$			
$-\mu_{zx}/E_z$	$-\mu_{yx}/E_y$	$1/E_x$			
			$1/G_{zy}$		
				$1/G_{yx}$	
					$1/G_{xz}$

The first letter in the subscript of μ represents direction of force applied and the second letter represents the transverse direction of E_{45}, which is the Young modulus along the 45° orientation in the plane zy, and μ_{45} is the Poisson ratio along the 45° orientation.

$$G_{zy} = E_{45}/2(1 + \mu_{45}) \tag{6.1}$$

All nine parameters can be predicted using the ultrasonic nondestructive method. This method is based on the measure of the time interval of ultrasonic oscillations in the longitudinal and the transverse directions. Gershberg [5] used semiconductors of model YKC-1 to measure the time interval. Semiconductors YKC-1 are provided with ultrasonic heads having frequencies ranging from 20 to 240 kHz. To maintain acoustic contact between the ultrasonic heads, the blades, and the blade surfaces, either viscous liquid or highly viscous oils can be used as the immersion medium. The acoustic heads are installed on the surface of the blade as shown in Figure 6.1. Along the longitudinal direction, a wave of 80 kHz was used to measure the time of interval. A 100-mm circle diameter is marked on the surface of the blade. The position of the acoustic heads is varied with each frequency and time interval of measurement (Figure 6.1).

The objective is to find the direction in which the ultrasonic wave can travel in a minimum time.

The velocity of the ultrasonic wave can be calculated using the following expression:

$$C = L/t * 10^3 \tag{6.2}$$

where C is the velocity of the ultrasonic oscillations, L is length between the two acoustic heads, and t is the time it takes for the ultrasonic oscillations to reach from one head to the other.

The equation to calculate the in-plane modulus of elasticity in any direction for an infinitely long plate is obtained for angle f. Correlation in an orthotropic body is needed to solve the different equations of motion and strain.

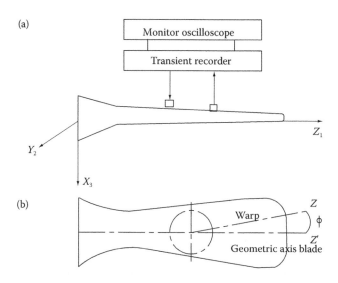

FIGURE 6.1
Schematic diagram showing (a) the position of acoustic heads installed on surface of turbine blades and (b) ultrasonic wave test method application.

$$E_\phi = C_\phi \rho * (1 - \mu_{1\phi}\mu_{2\phi}) \tag{6.3}$$

where ϕ is the angle in the direction of the zy plane, C_ϕ is the velocity in the direction of the angle ϕ, and $\mu_{1\phi}\mu_{2\phi}$ is Poisson's ratio in the direction of the angle ϕ.

The fiber composite has high strength, stiffness, and low weight that can be tailored to any application, and the reinforcing fibers can be oriented in the matrix to provide strength in any direction. From Equations (6.2) and (6.3), the material properties of an infinitely long plate are computed as follows:

$$\mu_{ZY} = \left(\frac{C_z}{C_y}\right)\mu_{YZ} \tag{6.4}$$

$$E_y = \frac{10\rho C_y^2}{g}\left(1 - \frac{C_z^2}{C_y^2} * \mu_{yz}^2\right) \tag{6.5}$$

$$E_z = \frac{C_z^2}{C_y^2} * E_y \tag{6.6a}$$

$$E_{45} = \left(-\frac{g}{10\rho C_{45}^2 A^2} - \frac{2}{A} \right) \tag{6.6b}$$

$$\mu_{45} = 1 + E_{45} * A \tag{6.7}$$

$$A = \frac{\mu_{ZY}}{E_z} - \frac{1}{2E_z} - \frac{1}{2E_y} \tag{6.8}$$

$$G_{zy} = \frac{E_{45}}{2(1 + \mu_{45})} \tag{6.9}$$

where C_z, C_y, and C_{45} are the velocities of longitudinal ultrasonic oscillations of the infinitely long plate in the warp, fill, and diagonal directions (m/s); g is the acceleration due to gravity (9.81 m/s²); and ρ is the density of the composites (1.998 g/cm³). The z direction set is called the warp and the y direction set is called the fill. To determine the elastic properties of a composite material, 12 plates having geometrical dimensions of 10 × 200 × 250 mm were fabricated containing 20% to 30% epoxy and 70% to 80% glass fiber and set under a pressure of 100 kg/cm², a temperature of 160°C, and a time of curing 3 to 6 min/mm. The fabricated plates, with defined warp and fill and diagonal directions, were cut into 10 × 10 × 200 mm and 10 × 30 × 200 mm specimens.

The elastic properties E_z, E_y, E_x, E_{45}, G_{zy}, G_{xz}, and G_{yz} for these specimens were determined using the ultrasonic method.

These properties are transferred from an infinitely long plate to finite dimensions of the turbine blade using appropriate coefficients given below:

$$K_z = \frac{E_z^t}{E_z^u}; \quad K_y = \frac{E_y^t}{E_y^u}; \quad K_x = \frac{E_x^t}{E_x^u} \tag{6.10}$$

where the superscript t denotes the standard destructive test and u denotes ultrasonic test. The elastic properties in the turbine blades fabricated from composites are given by the following equations:

$$\mu_{ZY} = \frac{K_z}{K_y} \left(\frac{C_z}{C_y} \right) \mu_{YZ} \tag{6.11}$$

$$E_y = \frac{K_y 10\rho C_y^2}{g} \left(1 - \frac{K_z}{K_y} x \frac{C_z^2}{C_y^2} * \mu_{yz}^2 \right) \tag{6.12}$$

$$E_z = \frac{K_z}{K_y} \frac{C_z^2}{C_y^2} * E_y \tag{6.13}$$

$$E_{45} = K_{45}\left(-\frac{G}{10\rho C_{45}^2 A^2} - \frac{2}{A}\right) \tag{6.14}$$

The coefficient values for the turbine blades were found to be $K_z = 0.885$, $K_y = 0.840$, and $K_{45} = 0.8$, where C_z, C_y, and C_{45} are velocity propagations of longitudinal oscillations in turbine blades in the warp, fill, and diagonal directions. Equations (6.7) to (6.9) can be used for calculating μ_{45}, A, and G_{zy} for a turbine blade. Application of the ultrasonic wave method for turbine blades is demonstrated in Figure 6.1. A circle drawn on the blade surface and ultrasonic heads was aligned with the circumference of the circle along the geometrical axis. The time taken for the wave to pass from one head to another was measured. Then, the heads were moved away from the geometrical axis by 5° and the propagation time was again measured. The results are shown in Table 6.2.

To achieve accurate results, it is necessary to take the mean $\bar{X}$ and the quadratic deviation $\bar{S}$ (Table 6.3).

From an analysis of the data for the 42 specimens, the values for $X = 3.56 \times 10^5$ and for $S = 0.25 \times 10^5$ were obtained. Figures 6.2, 6.3, and 6.4 show the experimental and theoretical distribution modulus curves of elasticity E_z, E_y, and G_{zy}.

We assumed that the empirical curve follows the law of normal distribution [8]. The normal law of distribution is given by:

$$p(x) = \frac{1}{(2\pi 6)^{1/2}} e^{-\frac{(\bar{x}-a)^2}{2\sigma^2}} \tag{6.15}$$

TABLE 6.2

Velocity Propagation of Longitudinal Waves in Blades

Blade No.	Base (mm)	Time (s)	Velocity C_z (m/s)	Time (s)	Velocity, C_y (m/s)	Time (s)	Velocity, C_{45} (m/s)
1	100	23.0	4350	25.7	3900	26.2	3800
2	100	23.0	4350	25.0	3950	26.0	3850
3	100	23.5	4250	26.0	3850	26.6	3750
4	100	23.0	4350	25.7	3900	26.2	3800
5	100	22.4	4450	25.7	3900	26.2	3800

TABLE 6.3

$E_z \times 10^3$ (kg/cm²)	Empirical Frequency	$\bar{x} \times 10^5$	$\bar{S}_x \times 10^5$	$\dfrac{\bar{x} - \bar{x}_1}{s}$ (Equation 6.15)	$p(x)$ (Equation 6.15)	m_1, m_1	χ^2 (Equation 6.16)
3.1		3.56	0.25				
3.15	5			1.63	0.0035	3, 2	0.133
3.2							
3.25							
3.3	7			1.04	0.139	6, 1	0.166
3.35							
3.4							
3.45	8			3.56	0.217	9, −1	0.101
3.5							
3.55							
3.60	10			0.016	0.240	10, 0	0
3.65							
3.70							
3.75	8			0.76	0.18	8, 0	0
3.80							
3.85							
3.90	4			1.35	0.096	4, 0	0
3.95							
4.0							

$N = \sum 42, \chi^2 = \sum 0.4.$

where X and S are substituted for "a" and "6," respectively. Using Equation (6.15) and substituting experimental results, we can check whether they follow the law of normal distribution. We assumed that the empirical curve follows the theory of normal distribution if the reliability testing is not less than 95%. Therefore, the probability is more than 5% (0.05). To compare the

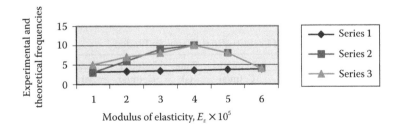

FIGURE 6.2
Experimental and theoretical curves distribution of modulus of elasticity E_z.

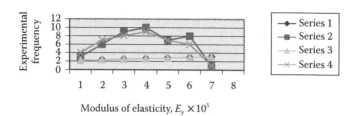

FIGURE 6.3
Experimental and theoretical curves distribution of modulus of elasticity E_y.

experimental and theoretical curves, the density of probability by Pearson's criterion is used, which is given by [9]

$$\chi^2 = \sum_{I=1}^{n} \left(\frac{m_i - m_{i'}}{m_i} \right)^2 \qquad (6.16)$$

where m_i are the experimental frequencies, and m_i' are the theoretical frequencies. After determining χ^2 from Equation (6.16), we determine coefficient K using the formula $k = n - r - 1$. Here, k is the number of degrees of freedom, n is the number of comparison frequencies, and r is the number of parameters in the theoretical function of distribution. The normal law of distribution has two parameters. In our case, we assume the worth variant when empirical frequencies are equal to 7, $\chi^2 = 0.166$ (see Table 6.3). Following the last equation, when $r = 2$, $n = 7$, and $k = 4$ from the table of probabilities [9], we find $P(x^2) = 0.9098$. The experimental data appears to agree with the theoretical curve. Similarly, the experimental values of E_y and G_{zy} specimens were obtained and checked as to whether they satisfy the Pearson's criterion. The Pearson's criterion for both E_y and G_{zy} are obtained as 0.9098, which implies that the experimental values agree by over 90% with the theoretical

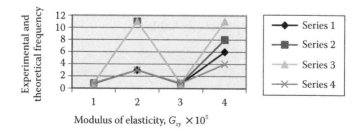

FIGURE 6.4
Experimental and theoretical curves distribution of modulus of elasticity G_{zy}.

values. The values of modulus of elasticity and modulus of shear for turbine blades are listed in Table 6.4.

The modulus of elasticity in an angle direction can be found as:

$$E_\phi = \frac{E_z \lambda}{\lambda \cos^4 \phi + 2B \cos^2 \phi + \sin^4 \phi} \tag{6.17}$$

where

$$\lambda = \frac{E_y}{E_z}; \quad 2B = 4\frac{E_y}{E_{45}} - (1 + \lambda)$$

The velocities of propagation of longitudinal waves in different angles with a step of 15° are determined using Equation (6.17). From these velocities, the modulus of elasticity is computed using Equations (6.7) through (6.9) and (6.11) through (6.14). The values of the coefficient introduced into Equation (6.2) are:

$$E_\phi = k_\phi C_\phi^2 \rho (1 - \mu_{1\phi}\mu_{2\phi}) \tag{6.18}$$

where k_ϕ is computed by

$$k_\phi = \frac{k_z}{\cos^4 \phi + b_0 \cos^2 2\phi + c_0 \sin^4 \phi} \tag{6.19}$$

where

$$c_0 = \frac{k_z}{k_y}; \quad b_0 = \frac{k_z}{k_{45}}\frac{c_0 + 1}{4}$$

The difference between the values computed by Equations (6.17) and (6.18) is only about 5%, which proves that the values obtained by the present non-destructive method are acceptable (see Table 6.5).

TABLE 6.4

Significances of Normal Modulus and Shear Elasticity in the Turbine Blades

Parameters	$E_z \times 10^5$ (kg/cm²)	$E_y \times 10^5$ (kg/cm²)	$E_{45} \times 10^5$ (kg/cm²)	$G_{zy} \times 10^5$ (kg/cm²)	$G_{zx,yx}$ (kg/cm²)	μ_{yz}	μ_{zy}	μ_{45}
Specimens	3.55	2.59	2.2	0.18	0.68	0.1	0.13	0.34
Blades	3.4	2.56	2.28	0.18	0.68	0.1	0.13	0.34

TABLE 6.5

Modulus of Normal Elasticity Depending on Angles

Angle (°)	μ_ϕ	k_ϕ	C_ϕ (m/s)	$E_\phi \times 10^5$ (kg/cm²) (Equation 6.17)	$E_\phi \times 10^5$ (kg/cm²) (Equation 6.18)	Error (%)
0	0.13	0.855	4350	3.4	3.3	3.0
15	0.20	0.86	4150	3.1	2.9	5.0
30	0.30	0.835	3950	2.6	2.5	4.0
45	0.34	0.80	3800	2.28	2.2	3.5
60	0.13	0.812	3900	2.29	2.38	4.0
75	0.128	0.824	3940	2.48	2.4	3.5
90	0.10	0.840	3900	2.56	2.46	4.0

6.1.3 Conclusions

Aerospace applications utilize advanced composites that replace traditional alloys in huge structures. Nondestructive test methods are necessary because all destructive testing is very expensive.

In this research, we developed a general nondestructive test method for determining the modulus of elasticity of high-speed gas and steam turbine blades. Reliability results of more than 90% prove that this method is very useful for aerospace applications.

6.2 Nondestructive Evaluation of Parts for Hovercraft and Ekranoplans

6.2.1 Introduction

The development of shipbuilding is connected with the solution to the crucial problem of increasing vessel speed. The maximum speed of displacement type vessels restricted by wave drag is 25 to 35 km/h for river vessels and 50 to 60 km/h for sea vessels. Using skimming and hydrofoils in order to lower wave drag on vessel motion made it possible to increase the speed up to 100 to 120 km/h. But because of their low seaworthiness, skimmers were not widely adopted and are now used mainly as sporting ships. Hydrofoils, on the contrary, became widespread. A Russian designer and scientist made the decisive contribution to hydrofoil creation: R. Alexceev (1918–1980). Thanks to his efforts in the 1940s and 1960s the Russian river and sea passenger hydrofoil fleet was created. This fleet continues to occupy leading positions in the world's high-speed transport systems.

Hovercrafts with dynamic principal support systems seem to be a very convenient form of transportation for this purpose. The regional aircraft market as well has currently developed a carbon fiber–epoxy fuselage section. Propeller blades utilize a hybrid of composites such as fiberglass skins, a carbon fiber–epoxy spar, and a polyurethane-foam core [10].

The ability of ultrasonic waves to travel in a web direction over a minimum time through curing fiberglass was advanced by this author in 1966. This effect can be used for communications such as radio wave propagation. However, only a combination of thermographs, ultrasound, and radiography techniques can predict physical parameters of material (density, thickness, moduli of elasticity) and strength. X-ray line sensor cameras have a wavelength of 400 to 600 nm and high sensitivity for capturing composite images. X-ray cameras with CCD chips have converted light into video signals also suitable for video microscopy, and laser light can measure the distance between a space station and planets very precisely.

Composites structures for space programs are mainly made of sandwich composites with aluminum honeycomb and graphite–epoxy face sheets and are located on the upper part of the launcher where the influence rate is the largest [11].

Influence rates of different stages vary on Ariane 5 from 7% for the lateral booster to 100% for upper parts, requiring that a strong effort be made in the design of structures in the launcher upper part. NASA's Marshall Space Flight Center (MSFC) is testing Raytheon's Radiance infrared camera to devise nondestructive evaluation (NDE) methods for assembling the space shuttle and other aerospace components [12].

Researchers have developed NDE techniques for any structural anomaly determination for composites on the space shuttle nose cap. They are using the Radiance system for the thermography component because the camera's 256×256 pixel InSb starting focal-plane array generates high-resolution images and is highly sensitive to slight temperature changes [12].

Thermoforming technology includes prepreg layup layers and a curing process with monitoring parameters: pressure, temperature, and polymerization time and a cooling process that had a temperature gradient of more than $10°F(12°C)$. It means that there was an irregularity field of temperature and thermal stresses.

The new development of braiding technology for aerospace components has used carbon–carbon or graphite–epoxy dry fabric, injected epoxy, where curing was shown in thermal cameras and cooling processes were not free of thermal stresses. Composite structures had low thermal conductivity during a faster heating process and a high gradient field of temperature changes.

An irregular field of temperature leads to significant thermal stresses, which could result in failure in structures in the process of fabrication. Curing and cooling processes for high thickness structures have a low speed of heatup rate that avoids significant thermal stress but increase time and labor cost.

The task of determining an optimal regime was sophisticated because in the process of curing epoxy matrix the exothermic reaction appeared in adiabatic conditions, which avoided contact with the outside environment. All the heat was used in preheating the epoxy resin.

The exponent of temperature increased the velocity of curing reaction. Self-heat in these conditions could arise from delaminating graphite–epoxy.

6.2.2 Designing Hovercraft Parts

The principal elements of hovercrafts are represented in Figure 6.5 [13].

The shell (pos. 1) is mainly made of sandwich composites with aluminum honeycomb and graphite–epoxy face sheets. Composite deck (pos. 2); four fans are accommodated in a deck (pos. 3) and split into three separate sections in the vertical trunk. Two air propellers are launched in a top deck at (pos. 4), which should be rotated in a horizontal position like commercial rotorcrafts. Two gas turbine engines with two transmissions are in (pos. 5). Two flexible wings are in (pos. 6), which can be moved from the shell and two ailerons in (pos. 7) stabilize the hovercraft in horizontal position.

Air pillows in (pos. 8) are stabilizing in a vertical position. The elevation could come from air system or rotors when air propellers change position to horizontal. Fans consist of straightening apparatus (pos. 3-1), wheel working in an air atmosphere (pos. 3-2) and guiding apparatus (pos. 3-3). The hovercraft's trunks were damaged in the process of exploitation under acting input shock forces and vibration, which could reduce reliability and term of service. Installed between the trunk walls and the ends of the blades are special polyurethane tubes that are implemented to dampen the roll from shock vibration [14]. Theoretical investigation was done in one study [15].

We know that gravity on the Moon is six times less than on the Earth, so the acceleration of the force gravity will be increased six times. What is important in the equation is motion input coefficient K reflecting the acceleration of force gravity.

Stress components for a rotating shell can be designated following Timoshenko and Goodier [16] as:

$$\sigma_{11} = \frac{3+\mu_{12}}{8}\rho\omega^2 \left(R_o^2 + R_i^2 \frac{R_o^2 * R_i^2}{r^2} - r^2 \right) \tag{6.20}$$

$$\sigma_{22} = \frac{3+\mu_{12}}{8}\rho\omega^2 \left(R_o^2 + R_i^2 + \frac{R_o^2 * R_i^2}{r^2} - \frac{1+3\mu_{12}}{3+\mu_{21}} - r^2 \right) \tag{6.21}$$

where
R_o, R_i, r = Outside, inside, and middle radii of the nose cap
μ = Density of material
C = Velocity of rotation and replaced in stress σ_{11}, σ_{22}

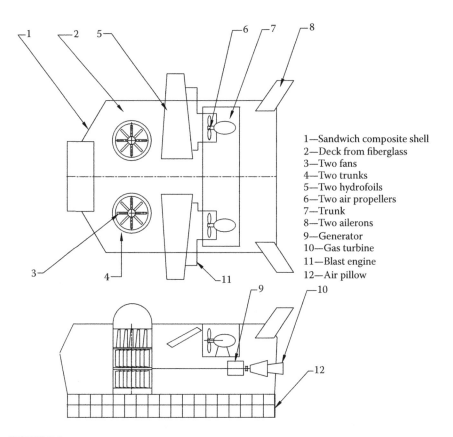

FIGURE 6.5
Principal elements of hovercrafts.

For a nose cap fabricated from orthotropic carbon–carbon or graphite–epoxy, the dynamic modulus of elasticity in the radial direction was determined as:

$$E_{11} = \frac{(3+\mu_{21})(1-\mu_{21}\mu_{12})}{8\alpha_{11}T} + \rho\omega^2(R_o^2 * R_i^2 - \frac{R_o^2 * R_i^2}{r^2} - r^2 \tag{6.22}$$

Dynamic modulus of elasticity in the tangential direction was determined as:

$$E_{22} = \frac{(3+\mu_{21})(1-\mu_{21}\mu_{12})}{8\alpha_{11}T} + \rho\omega^2\left(R_o^2 * R_i^2\right) - \frac{R_o^2 * R_i^2}{r^2} \frac{1+3\mu_{21}}{3+\mu_{21}} - r^2 \tag{6.23}$$

Figure 6.6 shows a nose cap with installed surface ultrasonic transducers. We used Panametrics Technology contact transducers with low frequencies of 50 kHz (X1021), 100 kHz (X1020), and 180 kHz (X1019).

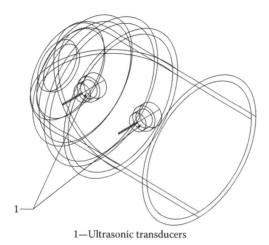

1—Ultrasonic transducers

FIGURE 6.6
Ultrasonic wave test method for nose cap.

High-voltage pulse receivers such as Panametrics 5058PR were used. For measuring temperature gradients we used infrared thermometers. The intensity of the radiation created by the infrared camera was a function of the temperature gradient. The infrared thermometers simply measured the intensity of radiation and thereby measured the temperature. The infrared cameras have 0.025°C sensitivity and detect anomalies such as delaminating. The influence of the velocity of curing on the rise of temperature gradient for three types of epoxy resin was shown in Figure 6.7.

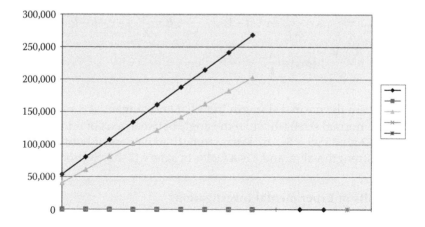

FIGURE 6.7
Correlation between thermal stresses and temperature gradient (compression zone). ▲ indicates a low viscosity epoxy resins; ■ indicates a high viscosity epoxy resins; *, ×, ♦ indicates a zero viscosity.

In an earlier study [17], we developed a strength theory and found that strength criteria in tensor form were:

$$a_{ikem}\sigma_{ik}\sigma_{ik} - \left[\frac{\left(\sigma_{ik}\delta_{ik}\right)^2 + \sigma_{ik}\sigma_{ik}}{2}\right]^{1/2} = 0 \qquad (6.24)$$

where a_{ikem} are components of tensor strength with different valence satisfying the following conditions of symmetry:

$$a_{ik} = a_{ki}; \ a_{ikem} = a_{emik}; \ a_{ikme} = a_{ikem}; \ a_{kiem} = a_{ikem}$$

where

   δ_{ik} = Kronecker's coefficient; $\delta_{ik} = 1$ if i equals k, $\delta_{ik} = 0$ if i does not equal k;
   σ_{ik} = stress component

These criteria can be used separately for tensile and compressive loads. Expanding Equation (6.24), criterion of thermal strength for biaxial stress conditions is obtained by:

$$\frac{\sigma_{11}^2 + c\sigma_{22}^2 + s\sigma_{11}\sigma_{22} + h\sigma_{22}\sigma_{11}}{(\sigma_{11}^2 + \sigma_{22}^2 + 2\sigma_{11}\sigma_{22})^{1/2}} \leq [\sigma_{br}] \qquad (6.25)$$

where σ_{11} are the thermal radial and tangential stresses and coefficients; c, s, d, h, and b are the variable relative strength depending on the history of loading and quality of materials, and are determined as:

$$c = \frac{X}{Y}; \ s = \frac{4X}{X_{12}^{45}} - c - d - 1; \ d - \frac{X}{S_{12}}; \ h = \frac{4Y}{X_{12}^{-45}} - c - b - 1;$$

$$b = \frac{Y}{S_{12}}; \ [\sigma_{br}] = \frac{X}{k}$$

where X, Y are the normal strength in radial and tangential directions; X_{12}^{45}, X_{12}^{-45} are the normal strength when the angle between axis of reinforced carbon fiber 1, 2 is ±45°; S_{12}, S_{21}, are shear strength in flatness of reinforced; σ_{br} is a complex strength value; and k is a factor of safety (2–2.5, Section 3.3).

6.2.3 Results of Experimental Investigations

For experimental data we selected a nose cap with geometrical dimensions $R_o = 10$ in, $R_i = 9$ in, and $r = 9.5$ in. The nose cap was fabricated from graphite–epoxy and had a density of material $\rho = 0.1497 \times 10^{-3}$ lb/in³, and Poisson's ratio was $\mu_{12} = \mu_{21} = 0.036$.

TABLE 6.6

Properties of Carbon–Carbon Composites

Modulus of Elasticity Acting in x,z Directions		Normal Strength (MPa/psi)		Shear Strength (MPa/psi)		Coefficient of Thermal Expansion (cm/cm °C)		
Significance	E_{11}	E_{22}		σ_{b1}	σ_{b2}	τ_{b12}	α_1	α_2
Compression	0.8×10^5/	0.65×10^5/		135.8/	116.5/	9.37/	0.118	0.115
	12.3×10^6	9.5×10^6		19.70	16.90	1360		
Tension	0.83×10^5/	0.71×10^5/		226/	172/	9.37/	0.118	0.115
	12.1×10^6	10.3×10^6		32.80	24.97	1360		

When we calculated modulus of elasticity E_1, E_2 for the nose cap, the velocity of rotation was changed from 3627 to 4800 rad/s and to 7000 rad/s. Modulus of elasticity in the radial direction was $E_1 = 25.1 \times 10^5$ psi. Modulus of elasticity in the tangential direction was $E_2 = 4.8 \times 10^5$ psi. Coefficients of thermal expansion were in the radial direction $\alpha_1 = 0.34 \times 10^{-3}$ in/in/°F, and in the tangential direction $\alpha_2 = 26.4 \times 10^{-3}$ in/in/°F.

The next nose cap was fabricated from carbon–carbon and had a density $\rho = 0.195 \times 10^{-3}$ lb/in³, modulus of elasticity in the radial direction was $E_1 = 12 \times 10^5$ psi, and modulus of elasticity in the tangential direction was $E_2 = 10 \times 10^5$ psi.

Value of Poisson's ratio was $\mu_{12} = \mu_2 = 0.036$, and coefficient of thermal conductivity for carbon–carbon β was 6.38 in/h in² °F.

Table 6.6 shows indicated properties for carbon–carbon composite that were used as a parameter relationship of strength in Table 6.7 for calculating complex strength values.

Coefficients of thermal expansion were in the radial direction $\alpha_1 = 0.118$ cm/cm/°C/3.8 × 10⁻³ in/in/°F, in the tangential direction $\alpha_2 = 0.115$ cm/cm °C/3.71 × 10⁻³ in/in/°F, and coefficients of thermal expansions in an arbitrary direction were determined as $\alpha_{12} = 2 \sin\theta*\cos\theta(\alpha_1 - \alpha_2)$.

Significance of the thermal stress components manufactured from carbon–carbon for the nose cap changed when the gradient of temperature changed from 1°F to 5°F for the compression zone (see Figure 6.3). Also, the significance of thermal stress components manufactured from carbon–carbon for

TABLE 6.7

Parameters' Relationship to Strength for Carbon–Carbon Composites

Significance	c	b	d	s	h
Compression	1.16	12.43	14.48	−12.34	−10.87
Tension	1.31	18.36	24.1	−21.86	−17.17

the nose cap changed when the gradient of temperature changed from 1°F to 5°F for the tension zone.

The correlation between thermal stresses and complex strength values in the compression zone are shown [18–20]. All the calculations were computerized to create a program in C language.

6.2.4 Conclusions

1. A new class of vessel—hovercraft and ekranoplans with composite structural parts—can be designed. Thermal stresses for orthotropic composite structures were calculated using parameters that were found by ultrasound and thermographic technologies.

2. Temperature gradients were calculated using an approximate solution (Equation 6.6). Temperature fluctuated from 0.5°F to 3°F. NDE thermographic cameras showed that a temperature gradient can be very useful in the process of curing and cooling composite structures.

3. In spite of the significance of thermal stresses whose levels can be 2 to 3×10^5 psi, the complex strength value will be on a level of 1.4 to 1.7×10^4 psi in the compression zone, which is less than the critical strength that we found in the samples.

6.3 Dynamic Local Mechanical and Thermal Strength Prediction Using NDE for Material Parameters for Evaluation of Aerospace Components

6.3.1 Introduction

The overall approach is to determine the local elastic constant and material parameters using nondestructive methods, which characterize strength in the domain point and also determine the local stresses from the temperature profile.

The purpose of this research is to present and discuss the developed NDE methods used to increase the reliability estimation of dynamic strength.

The thermal, radial, and tangential stresses for an orthotropic nose were used as an example, while the dynamic modulus of elasticity for stress analysis was used and all the stress components for strength criteria were established [21–23].

Agfa Nondestructive Testing, Inc., has become a leading supplier of NDT systems after successfully testing caps for three years using the ultrasound method [24].

Composite structures for space programs have been mainly manufactured as sandwich composites with aluminum honeycomb and graphite–epoxy face sheets, and are located on the upper part of the space vehicle launcher where their influence has the largest effect [25].

For example, the influence rates of composites on different stages vary on the Ariane 5 from 7% for the lateral booster to 100% for the upper parts. They require that a strong effort be made in the design of structures in the launcher upper part, and the demand for NDT equipment and procedures becomes imperative.

Launched in April 2001, NASA's Mars Odyssey is now prepared to collect data that will offer insights into the makeup and history of the red planet.

With the spacecraft's 20-ft boom successfully deployed, two neutron detectors and a gamma ray spectrometer (GPS) mounted on its end, the spacecraft can measure the quantity and distribution of primary elements located at or near the planet's surface and also the modulus of elasticity of the primary elements. Additionally, silicon, oxygen, iron, magnesium, potassium, aluminum, calcium, sulfur, and carbon are among the 20 primary elements being measured.

NASA's MSFC has tested Raytheon's Radiance infrared camera to devise nondestructive evaluation (NDE) methods for assembling the space shuttle and other aerospace components. These NDE techniques have been used for any structural anomaly determinations for composites on the space shuttle nose cap, and the Radiance system is used for the thermography component because the camera's 256 × 256 pixel InSb standing focal plane array generates high-resolution images and is highly sensitive to slight temperature changes [26].

Thermoforming technology includes prepreg layup layers, a curing process with monitoring parameters, pressure, temperature of polymerization time, and a cooling process with a temperature gradient of more than 10°F (12°C). This means that there is an irregular field of temperature and thermal stresses that can be detected by NDT methods.

Newly developed braiding technology for aerospace components has used carbon–carbon or graphite–epoxy dry fabric, injected epoxy, and curing was shown by thermal cameras. Cooling processes were not free of thermal stresses and were also seen by NDT methods. Composite structures had a low thermal conductivity during the faster heating process and a high gradient field of temperature changes. Additionally, an irregular field of temperature had significant thermal stresses, which could result in failure in structures in the process of fabrication. These also were seen by NDT.

Curing and cooling processes for high-thickness structures with a low speed of heatup rate can avoid a significant thermal stress, but they increase time and labor costs. The task of determining the optimal regime is sophisticated because in the process of an curing epoxy matrix, the exothermic reaction appears in adiabatic conditions, which avoid contact with the outside environment. All the heat is used in preheating the epoxy resin.

Finally, the velocity of reaction increases by the exponent of temperature and self-heat in these conditions and could delaminate graphite–epoxy composites. Theoretical investigation had been done in one study [27].

Dynamic stresses for orthotropic components could be described as:

$$\sigma_x = \frac{d^2\phi}{dx^2}; \quad \sigma_y = \frac{d^2\phi}{dy^2}; \quad \tau_{xy} = \frac{d^2\phi}{dxdy} \tag{6.26}$$

Here, x, y are the vectors describing the directions in which the dynamic stresses act, and ϕ is the stress function that can be shown as:

$$\phi = Q_{ij}\varphi \tag{6.27}$$

Q_{ij} are the nine stiffness constants described in Equation (6.26), and φ is the contour of profile for the orthotropic parts.

Thus, Equation (6.27) is:

$$\sigma_x = \frac{d^2\phi}{dx^2} = h\rho C_x^2$$

$$\sigma_y = \frac{d^2\phi}{dy^2} = h\rho C_y^2 \tag{6.28}$$

$$\tau_{xy} = \frac{d^2\phi}{dxdy} = h\rho C_{xy}^2$$

where
  h = Parameter of length propagation of the ultrasonic wave
  ρ = Density of composites (for fiberglass, 1.998 g/cm³)
  C = Velocity of ultrasonic propagation (m/s)

By replacing the stress function from Equation (6.27), the results are shown in Equation (6.28):

$$\sigma_x = Q_{11}\frac{d^2\varphi}{dx^2} = Q_{11}h\rho C_x^2$$

$$\sigma_y = Q_{22}\frac{d^2\varphi}{dy^2} = Q_{22}h\rho C_y^2 \tag{6.29}$$

$$\tau_{xy} = Q_{12}\frac{d^2\varphi}{dy^2} = Q_{12}h\rho C_{xy}^2$$

If the mechanical deformation is equal to zero only considered under thermal stresses [23], the equation for thermal stresses will become:

$$\sigma_{ij} = Q_{ij}\alpha_{ij}T \tag{6.30}$$

where
α = Coefficient of temperature expansion
T = Temperature gradient
Q_{ij} = Stiffness constants can be determined using an NDE method for velocity propagation [21]

For the orthotropic nose, thermal radial stresses are:

$$\sigma_{11} = -Q_{11}\alpha_{11}T = \frac{E_1}{1-\mu_{12}\mu_{21}}\alpha_{11}T$$

while the thermal tangential stresses are:

$$\sigma_{22} = -Q_{22}\alpha_{22}T = \frac{E_2}{1-\mu_{12}\mu_{21}}\alpha_{22}T \tag{6.31}$$

where
E_1, E_2 = Moduli of normal elasticity in the radial and tangential directions
μ_{12}, μ_{21} = Poisson's ratio of material: the first symbol designates the direction of force, and the second symbol designates the direction of transverse deformation
α_{11}, α_{22} = Coefficients of thermal expansion in the radial and tangential directions
T = Temperature gradient

The differential equation of heat conductivity without the exothermic reaction of curing of the nose cap is as follows:

$$\frac{dT}{dt} = \beta\left(\frac{d^2T}{dr^2} + \frac{1}{r}*\frac{dT}{dr}\right) \tag{6.32}$$

where t is the time of curing, and R, r are the outside and middle radius of the nose cap. β is the coefficient of thermal conductivity. Thus, in selecting the boundary conditions for Equation (6.32), $T(r,0) = 0$, $T(R,t) = bt$, and b is the velocity of the curing (cooling) process.

The first approach to the solution of Equation (6.33) is:

$$T(r,t) = \frac{bR^2}{\beta}*\left[\frac{\beta t}{R^2} - \frac{1}{4}\left(1 - \frac{r^2}{R^2}\right)\right] \tag{6.33}$$

The gradient of temperature T during the period of curing (cooling) may be responsible for the geometrical parameters of the nose cap, the thermal conductivity of the epoxy resins, and the velocity of curing b.

We must also look at the stress components for a rotating shell that can be designated following Timoshenko's [16] studies and results. Here, R_o, R_i, and r are the outside, inside, and middle radius of the nose cap, ρ is the density of material, ω is the velocity of rotation and replaced in Equation (6.33) from Equations (6.34) and (6.35).

Figure 6.6 shows a nose cap with the installed surface ultrasonic transducers. We used Panametrics Technology contact transducers with low frequencies of 50 kHz (X1021), 100 kHz (X1020), and 180 kHz (X1019). The high-voltage pulse receivers such as Panametrics 5058PR were also used.

In measuring the temperature gradients, infrared thermometers were used and the intensity of the radiation created by the infrared camera was a function of the temperature gradient. The infrared thermometers simply measured the intensity of radiation and thereby measured the temperature. The infrared camera had a 0.025°C sensitivity and was able to detect anomalies such as delaminations. The influence of the velocity of curing on the rise of the temperature gradient for three types of epoxy resins is shown in Figure 6.8.

The strength of composites can be predicted using a second-order polynomial (Tsai, Wu, Hoffman, Hill [28–31]).

A strength criterion of the second order is not capable of handling airstream load, particularly for strong anisotropic materials such as carbon–carbon or graphite–epoxy. In our work [17], we developed a strength theory and found that the strength criteria are in tensor form. These criteria can be used separately for tensile and compressive loads.

The probability of local cracking can be predicted using mathematical models, which include the first and second approach in the law of normal distribution.

$$P(t) = \frac{1}{(2\pi)^{1/2}} e^{-t^2/2} \qquad (6.34)$$

Here,

$$\text{parameter } t = \frac{\sigma_{bi} - \sigma_{bm}}{S_j}.$$

where
σ_{bi} = Current strength in x, y, z directions
σ_{bm} = Middle strength in x, y, z directions
S_j = Sample of the standard deviation for each environment via

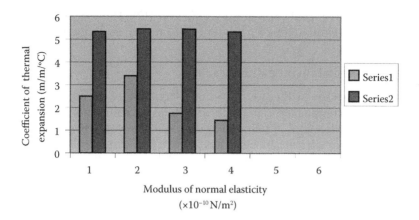

FIGURE 6.8
Stiffness of the nose cap aviation parts.

$$S_j^2 = \frac{1}{n_j} \sum_1^{n_j} (\sigma_{bi} - \sigma_{bm})^2$$

where n_j is a number of test samples.
    We calculated the middle strength σ_{bm} as:

$$\sigma_{bm} = \frac{1}{n_j} \sum_1^{n_j} \sigma_{bi}$$

Thus, for a single test condition (such as compression strength in the fiber direction), the data was collected for each environment being tested. The number of observations in each environmental condition was n_j, where the j subscript represented the total number of environments being pooled. If the assumption of normality is significantly violated, the other statistical models should be investigated to fit the data.
    Finally, the dynamic strength of construction could be predicted as:

$$\sigma_d = 1 - P(t) \tag{6.35}$$

The matrix of material parameters was described as a fourth-rank polynomial equation. The matrix of strength properties was described in the same manner as a fourth-rank polynomial equation [17]. The load of dynamic response does not follow Hook's law and has a nonlinear character.

6.3.2 Experimental Investigation Results

The thermal stresses in the nose cap manufacturing of graphite–epoxy or carbon–carbon composites can reach a significant value; however, it can never reach the threshold of failure. For the experimental data we selected a nose cap with geometrical dimensions $R_o = 10$ in (0.254 m), $R_i = 9$ in (0.228 m), $r = 9.5$ in (0.241 m). The nose cap was fabricated from graphite–epoxy and the density of the material had a density of $\rho = 0.420 \times 10^3$ kg/m^3. The Poisson ratio was $\mu_{12} = \mu_{21} = 0.036$.

When the modulus of elasticity was calculated E_1, E_2 for the nose cap, the velocity of rotation changed from 3627 to 4800 rad/s and to 7000 rad/s. The modulus of elasticity in the radial direction was $E_1 = 3.4 \times 10^{10}$ N/m^2, while the modulus of elasticity in the tangential direction was $E_2 = 2.5 \times 10^{10}$ N/m^2 (see Table 6.5). The coefficient of thermal expansion (CTE) in the radial direction was $\alpha_{11} = 5.45 \times 10^{-6}$ m/m/°C, while in the tangential direction it was $\alpha_{22} = 5.34 \times 10^{-6}$ m/m/°C.

A second nose cap was fabricated from carbon–carbon and had a density of $\rho = 0.548 \times 10^{-3}$ kg/m^3. The modulus of elasticity in the radial direction was $E_1 = 1.74 \times 10^{10}$ N/m^2 and the modulus of elasticity in the tangential direction was $E_2 = 1.45 \times 10^{10}$ N/m^2. The Poisson ratio was $\mu_{12} = \mu_{21} = 0.036$, while the coefficient of thermal conductivity for carbon–carbon β was 0.903 $\times 10^{-4}$ m/hrm^2/°C.

Table 6.8 shows the properties for graphite–epoxy composite that were used to calculate the thermal, radial, and tangential stresses in Equation (6.31) (see Table 6.9).

The correlations between thermal stresses radial σ_{11} and tangential σ_{22} and temperature gradient T are presented in Figure 6.7.

For carbon–carbon composites, the coefficients of thermal expansion that were in the radial direction are $\alpha_{11} = 5.45 \times 10^{-6}$ m/m/°C and in the tangential

TABLE 6.8

Properties of Elasticity for Graphite–Epoxy Composites

Description	Values of Characteristics (N/m²)							
	E_{11}	E_{22}	E_{45}	G_{12}	G_{21}	μ_{12}	μ_{21}	μ_{45}
Properties of elasticity on the patterns	3.56×10^{10}	2.59×10^{10}	2.24×10^{10}	0.818×10^{10}	0.68×10^{10}	0.13	0.10	0.34
Properties of elasticity on the nose cap	3.4×10^{10}	2.5×10^{10}	2.2×10^{10}	0.818×10^{10}	0.68×10^{10}	0.13	0.10	0.34

TABLE 6.9

Significance of Thermal Stresses on Nose Cap Manufacturing from Graphite–Epoxy Composites

| Temperature (°C) | Stresses Acting in Radial and Tangential Directions (N/m²) | |
	$\sigma_{11} \times 10^{-5}$	$\sigma_{22} \times 10^{-5}$
20	2237.0	1611.6
17	2072.7	1493.0
14	1908.2	1374.0
12	1743.7	1256.0
9	1579.2	1137.6

direction $\alpha_{22} = 5.34 \times 10^{-6}$ m/m/°C. The CTE in an arbitrary direction was determined as $\alpha_{12} = 2 \sin\alpha * \cos\alpha(\alpha_{11} - \alpha_{22})$.

The significance of the thermal stress components manufactured from carbon–carbon for the nose cap changed when the gradient of temperature changed from −14°C to 17°C for the compression zone. Also, the same is true for the tension zone.

Future developments of NDE methods consist of the elimination of thermal stresses, and to compare with strength parameters see Table 6.10.

Failure criteria are also needed for design and for guiding materials improvement [32,33]. The surface of the equally dangerous biaxial stress conditions for graphite–epoxy composite is shown in Figure 6.9.

All the experimental points are found inside this surface, and for the nose cap there are no dangerous conditions. However, if thermal stresses reach a threshold of failure the nose cap can collapse, and this means that there are points outside of the surface of strength. It is recommended that the surface of strength be drawn for all the biaxial and triaxial stress conditions.

The complex strength values can be calculated if we know all the thermal stresses and all the strength coefficients c, a, d, e (see Table 6.11).

All the calculations in this program were computerized to create a program in C language.

TABLE 6.10

Parameters of Strength for Graphite–Epoxy Composites (N/m²)

Description	Normal Strength, ×10⁻⁵	In-Plane Shear Strength, ×10⁻⁵	Interlaminar Shear Strength, ×10⁻⁵	Normal Strength Acting in Diagonal Directions, ×10⁻⁵
Significance	$\sigma_{11}, \sigma_{22}, \sigma_{33}$	τ_{12}	τ_{13}	$\sigma_{12}^{45}, \sigma_{13}^{45}, \sigma_{23}^{45}$
Compression zone	3448, 827, 3448	34.4	68.9	2068, 2068, 3448
Tension zone	6896, 1241, 34.4	34.4	68.9	4068, 827, 3620

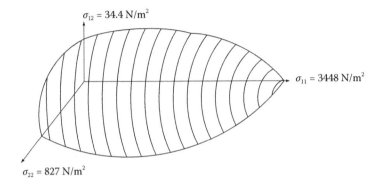

σ_{11}—Normal strength in x direction

σ_{22}—Normal strength in y direction

σ_{12}—Shear strength in 12 directions

$\sigma_{12} = 34.4$ N/m^2

$\sigma_{11} = 3448$ N/m^2

$\sigma_{22} = 827$ N/m^2

FIGURE 6.9
Surface of equally dangerous biaxial stress conditions for graphite–epoxy composites.

6.3.3 Conclusions

1. A methodology for predicting dynamic strength using the nondestructive evaluation of aerospace components by ultrasound, x-ray, digital radiography, and thermography technologies was established.

2. Thermal stresses for orthotropic composite structures were calculated using parameters, which were found by ultrasound and thermography technologies.

3. Temperature gradients were calculated using the approximate solution of Equation (6.31) fluctuating from −8°C to −16°C. NDE thermography cameras that can observe and record temperature gradients can be very useful in the process of curing and cooling composite structures.

4. Probability of local cracking can be predicted using statistical models such as the first and second approach in the law of normal distribution.

TABLE 6.11

Parameters Relationship to Strength for Graphite–Epoxy Composites

Significance	c	a	d	e
Compression zone	4.17	0.40	100	103.56
Tension zone	5.56	1.7	200	−198.8

5. In spite of the visibility of cracks, the samples tested reflect 90% of initial strength. However, it is very important to use nondestructive evaluation methods in the determination of cracks that are not visible.

6.4 Nondestructive Evaluation of Lightweight Nanoscale Structural Parts for Space Shuttle and Satellites

6.4.1 Introduction

The development of novel multifunctional nanoscale structural materials with high strength/weight or stiffness/weight ratios and the related technological process is a significant problem in space communication technology. In recent years, carbon-reinforced epoxy composites have been used extensively in the upper-stage structures of satellite vehicles to improve payload performance and reduce costs. Lattice structures are the choice of designers and they hold promise for interstage structures in launched space vehicles. The prediction of lattice structural stiffness and strength using the nondestructive digital method is the subject of this research. NASA's MSFC has tested Raytheon's Radiance infrared camera to devise nondestructive evaluation (NDE) methods for assembling the space shuttle and other aerospace components.

Thermal management materials are recognized as a way to improve mechanical properties. Hybrid carbon–boron/cyanate ester hybrid prepregs have already been successfully tested in space structures. NDE methods have been described in several studies [34–36] and earlier in this chapter. The ultimate objective consists of developing nondestructive evaluation of nanoscale lightweight materials and advanced polymer matrix composites that will critically contribute to space vehicle weight reduction [37].

The new nanoscale hybrid material would provide engineers and spacecraft designers with a superlightweight material, with superior stiffness and strength, near-zero CTE, and minimal microcracking. Additionally, all of the benefits of processing thermoplastics would become available, such as molding and thermoforming, welding, roll wrapping, in situ automated tape laying, and so forth. As a result of the fact that polyetheretherketone (PEEK) and polyphenylene sulfide (PPS) are already space qualified, there would be broader opportunities for accelerated commercial development and rapid acceptance of new material.

The structural elements of space vehicles represent a significant portion of the total weight. Therefore the appropriate selection of lightweight materials permits increased structural efficiency overall and allows for higher payload weight.

Contemporary carbon–carbon structures displaced carbon–aluminum frameworks providing 20% to 30% weight reduction. However, ubiquitous epoxy matrices typically used in such frameworks remain to be underperforming components. For example, Khassanchine et al. [34] investigated an outgassing process of polymeric composites subjected to ultraviolet radiation.

Critical material failures were attributed to desorption of the volatile products absorbed or generated by the epoxy polymer. The experimental data showed that the outgassing rate depends on the temperature of the material and the volume ratio of the polymeric binder/filling agent. Another frequent cause of failure is associated with microcracking that results from the extreme thermal cycling [35]. The proton irradiation test was shown to produce the significant changes in the mechanical properties of epoxy polymers [36].

Carbon–cyanate ester composites were proposed as a replacement for the carbon–epoxy composites with improved outgassing, microcracking, and moisture absorption characteristics [38]. The effect of thermal cycling on the microcracking behavior and dimension stability on the comparative panel of the composite material was investigated by Lawrence [39]. Thornel PAN T50 fibers and three pitch-based fibers, P55, P7, and P120 were selected for their good fiber stiffness and negative CTE. Thornel fibers were impregnated by the Fiberite epoxy, Amoco ERL 1962 toughened epoxy, and YLA R53 cyanate ester. After curing, all the materials possessed positive CTE and compromised dimension stability. Following thermal cycling at ±50°F, ±150°F, and ±250°F, every system produced microcracking though the cyanate ester performed better than epoxy. To further control microcracking, Roy et al. [40] used cyanate ester resin filled with a 1% to 5% dispersion of the silicate nanoclay particles in polypropylene. This new family of materials exhibits enhanced stiffness and compression strength improved by 12% to 20%.

PEEK and PPS sulfide, new robust high-temperature polymers currently finding broader uses in the aerospace industry, may potentially address a number of limitations inherently pertinent to both epoxy and cyanate resins. PEEK is particularly promising to significantly reduce microcracking damage. However, related PEEK and PPS carbon composites and prepregs have yet to be better proven and broader tested in that respect.

For reinforcement purposes in the space and avionics structures, carbon fibers remain the material of choice to achieve higher safety margins for both stiffness and strength. For example, the IM7 (HS-CP-6000) filaments with high tensile strength and modulus as well as good shear strength is one of the most popular materials for critical applications. Recently, boron fibers gained significant recognition, too, as a way to improve mechanical properties. Hybrid carbon–boron–cyanate ester hybrid prepregs have already been successfully tested in space structures.

6.4.2 Advantages of the Carbon–Boron Nanoscale Fibers

Dynamic aspects of the behavior of nanoscale carbon fiber–boron structures in the manufacturing of carbon fiber–epoxy composites for interstage structures in launch vehicles have been developed. The mechanism of the deposition of boron, carbon, and silicon was described by Thomas [41]. The silicon and boron forms a stronger bond than carbon and boron, so the silicon–boron bond strength is 1.5 times greater than that of the carbon–boron bond.

In our own work [42,43], we demonstrated that carbon fibers encapsulated by deposits of the core glass silica and germanium can be reinforced by boron fibers in the vapor deposition process where borane (BH_3) reacts with glass silica ($SiCl_4$) to form silicon–boron filaments on the nanoscale level. Boron fibers have twice the stiffness and five times the strength of steel.

Boron fibers are typically made using the chemical vapor deposition process and precipitated onto a fine tungsten or carbon filament [44]. The resulting boron fiber is strong, stiff, light in weight, and possesses excellent compressive properties as well as buckling resistance. For example, Special Materials, Inc. (formerly Textron), uses chemical vapor deposition (CVD) for creating the boron layers. The process uses fine tungsten wire for the substrate and boron trichloride gas as the boron source [45]. The boron manufacturing process is precisely controlled and constantly monitored to assure consistent production of boron filaments with diameters of 4.0 and 5.6 mil (100 and 140 mm). Combining the boron fiber with graphite prepreg, a high-performance material Hy-Bor, resulted in a Hy-Bor–laminate material with exceptional mechanical properties, shown in Table 6.12.

Other types of materials, such as low thermal conductivity ceramics, can offer advantages for protective coatings of the satellites. Thermal barrier coatings (TBCs) have thin ceramic layers that are generally applied by plasma spraying or by physical vapor deposition, and are used to insulate air-cooled

TABLE 6.12

Mechanical Properties of the Hy-Bor–Laminate (4.0 mil) [45]

Mechanical Property	Values
Tensile strength	275 ksi (1896 MPa)
Tensile modulus	35 msi (241 GPa)
Flexural strength	350 ksi (2413 MPa)
Flexural modulus	31 msi (214 GPa)
Compression strength	400 ksi (2756 MPa)
Compression modulus	35 msi (241 GPa)
Interlaminar shear strength	15 ksi (103 MPa)
Strain	0.86%
Short beam shear strength	17 ksi (117 MPa)

metallic components from hot gases in gas turbine and other heat engines [46]. The ceramic layer consists of 95.4% ZrO_2 and 4.6% Y_2O_3. However, these coatings have porous and microcracked structures. Recently, scandium was identified as a stabilizer that could be used in addition to yttrium [47]. A composition of 3% scandium and 2.5% yttrium may confer the desired phase stability at 1400°C.

6.4.3 NDE Technique

Apparently, the materials with a perfect combination of high modulus/ strength and low CTE combined with a structure's reduced weight are the most desirable composites to address many issues related to space flight requirements.

The most important issue in selecting an NDE technique involves the number and types of flaws contained within a material. Generally, monolithic ceramics have some degree of porosity and dislocations; ceramic–matrix composites contain several types of defects, including interlaminar porosity and processing-induced voids.

To consider when and how various NDE techniques can be best applied to the examination of ceramic materials, this section reviews the following approaches: ultrasonic testing (UT), radiography, x-ray computed tomography (CT), and acoustic emission (AE). Table 6.13 provides a comparison of these NDE techniques as applied to the realm of advanced ceramics [48].

6.4.4 Ultrasonic Testing

Ultrasonic testing is one of the most widely used NDE techniques for quality control and service integrity evaluation because of its relatively inexpensive cost and the convenience of data acquisition. Generally, UT can be used to detect flaws, determine the size, shape, and location of defects, and identify discontinuities of materials. Also, the determination of ultrasonic velocities can be used to measure the modulus of elasticity for advanced ceramics and liquid polymers [49–52]. Sound that possesses frequencies so high that it cannot be heard is called ultrasound (the frequency range is typically greater than 20 kHz). In ultrasonic testing, beams of high-frequency sound waves are introduced into materials so as to detect both surface and internal flaws [53]. The sound waves travel through the material (with some attendant loss of energy) and are deflected at interfaces and/or defects. The deflected beam can be displayed and analyzed to assess the presence of flaws or discontinuities. Most ultrasonic inspections are performed at frequencies between 0.1 and 25 MHz.

A number of ultrasonic evaluation methods—such as A-, B-, and C-scans— have been used to study various types of flaws in ceramic materials [48–52]. The UT A-scan presents one-dimensional defect information. In the oscilloscope view, the A-scan signal displays the pulse and amplitude against

TABLE 6.13

Key NDE Techniques for Analyzing Advanced Ceramics and Composites

Characteristics	Ultrasonics	X-ray Computed Tomography
Principles	Sonic transmission	X-ray transmission
Variables	Scattering, attenuation, and velocity	Absorption and attenuation coefficients
Advantages	Suitable for thick materials; relatively quick testing time	Creates cross-sectional view of the entire transmitted thickness
Limitations	Requires water immersion or acoustic coupling	Expensive; limited specimen size; radiation hazard
Detectable flaw	Voids, delaminations, porosity, and inclusions	Voids, delaminations, porosity, and inclusions
Characteristics	**Radiography**	**Acoustic Emission**
Principles	X-ray, gamma-ray, and neutron transmission of penetrating radiation	Stress wave emission
Variables	Absorption and attenuation coefficients	Amplitude, counts, and number of events
Advantages	Extensive available database	Real-time monitoring
Limitations	Expensive; depth of defect not indicated; radiation hazard	Requires a prehistory of stresses for flaw detection
Detectable flaw	Voids, delaminations, porosity, and inclusions	Delaminations and inclusions

time. The A-scan display is commonly used to measure material thickness. The UT B-scan displays a parallel set of UT A-scans with two-dimensional data (i.e., the B-scan presents defect distribution through the material's cross section). The B-scan can also be used to inspect rotating tubes and pipes, because it provides a cross-sectional view of defect distribution. The UT C-scan is the most widely used scan mode, as it provides a two-dimensional presentation of defect distribution. A C-scan displays the size and position of flaws in an area parallel to the surface through the raster scan of two axes. A C-scan presentation is a very effective way to investigate flaw distribution, since the presence of the flaw as well as its severity can be readily indicated on a drawing of the part being inspected.

To measure the time of flight or attenuation of the UT signal, the UT scan mode may be employed in either a through-transmission mode (using both a transmitting transducer and a receiving transducer) or a pulse-echo mode (using a single pulser/receiver transducer). Figure 6.10 shows a pulse-echo mode setup with a pulser/receiver transducer and the through-transmission ultrasonic (TTU) setup using a pair of focused transducers. The scan is performed in an immersion tank. In Figure 6.10a, the pulser/receiver transducer is used to generate ultrasonic sound waves and receive the reflected beams. The transducer obtains the traveling sound-wave signals, which are displayed on the oscilloscope with amplitudes against the traveling time.

Finally, a computer gathers the amplitudes and forms a scan image. In the TTU mode, the transmitting transducer (pulser) and receiving transducer (receiver) are aligned, and the beam path is kept perpendicular to the test specimen during the scan (Figure 6.10b). The two transducers are ganged using a yoke arrangement that maintains the alignment of the focused beam (see Figures 6.10 and 6.11).

Thermal management and thermal stresses are critical issues in many ceramics applications [53].

6.4.5 Composite Micromechanical Model

Micromechanic techniques are used to predict the effective properties and deformation response of the individual plies in the composite laminates [54]. Laminate theory is then used to compute the effective stress and deformation response of the entire composite. The composites are assumed to have

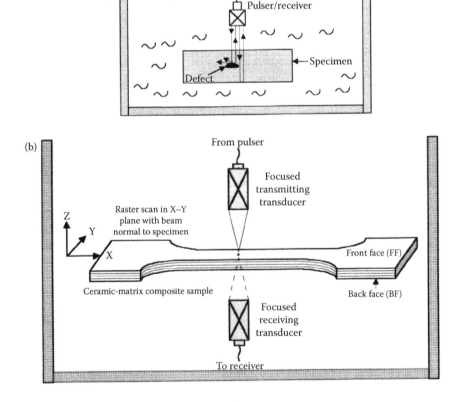

FIGURE 6.10
Setups for ultrasonic scanning: (a) pulse-echo mode with a pulser/receiver transducer and (b) through-transmission using a pair of focused transducers in an immersion tank.

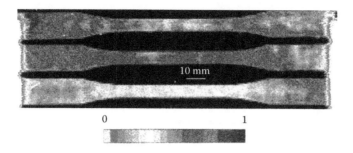

FIGURE 6.11
Ultrasonic C-scans for woven Nicalon/SiC tensile specimens [15–18]. Relative amplitudes range from 0 (least relative amplitude) to 1 (greatest amplitude).

periodic, square fiber packing and a perfect interfacial bond is assumed. Thermal effects will also be examined both experimentally and analytically. The fibers are assumed to be transversely isotropic and linearly elastic with a circular cross section. The matrix is assumed to be isotropic, with a rate-dependent, nonlinear deformation response computed using the equations described below. Experimentally, we can determine deformation by optical interferometry. In double exposure holographic interferometry is applied to determine displacement under load [55].

High strength/light weight or high stiffness/light weight ratios for aniso-tropic composites like carbon–boron–PEEK, or carbon–boron–PPS under different combinations of applied stress components (biaxial and triaxial stress conditions) pose a challenge to designers for establishing reliable failure criteria.

The ability to predict the strength of high performance composite materials under complex loading conditions like atmospheric and cryogenics is a necessary ingredient for rational design. Information that can be used for predicting strength stems on the micromechanics level. As soon as the main principles are established, the approach can then be used separately for tensile (cryogenic) and compressive (atmospheric) loads. Some possible methodology postulates to be taken into account include having strength criteria invariant with respect to coordinate transformation and satisfying Drucker's postulate stating that the strength surface (a plot of the limiting values of strength in a nine-dimensional stress space) must be convex. Elements of the lattice structure are shown in Figure 6.12 and a model of satellite lattice structure is shown in Figure 6.13.

6.4.6 Micromechanics Equations

We derived Equation (6.36) as a criterion of strength for triaxial stress conditions [17]:

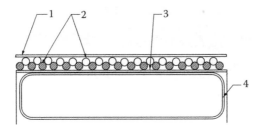

1. Protection layer of
   boron spray from radiation,
   microcracking
2. Helical winding
   type K63712 carbon–boron
   peek or carbon–boron–PPS
3. Circumferential prepreg
   K63712 carbon–boron–PEEK
4. Pultrusion PP cylinder

FIGURE 6.12
Elements of the lattice structure.

$$\frac{\sigma_x^2 + c\sigma_y^2 + b\sigma_z^2 + d\tau_{xy}^2 + p\tau_{yz}^2 + r\tau_{zx}^2 + s\sigma_x\sigma_y + t\sigma_y\sigma_z + f\sigma_z\sigma_x}{\left(\sigma_x^2 + \sigma_y^2 + \sigma_z^2 + \tau_{xy}^2 + \tau_{yz}^2 + \tau_{zx}^2 + \sigma_x\sigma_y + \sigma_y\sigma_z + \sigma_z\sigma_x\right)^{1/2}} \leq \left[\sigma_{bx}\right] \quad (6.36)$$

where σ_x, σ_y, σ_z, τ_{xy}, τ_{yz}, τ_{zx} are normal and shear microstresses, respectively.

Coefficients c, b, d, p, r, s, and t are the relative variable strengths that depend on the loading history, and quality of materials—carbon–boron–fiber hybridization, impregnation by liquid polymers, and so forth. They are determined as:

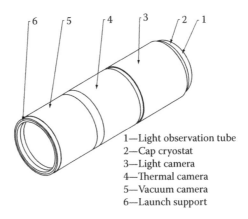

1—Light observation tube
2—Cap cryostat
3—Light camera
4—Thermal camera
5—Vacuum camera
6—Launch support

FIGURE 6.13
Model of the satellite lattice structure.

$$c = \frac{X}{Y}; \quad b = \frac{X}{Z}; \quad d = \frac{X}{S_{bxy}}; \quad p = \frac{X}{S_{byz}}; \quad r = \frac{X}{S_{bzx}};$$

$$s = \frac{4X}{X_{bxy}^{45}} - c - d - 1; \quad t = \frac{4X}{Y_{byz}^{45}} - c - b - p; \quad f = \frac{4X}{X_{bzx}^{45}} - b - r - 1$$

(6.37)

where X, Y, Z are the empirical microtensile (compressive) strengths in x, y, and z directions; S_{bxy}, S_{byz}, S_{bzx} are the empirical microinterlaminar shear strengths in xy, yz, and zx directions; X_{bxy}^{45}, Y_{byz}^{45}, Z_{bzx}^{45} are the empirical micro-tensile (compressive) strengths in diagonal directions xy, yz, and zx.

In triaxial stress conditions ($\sigma_x = \sigma_y = \sigma_z$) with all shear stresses at zero, we get a value of the hydrostatic pressure p as:

$$p = \frac{\delta(6.0)^{1/2} |X|}{R(1 + c + b + s + t + f)}$$

(6.38)

In Equation (6.39), we then substitute σ_x from Equation (6.38)

$$\sigma_x = \sigma_y = \sigma_z = \frac{pR}{\delta}$$

(6.39)

where p is hydrostatic pressure, R is the middle radius of the lattice structure, and δ is middle thickness of the lattice structure.

Failure load will be determined as the relationship of hydrostatic pressure to square p/S.

$$p = \frac{\delta(6.0)^{1/2} |X|}{R4 \left\{ \left[\dfrac{X}{X_{bxy}^{45}} + \dfrac{2X}{Y_{byz}^{45}} + \dfrac{2X}{Z_{bzx}^{45}} \right] \dfrac{X}{Y} \left[\dfrac{X}{Z} + \dfrac{X}{S_{bxy}} + \dfrac{X}{S_{bzx}} \right] \right\}}$$

(6.40)

After the prepreg was pultruded to form a composite laminate, uniaxial tensile stress tests were performed to determine tensile strength of the laminate. Dramatic improvements in tensile strength and tensile modulus were observed using transmission electron microscopy (TEM).

In biaxial stress conditions, every layer has acted as a normal and shear microstress: σ_{11}, σ_{22}, and τ_{12} [56,57].

$$|\sigma_{11}| = |Q_{11} \quad Q_{12} \quad 0| |\varepsilon_{11}|$$
$$|\sigma_{22}| = |Q_{21} \quad Q_{22} \quad 0| |\varepsilon_{22}|$$
$$|\tau_{12}| = |0 \quad 0 \quad Q_{66}| |\gamma_{12}|$$

(6.41)

where

$Q_{11}, Q_{22}, Q_{12}, Q_{21}$ = Stiffness of every layer
$\varepsilon_{11}, \varepsilon_{22}, \gamma_{12}$ = Elastic normal and shear strains

If we include the thermal effect, Equation (6.36) will be transformed as:

$$|\sigma_{11}| = |Q_{11}\ Q_{12}\ 0|\,|\varepsilon_{11}|\,|Q_{11}\ Q_{12}\ 0|\,|\alpha_{11}\ \alpha_{12}\ 0|\,\Delta T$$
$$|\sigma_{22}| = |Q_{21}\ Q_{22}\ 0|\,|\varepsilon_{22}| + |Q_{21}\ Q_{22}\ 0|\,|\alpha_{21}\ \alpha_{22}\ 0|\,\Delta T$$
$$|\tau_{12}| = |0\ \ 0\ \ Q_{66}|\,|\gamma_{12}|\,|0\ \ 0\ \ Q_{66}|\,|0\ \ 0\ \ \alpha_{66}|\,\Delta T$$

where, $\alpha_{11}, \alpha_{12}, \alpha_{21}, \alpha_{22}, \alpha_{66}$ are the coefficients of thermal expansion and ΔT is the temperature gradient.

The total in-plane stresses for the lamina are assumed to be equal to the volume average of the in-plane stresses for each slice, in tensor form as follows:

$$|\sigma_{ij}| = \sum \sigma_{ij}(h_i - h_d) + \sum Q_{ij}(h_i - h_d)\alpha_{ij}\Delta T \tag{6.42}$$

where

σ_{ij} = Stresses of each layer
Q_{ij} = Stiffness of each layer
h_{ij} = Thickness of every slice
h_d = Number of slices with defects
α_{ij} = Coefficient of thermal expansion
ΔT = Temperature gradient

The number of defects are related to quality of resin and air impacts in adhesion slides. In double-exposure interferometry, one makes two successive holograms on the same film. For the first exposure the cantilever bar is fixed in the beginning position, and for the second position bending and deflection are detected in the bar.

This deflection can be found by Equation (6.43) [55]:

$$\Delta y = \frac{n\lambda}{2(\cos\alpha + \cos\beta)} \tag{6.43}$$

where n counts both dark and light fringes from some undeflected reference point, λ is a length of light, α is the angle of incident light, and β is the angle of reflection light.

In classical theory, cantilever bar deflections are:

$$\Delta y = \left[\frac{-W}{6}(1 - x)^3 - \frac{Wl^2 x}{2} + \frac{Wl^3}{6}\right]\frac{1}{EI} \tag{6.44}$$

where W is the total weight of the bar, l is a length of the bar, I is a moment of area, and E is Young's modulus for the material.

Comparing Equations (6.43) and (6.44), we see that deflection comes from double-exposure interferometry and classical laminate theory.

6.4.7 Experimental Results

We determined deflection using a Nikon digital camera (see Figure 6.14) in the cantilever bar. The picture of deflection will be transformed to AutoCAD.

The stiffness was determined using correlation between velocity of propagation of the ultrasonic waves and modulus of elasticity [19].

The ultrasonic test of the lattice helical structure is shown in Figure 6.15.

The temperature gradient was determined using a Radiance infrared camera [53].

Our NDE analysis was performed for the IM7 carbon fibers, whose mechanical properties are shown in Tables 6.14 and 6.15. However, no combination of the IM7 material with boron fiber and PEEK or PPS polymer afforded targeted near-zero CTE. Some results of our estimations are summarized in Tables 6.14 and 6.15. Consequently, we were looking for carbon fibers with maximally negative CTE and also improved modulus. Driven by these selection criteria, we identified as our leading candidate the Pitch Carbon Fiber Dialead K63712 from Mitsubishi (CTE = −0.6 PPM/F, modulus of elasticity 93 msi). According to our model, it allows improving tensile modulus of the laminate and approaches near-zero CET in a certain combination with boron fibers, in the 10% to 20% range of volume fractions (see Tables 6.16 and 6.17).

Other types of materials, such as low thermal conductivity ceramics, can offer advantages for protective coatings of satellites. Thermal barrier coatings

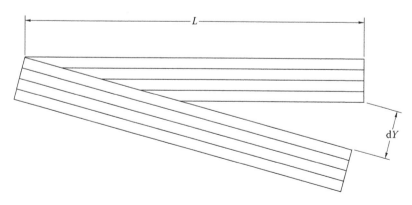

FIGURE 6.14
Determining deflection on the cantilever bar using a Nikon digital camera.

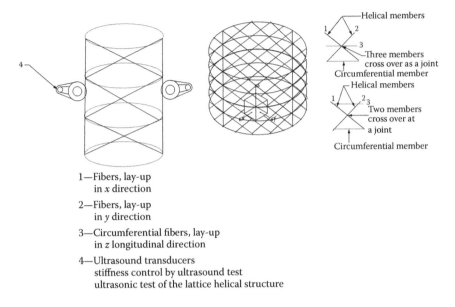

1—Fibers, lay-up
   in *x* direction

2—Fibers, lay-up
   in *y* direction

3—Circumferential fibers, lay-up
   in *z* longitudinal direction

4—Ultrasound transducers
   stiffness control by ultrasound test
   ultrasonic test of the lattice helical structure

FIGURE 6.15
Ultrasonic test of the lattice helical structure.

(TBCs) have thin ceramic layers that are generally applied by plasma spraying or by physical vapor deposition, and are used to insulate air-cooled metallic components from hot gases in gas turbines and other heat engines [58]. The ceramic layer consists of 95.4% ZrO_2 and 4.6% Y_2O_3. However, these coatings have porous and microcracked structures. Recently, scandium was identified as a stabilizer that could be used in addition to yttrium [57]. A composition of 3% scandium and 2.5% yttrium may confer the desired phase stability at 1400°C.

TABLE 6.14

Mechanical Properties of the IM7 Carbon Fibers

Mechanical Property	Values
Tensile strength	2760 MPa
Tensile modulus	168 GPa
Compression strength	1655 MPa
Compression modulus	148 GPa
Short beam shear strength	100 MPa
Fiber density	1770 kg/m³
Fiber volume fraction	62%

TABLE 6.15

Modulus of Elasticity and Coefficient of Thermal Expansion for the Prepreg Components

Material	IM7/Carbon	Boron/Fibers	K63712/Carbon	PEEK	PPS
Modulus of elasticity, E (GPa/msi)	273.8/40	342.3/50	636.6/93	3.49/0.51	3.76/0.55
Coefficient of thermal expansion, CTE (ppm/°F)	−0.2	2.5	−0.6	26	28.9

6.4.8 Physical Characteristics of Prepreg

The typical prepreg tapes manufactured by Phoenixx TPC have a 25-mm width and thickness of 0.15 mm. We estimated that our final laminate would include 10 to 12 and 6-8 boron-based layers, with a total thickness of 2.0–3.0 mm. The lamination sequences of the prepreg layout would include a cross ply configuration, $[0,90]_{2s}$ and two quasi-isotropic configurations, $[0/+45/−45/90]_s/[0/+45/90/−45]_s$, in alternating order. The curing process will be carried out and optimized using a hot hydraulic press, within a pressure range of 20 to 50 psi and curing temperature 625°F to 725°F (330–385°C).

6.4.9 Enhancing Microcracking Resistance Using Montmorilonite Dispersion in PP

Intercalated montmorillonite nanocomposites have been shown to reduce microcracking and improve mechanical characteristics of carbon–cyanate laminates through better stress distribution and thermocycle structural relaxation. We expect similar effects to take place in the carbon–PEEK, and particularly, somewhat less robust carbon–PPS prepregs. One of the

TABLE 6.16

Modulus of Elasticity and Coefficient of Thermal Expansion for the Prepreg Laminates Based on the IM7 Carbon Fibers

Material	IM7/PEEK	IM7/PPS	IM7/Boron/PEEK[a]	IM7/Boron/PPS[b]
Modulus of elasticity, E (GPa/msi)	165.7/24.2	383.4/56.0	384.7/56.2	384.7/56.2
Coefficient of thermal expansion, CTE (ppm/°F)	0.02	0.06	0.57	0.57

[a] IM7, 60, matrix, 40% volume fractions.
[b] 10% boron fibers volume fraction.

TABLE 6.17

Estimated Modulus of Elasticity and Coefficient of Thermal Expansion for the Prepreg Laminates Based on the K63712 Carbon Fibers Using NDE Methods

Material	K63712–PEEK[a]	K63712–PPS[a]	K63712–Boron–PEEK[b]	K63712–Boron–PPS[c]
Modulus of elasticity, E (GPa/msi)	383.4/56.0	383.4/56.0	384.7/56.2	384.7/56.2
Coefficient of thermal expansion, CTE (ppm/°F)	−0.50	−0.48	0.00	0.00

[a] K63712, 60, matrix, 40% volume fractions.
[b] 16% boron fibers volume fraction.
[c] 19% boron fibers volume fraction.

tested compositions is based on the material prepared from the organoclay Nanocor® 1.34TCN, a maleic anhydrid grafted polypropylene Epolene® G-3003 and regular polypropylene, by reacting a three-component suspension in xylene. The resulting powder in an optimized ratio of PP/EpolenClay 85:10:5 was then extruded to ensure uniform thermal stability for all the materials and used as an additive component in 1% to 5% quantity [58–60].

6.4.10 Testing Mechanical and Thermal Properties of the Prepreg Laminates

We tested the key mechanical properties of the laminated plates (25×2.4 mm) of different lengths including tensile and compression strength (modulus) at $0°$, $\pm45°$, and $90°$ configurations; shear strength/modulus, and interlaminar shear strength, as described by Liaw et al. [61]. Thermal properties of interest, thermal expansion coefficient (CTE), and thermal conductivity were examined following Liaw et al. [62]. The combined test panel was based on the following ASTM standards: ASTM D638 (ref. D3039/D3039M), "Test Method for Tensile Properties of Polymer Matrix Composite Materials"; ASTM D696 (ref. D3410), "Test Method for Compression Properties of Polymer Matrix Composite Materials"; ASTM D732, "Shear Strength of Plastics by Punch Tool"; ASTM D903, "Peel or Stripping Strength of Adhesive Bonds"; and ASTM D696 "Coefficient of Linear Thermal Expansion of Plastics." We also examined the microscopy of the laminate surface after thermal cycling for presence of microcracking.

The thermal stresses in the nose cap manufacturing of graphite–epoxy or carbon–carbon composites can reach a significant value; however, it can never reach the threshold of failure [20].

For the experimental data we selected a satellite lattice structure model with geometrical dimensions $R_o = 10$ in (0.254m), $R_i = 9$ in (0.228 m), $r = 9.5$ in

(0.241 m). This model was fabricated from graphite–epoxy and the density of the material had a density of $\rho = 0.420 \times 10^3$ kg/m³.

The Poisson ratio was $\mu_{12} = \mu_{21} = 0.036$

When the modulus of elasticity was calculated E_1, E_2 for the satellite model, the velocity of rotation changed from 3627 to 4800 rad/s and to 7,000 rad/s. The modulus of elasticity in the radial direction was $E_1 = 3.4 \times 10^{10}$ N/m², while the modulus of elasticity in the tangential direction was $E_2 = 2.5 \times 10^{10}$ N/m² (see Table 6.8). The CTE in the radial direction was $\alpha_1 = -5.45 \times 10^{-6}$ m/m/°C, while in the tangential direction it was $\alpha_2 = -5.34 \times 10^{-6}$ m/m/°C.

A second satellite model was fabricated from carbon–carbon and had a density of $\rho = 0.548 \times 10^{-3}$ kg/m³. The modulus of elasticity in the radial direction was $E_1 = 1.74 \times 10^{10}$ N/m² and the modulus of elasticity in the tangential direction was $E_2 = 1.45 \times 10^{10}$ N/m². The Poisson ratio was $\alpha_{12} = \alpha_{21} = 0.036$, while the coefficient of thermal conductivity for carbon–carbon α was -0.903×10^{-4} m/hrm²/°C.

Table 6.8 shows the properties for graphite–epoxy composite that were used to calculate the thermal, radial, and tangential stresses in Equation 6.31 (see Table 6.9).

The correlation between thermal stresses radial σ_{11} and tangential σ_{22} and temperature gradient T are determined by Golfman [27].

For carbon–carbon composites, the coefficients of thermal expansion that were in the radial direction are $\alpha_1 = -5.45 \times 10^{-6}$ m/m/°C and in the tangential direction $\alpha_2 = -5.34 \times 10^{-6}$ m/m/°C.

The CTE in an arbitrary direction was determined as $\alpha_{12} = 2 \sin\alpha^* \cos\alpha$ $(\alpha_1 - \alpha_2)$.

The significance of the thermal stress components manufactured from carbon–carbon for the satellite model changed when the gradient of temperature changed from −14°C to −17°C for the compression zone. Also, the same is true for the tension zone.

Future developments of NDE methods consist of the determination of the thermal stresses and strength parameters (see Table 6.18), and transfer of images by optical cameras. Different aspects of using NDE are discussed in Refs. [63,64].

TABLE 6.18

Parameters of Strength for Graphite–Epoxy Composites (N/m²)

Description	Normal Strength, ×10⁻⁵			In-Plane Shear Strength, ×10⁻⁵	Interlaminar Strength, ×10⁻⁵	Normal Strength Acting in Diagonal Directions, ×10⁻⁵		
Significance	σ_{11}	σ_{22}	σ_{33}	τ_{12}	τ_{13}	σ_{12}^{45}	σ_{13}^{45}	σ_{23}^{45}
Compression zone	3448	827	3448	34.4	68.9	2068	2068	3448
Tension zone	6896	1241	6896	34.4	68.9	4068	827	3620

TABLE 6.19

Parameter Relationship to Strength for Graphite–Epoxy Composites

Significance	c	a	d	e
Compression	4.17	0.40	100	103.56
Tension	5.56	1.7	200	−199.8

The parameter relationship to strength for graphite–epoxy composites is shown in Table 6.19.

The significance of the thermal stresses on nose cap manufacturing from graphite–epoxy composites is indicated in Table 6.9.

6.4.11 Concluding Remarks

1. A methodology for predicting strength and stiffness of advanced carbon–boron fibers and liquid polymer composites manufacturing using vapor deposition nanoscale technology and the nondestructive evaluation of space satellite components by ultrasound, x-ray, digital radiography, and thermography technologies was established.

2. A holographic optical interferometer technique was used in composite lattice structures. As an example, we demonstrated deflection on the cantilever bar by using a Nikon digital camera.

3. Thermal stresses for lattice composite structures were calculated using parameters that were found by ultrasound and thermography technologies. We used NDE thermography cameras with high-resolution images.

4. Probability of local cracking can be predicted using micromechanics models, by computing slices of composite layers.

5. In spite of the visibility of cracks, the samples tested reflect 90% of initial strength. However, it is very important to use nondestructive evaluation methods in the determination of cracks that are not visible.

6.5 Noncontact Measurement of Delaminating Cracks Predicts the Failure in Fiber Reinforced Polymers

6.5.1 Introduction

Noncontact measurement of delaminating cracks in fiber reinforced polymers (FRPs) is of great importance in predicting fatigue failure of these polymers. Currently used FRP failure prediction methods lack multifunctional self-diagnostic capabilities. The previous section established the correlation between crack delaminating of the FRP matrix resin and the strain on it. However, laminate stress–strain curves are highly nonlinear, especially at elevated temperatures.

This section examines the correlation during fatigue testing between temperature gradients and the appearance of nonlinear deformation cracks on the test object. This section also shows that measuring the boundary surface temperature gradients of an FRP test package can directly self-diagnose and predict the appearance of delaminating cracks, and that injecting a resin agent and solid chemical catalyst can create an automatic self-healing process that increases the durability of the FRP object.

6.5.2 Fatigue Strength Prediction

Recent work [65,66] described the fatigue damage mechanisms of carbon FRPs. During fatigue tests, FRPs formed delaminating cracks [67]. These authors found a correlation between crack length in a delaminated matrix resin and the applied strain.

In certain composites, such as carbon fiber IM7 in polyetheretherketone (PEEK) polyamide and carbon fiber IM7 in polyphenylene sulfide (PPS) polyamide, layups laminate stress–strain curves are highly nonlinear, especially at elevated temperatures.

We predict FRP fatigue stress (σ_s) from noncontact measurement of delaminating cracks using Equation (1.5):

$$\sigma_s = \int_1^n E_{11} \Delta \varepsilon^2 e^n \partial n + \int_1^n E_{11} \alpha T e^n \partial n$$

where

σ_s = Fatigue stress
E_{11} = Modulus of elasticity in the fiber-reinforced direction
$\Delta \varepsilon^2$ = Strain applied on the delaminating layer
n = Number of stress cycles per minute (changes from 1 to 1000 cycles)
T = Temperature gradient
α = Coefficient of thermal expansion
e^n = Exponential function of natural logarithm

Previous work described how to measure strain ($\Delta \varepsilon^2$) in a resin matrix using embedded fiber optic sensors [67], and how to measure temperature on the boundary surface of an FRP object with infrared thermography cameras. Our previous work described how to measure the modulus of elasticity using an ultrasonic NDE method. In this section, we replace the integrals of Equation (1.5) with summations in Equation (1.6):

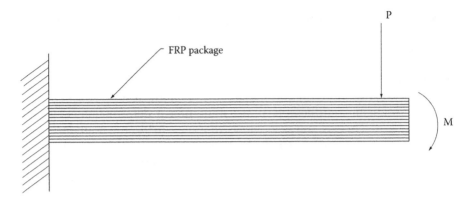

FIGURE 6.16
Carbon fiber reinforced matrix model.

$$\sigma_s = \sum_1^n E_{11}\Delta\varepsilon^2 e^n \sum_1^n E_{11}\alpha T e^n$$

Thermography is the use of an infrared imaging and measurement camera to "see" and "measure" thermal energy emitted from an FRP object. Thermal, or infrared, energy is light that is not visible because its wavelength is too long to be detected by the human eye; it is the part of the electromagnetic spectrum that we perceive as heat. Unlike visible light, in the infrared world everything with a temperature above absolute zero emits heat. The higher infrared thermography cameras produce images of invisible infrared, or "heat," radiation and provide precise noncontact temperature measurement capabilities.

TABLE 6.20

K_{-1} values of IM7/PEEK [Thermo-Lite™ Prepreg, Made from Carbon Fiber IM7 and PEEK Polyamide (40% and 60% by Volume, Respectively)]

Laminate Property	Fiber Orientation	Test Data	K_{-1} Ratio
Fatigue strength, σ_{-1} (ksi)	0°	285	0.0117
	90°	11.2	0.086
Tensile modulus, E_{-11} (msi)	0°	24.2	
	90°	1.3	
Compression strength (ksi)	0°	136	0.0093
	90°		

TABLE 6.21

K_{-1} values of IM7/PPS [Thermo-Lite™ Prepreg, Made from Carbon Fiber IM7 and PPS Polyamide (40% and 60% by Volume, Respectively)]

Laminate Property	Fiber Orientation	Test Data	K_{-1} Ratio
Fatigue strength (ksi)	0°	285	0.0117
	90°	12.5	0.083
Tensile modulus (msi)	0°	24.2	
	90°	1.5	
Compression strength (ksi)	0°	185	0.0108
	90°		

The internal heat of a tested FRP object increases during fatigue testing [68]. If the test frequency is above 1000 cycles per minute (cpm), the internal heat of the sample typically rises to 50°C to 70°C. For test frequencies below 300 cpm, the internal heat decreases to 25–30°C. The effect of thermal cycling-induced microcracking in fiber-reinforced polymer matrix composites is studied on the console loading model cantilever beam (see Figure 6.16). P is an impulse load and M is a bending moment.

On the level of 1000 cpm, fatigue stress must be less than fatigue strength $(\sigma_s < \sigma_{-1})$.

6.5.3 Temperature Measurement of the Surface of an FRP

By defining the ratio σ_s/E_{11} as coefficient K_{-1}, Equation (1.6) becomes:

$$K_{-1} = e^n \left(\sum \Delta\varepsilon^2 - \sum \alpha T \right) \qquad (6.50)$$

Tables 6.20 and 6.21, based on test data from one study [69], give K_{-1} values for two orthogonal fiber orientations of two FRPs.

In Equation (6.50), when $n = 1000$ cpm, e^n is greater than 2^{1000}, making K_{-1}/e^n extremely small. If we set this ratio to zero, we obtain this correlation between temperature gradient in Equation (6.51) and strain.

TABLE 6.22

Modulus of Elasticity and Coefficient of Thermal Expansion for Prepreg Laminates

Material	IM7/PEEK	IM7/PPS
Modulus of elasticity (msi)	24.2	24.2
Coefficient of thermal expansion (ppm/°F)	−0.02	−0.06

TABLE 6.23

Correlation between Length Delaminating and Strain

Length Delaminating, C (mm)	Number of Cracks (n)	Parameter $C*n$	Strain, $\Delta\varepsilon^2$ (%)
6	1	6	2.4
12	2	24	4.9
18	3	54	7.3
24	4	96	9.8
30	5	150	12.4

$$T = \frac{\sum \Delta\varepsilon^2}{\sum \alpha} \tag{6.51}$$

Table 6.22 shows the modulus of elasticity and coefficient of thermal expansion for the prepreg laminates based on IM7 carbon fibers shown in Tables 6.20 and 6.21 [69].

Ten percent of boron fibers by volume was added to IM7 fiber. Strain in different layers of the FRP object correlates with each layer. We assume that correlation between strain $\Delta\varepsilon^2$ during delaminating and number of cracks is:

$$\Delta\varepsilon^2 = (n * C)^{1/2} \tag{6.52}$$

where n is a number of cracks and C is the length of cracks. The values of parameters are given in Table 6.23.

Figure 6.17 shows the measured correlation between temperature gradient and strain propagation.

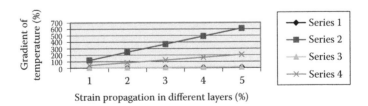

FIGURE 6.17

Correlation between temperature gradient and strain.

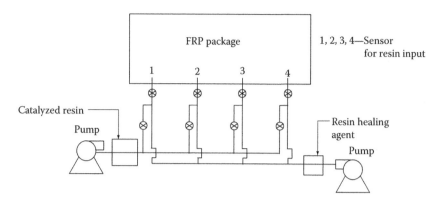

FIGURE 6.18
Automatic resin input.

Figure 1.4 shows the test stand used to measure delaminating cracks. Fiber-optic sensors are embedded in FRP package and detect laser light passing through the material. The infrared camera measures temperature at the side surface boundary of the package. P is the impulse load and M_x is the bending moment.

The coefficient of thermal expansion α_y has a back correlation with the coefficient of thermal conductivity α, determined in Equation (6.53).

$$\alpha_y = 1/\alpha \Delta L/L \Delta Q/A \Delta t h \tag{6.53}$$

where
ΔL = change of length of sample
ΔQ = total heat energy conducted
A = area through which conduction takes place
Δt = time during which the conduction occurred
h = thickness of sample material

Ceramics have a coefficient of thermal conductivity α equal or less than the coefficient of thermal diffusivity β, which for adiabatic conditions is:

$$\beta = 0.1388 \, h^2/t_{0.5} \tag{6.54}$$

where
h = sample thickness
$t_{0.5}$ = time's at 50% of the temperature increase

The Netzsch Company used CMC NDE laser flash apparatus LFA 427 to determine thermal diffusivity.

6.5.4 Fatigue Strength Improvement

Figure 6.18 shows the simultaneous injection of healing agent and catalyst into the test package during stress testing. Injection occurred when a temperature gauge indicated to the injection pumps when the surface temperature of the package increased. The healing agent and solid catalyst are first dispersed [70]. Valves control the four pumps as indicated in Figure 6.18.

6.5.5 Conclusions

1. Laser light, detected by embedded sensors, is a good measure of the formation of delaminating cracks during fatigue testing.

2. An infrared radiation camera accurately measures the surface temperature of test objects during fatigue testing.

3. Surface temperature is a good predictor of fatigue strength as measured by fatigue testing.

4. Injecting a healing agent and a solid catalyst can increase fatigue strength by up to 30%.

References

1. Golfman, Y. 1966. Influence loops on wire sensors on the index of deformations. Science Conference of Forest Academy, Leningrad, USSR, 5, 15–18.
2. Rose, J. L. et al. 1948. A comb transducer model for guided wave NDE ultrasonic. *Ultrasonics* 36 (1–5): 163–168.
3. Rose, J. L. 1998. *Ultrasonic Waves in Solid Media*. Cambridge University Press.
4. Golfman, Y. 1993. Ultrasonic non-destructive method to determine modulus of elasticity of turbine blades. *SAMPE Journal* 29 (4): 31–35.
5. Gershberg, M. 1971. Nondestructive methods for control shipbuilding materials. Leningrad, USSR: *Shipbuilder*, 1–5.
6. Golfman, Y., and M. Gershberg. 1971. Estimate strength of fiberglass. *Journal of Shipbuilding Technology* 1968 (76): 48–55.
7. Lekhnitskii, S. G. 1981. *Theory of Elasticity of Anisotropic Body* (translated in English). Moscow: Mir Publishers.
8. Miller, I. K., J. E. Freund, and R. Jonson, 1989. *Probability and Statistics for Engineers*, 4th ed. Englewood Cliffs, NJ: Prentice-Hall.
9. *Methods of Statistical Treatment of the Experimental Datum*. 1966. Moscow: Committee of Standard Measures and Devices.
10. Raasch, J. E. 2007. Scientific and technical aerospace reports, NASA, Langley Research Center, No. 18, 1–73.
11. Prel, Y. 1999. Composite materials on Ariane 5 launcher. *SAMPE Journal* 8, http://www.sampe.org.gov.

12. Nicolas, J. M. 2006. Implementation of infrared nondestructive evaluation in fiber reinforced polymer double web beams, http://sholar.Lb.vt.edu/theses.
13. Golfman, Y. 1978. Composites application in hovercrafts. Leningrad, USSR: *Shipbuilder.*
14. Golfman, Y. 1978. Hovercrafts with composites fans. Russian patent # 249058.
15. Golfman, Y. 2002. Nondestructive evaluation of parts for hovercrafts and ekranoplans. *Journal of Advanced Materials* 34 (4): 3–7.
16. Timoshenko, S., and J. N. Goodier. *Theory of Elasticity.* New York: McGraw-Hill Book Company.
17. Golfman, Y. 1991. Strength criteria for anisotropic materials. *Journal of Reinforced Plastics and Composites* 10 (6): 542–556.
18. Whitney, J. M., I. M. Daniel, and R. B. Pipes. 1982. *Experimental Mechanics of Fiber Reinforced Composite Materials.* Englewood Cliffs, NJ: Prentice-Hall.
19. Golfman, Y. 1994. Effect of thermoelasticity for composite turbine disk. 26th International Conference, Atlanta, GA.
20. Tsai, S. W. 1963. Composite moduli of unidirectional fiber reinforced media. www.ntrs.nasa.gov/search.jsp.
21. Tsai, S. W. 1964. Dynamic of composite structures and materials, NASA Center. www.ntrs.nasa.gov/search.jsp.
22. Golfman, Y. 2001. Nondestructive evaluation of aerospace components using ultrasound and thermography technologies. *Journal of Advanced Materials* 33 (4): 21–25.
23. Golfman, Y. 1994. Effect of thermoelasticity for composite turbine disk. 26th International SAMPE Technical Conference, Atlanta, GA.
24. Agfa Integrates NDT Technologies. 2002. High-Performance Composites, May 2002.
25. Golfman, Y. 2003. Dynamic aspects of the lattice structures behavior in the manufacturing of carbon–epoxy composites. *Journal of Advanced Materials* 35 (2): 3–8.
26. NASA uses radiance infrared camera for component evaluation. *NASA Tech Briefs* 24 (1).
27. Golfman, Y. 2005. Dynamic local mechanical and thermal strength prediction using NDE for material parameters evaluation of aerospace components. *Journal of Advanced Materials* 35 (1): 61–66.
28. Wu, E. M. 1974. Phenomenological anisotropic failure criteria. In *Mechanics of Composite Materials*, vol. 2, ed. G. P. Sendeckyj. New York: Academic Press.
29. Tsai, S. W., and Wu, E. M. 1971. A generalized theory of strength for anisotropic materials. *Journal of Composites Materials* 5 (1): 58–80.
30. Hoffman, O. 1967. The brittle strength of orthotropic materials. *Journal of Composite Materials* 1 (2): 200–206.
31. Hill, R. 1963. Elastic properties of reinforced solids, some theoretical principles. *Journal of the Mechanics and Physics of Solids* 11: 357–372.
32. Icardi, U., and L. Ferrero. 2008. A comparison among several recent criteria for the failure analysis of composites. *Journal of Advanced Material* 40 (4): 73–103.
33. Tsai, S. W. 1986. *Composites Design.* Dayton, OH: U.S. Air Force Materials Laboratory.
34. Khassanchine, R. H. et al. 2006. Influence of UV radiation on outgassing of polymer composites. *Journal of Space and Rockets* 43 (2): 410–413.

35. Mallick, K. et. al. 2003. Thermo-micromechanics in a cryogenic pressure vessel. A1AA-2003-1765, 44th A1AA/ASME/ASCE/AHS/ASC Structures, Structural Dynamics and Material Conference, Norfolk, VA, Apr. 7–10, 2003.

36. Gao Yu et al. 2006. Effect of proton irradiation on mechanical properties of carbon/epoxy composites. *Journal of Spacecraft and Rockets* 43 (3): 505–508.

37. Golfman, Y. 2000. Non-destructive evaluation of lightweight nanoscale structural parts for space satellite vehicles. *JEC Composites* 35: 1–18.

38. McConnell, V. P. 1992. Tough promises from cyanate esters. *Advanced Composites* 7 (3): 28–37.

39. Lawrence, B. T. The Effect of Long-Term Thermal Cycling on the Microcracking Behavior and Dimensional Stability of Composite Materials. http:/scholar .lib.vt.edu/theses/available/end-11597-95721.

40. Roy, S. et al. 2005. Characterization and modeling of strength enhancement mechanisms in a polymer/clay nanocomposite. A1AA-2005-1853, 46th A1AA/ ASME/ASCE/AHS/ASC Structures, Structural Dynamics and Materials Conference, Austin, TX, Apr. 18–21, 2005.

41. Thomas, E. 1991. *Organic Synthesis. The Roles of Boron and Silicon*. Oxford, UK: Oxford University Press.

42. Golfman, Y. 2001. Fiber draw automation control. *Journal of Advanced Materials* 34 (2): 35–40.

43. Golfman, Y. 2007. Vapor-phase deposition for the thermoprotective layers for the space shuttle. *Journal of Advanced Materials, Special Edition* 3: 58–64.

44. Kozoil, K. et. al. 2007. High performance carbon nanotubes fiber. *Science* 318 (5858): 1892–1895.

45. Specialty Materials, Inc. Manufacturing of Boron SCS Silicon Carbide Fibers. http://www.specmaterials.com.

46. Mess, D. 2003. Low-conductivity thermal-barrier coatings. *Tech Briefs* 27 (6).

47. Mess, B. 2003. Scandia and yttria stabilized zirconia for thermal barriers. *Tech Briefs* 27 (10).

48. Cartz, L. 1995. *Nondestructive Testing*. Materials Park, OH: ASM.

49. Schwartz, M. M. 1997. *Composite Materials*, vol. 1. Upper Saddle River, NJ: Prentice-Hall.

50. Bray, D. E., and D. McBride. 1992. *Nondestructive Testing Techniques*. New York, NY: John Wiley & Sons.

51. Mix, P. E. 1987. *Introduction to Nondestructive Testing*. New York, NY: John Wiley & Sons.

52. *ASM Handbook: Nondestructive Evaluation and Quality Control*, vol. 17. 1992. Materials Park, OH: ASM, p. 231.

53. http://rst.gsfc.nasa.gov/intor/Part_2_24html.

54. Goldberg, R. K., G. D. Roberts, and A. Gilat. 2004. Analytical studies of the high strain rate tensile response of a polymer matrix composites. *Journal of Advanced Materials* 36 (3): 14–24.

55. Rice University, Physics 332, Holographic Interferometry, www.ruf.rice.edu/ dods/Files332/halography.pdf.

56. Azzi, V. D., and S. W. Tsai. 1965. Anisotropic strength of composites. *Experimental Mechanics* 5: 283–288.

57. Hashin, Z. 1987. Analysis of orthogonally crack laminate under tension. *Journal of Applied Mechanics* 54: 872–879.

58. Zweben, C., 2007. Advances in high-performance thermal management materials—a review. *Journal of Advanced Materials* 39 (1): 3–10.
59. Kim, J. et al. 1997 Nondestructive evaluation of continuous nicalon fiber reinforced silicon carbide composites. In *Nondestructive Evaluation and Materials Properties III*, ed. P. K. Liaw et al., 55–63. Warrendale, PA: TMS.
60. Kim, J. et al. 1997. Nondestructive evaluation of Nicalon/SiC composites by ultrasonics and x-ray computed tomography. *Ceramic Engineering and Science Proceedings* 18 (4): 287–296.
61. Liaw, P. K. et al. 1998. Nondestructive evaluation of woven fabric reinforced ceramic composites. In *Nondestructive Evaluation of Ceramics, Ceramic Transactions*, vol. 89, ed. C. Schilling et al., 121–135. New York.
62. Liaw, P. K. et al. 1996. Investigation of metal and ceramic-matrix composites moduli: Experiment and theory. *Acta Metallurgica* 44 (5): 2101–2113.
63. George, A. R., and A. B. Strong. 2008. A new spectroscopic method for the non-destructive Characterization of weathering damage in plastics. *Journal of Advanced Materials* 40 (2): 41–56.
64. Sharma, S. K. et al. 2010. Investigation of intermediate cracks debonding in reinforced concrete beams strengthened with fibre reinforced polyner plates. *Journal of Advanced Materials* 42 (2) 17–33.
65. Kelly, A. 1989. *Concise Encyclopedia of Composite Materials*. New York, NY: Pergamon, p. 199.
66. Gamstedt, E. K., and R. Talrega. 1999. Fatigue damage mechanisms in unidirectional carbon-fiber-reinforced plastics. *Journal of Materials Science* 34: 2535–2546.
67. Silva Munoz, R., and A. Lopez Anido. 2008. Monitoring of marine grade composite doubler plate joints using embedded fiber optic strain. *Journal of Advanced Materials*, (4): 52–72.
68. Golfman, Y. 2004. The fatigue strength prediction for aerospace components using reinforced fiber glass or graphite/epoxy. *Journal of Advanced Materials* 36 (2): 39–43.
69. Phoenixx TPC. 2006. Thermo-Lite™ Thermoplastic Composites. SAMPE Meeting, Boston, Massachusetts.
70. Kessler, M. R., N. R. Sottos, and S. R. White. 2004. Self-healing structural composite materials. *Composite Part A* (34): 744–747.

7

Protective Coating Process for Aviation Parts

7.1 Developing a Nonthermal-Based Anti-Icing/ Deicing of Rotor Blade Leading Edges

7.1.1 Introduction

Naval aircraft regularly operate in hostile environments that include sandy or dusty landing zones and severe sand, rain, or ice storms. Helicopters and other vertical/short takeoff and landing (VTOL/STOL) aircraft such as the V-22 Osprey are expected to endure these severe environments without rapid erosion to the leading edge of their rotor blade. To avoid rapid deterioration of the rotor blade and potential irreparable damage, the leading edge is typically protected with erosion-resistant materials.

The purpose of this chapter is to develop a nonthermal-based anti-icing/ deicing system compatible with heavy and large electrical power for both metallic and polymer-based composite leading edges.

Thermal deicing systems are heavy and require a large amount of electrical power. As a result, the deicing system is only run periodically, allowing ice accretion on the rotor. Furthermore, melted ice may flow and refreeze further aft. Debounded pieces of ice could impact sensitive parts of the aircraft [1–3].

The proposed technology should reduce the overall power needed for anti-icing/deicing the leading edges of rotor blades and improve safety of the aircraft in severe icing conditions. Specifically, the developed technology should demonstrate the anti-icing/deicing capability through a 0.15-in-thick leading edge layer within 20 s. Suitability of the appropriate solution for rotor blade applications should be demonstrated via subscale bench tests. We need to demonstrate the proof of concept through an initial development effort that indicates the scientific merit and feasibility of the anti-icing/deicing mechanism for metallic and polymer-based leading edge materials.

The dynamic analysis and experimental testing of thin-walled structures driven by shear tube actuators was described by Palacios and Smith [4].

The work of Kandagal and Venkatraman [5] is a proof-of-concept study using piezoceramic actuators to induce vibration in a host structure such that the induced out-of-plane shear stresses can shear off the ice.

It is shown experimentally that vibratory deicing is successful at those resonance frequencies at which the peak values of the out-of-plane shear stresses are greater than the adhesive shear strength values of ice on an aluminum surface. Numerical simulations support the experimental observations.

Adhesive shear strength analysis of impact ice was done by Chu and Scavuzzo [6] and Gent et al. [7], and an ultrasonic shear wave anti-icing system for helicopter rotor blades was described by Smith et al. [8]. Experimental tests were performed in a fabricated deicing and anti-icing prototype formed by piezoelectric actuators and an aluminum plate. As the driving amplitude of the shear actuator increases, the ice shear adhesion strength decreases.

As the driving frequency approaches the ultrasonic resonance frequency of the system, the adhesion strength of the attached ice layer decreased. A 100% reduction in accreted ice shear adhesive strength was reached when the actuator was driven at its resonance frequency and at an amplitude of 450 V for a period of 90 s.

Resonance frequencies will be different for aluminum and carbon fiber blades because critical damping coefficients depend on the nature of materials.

7.1.2 Concept of Ultrasonic Shear Wave

The correlation between ultrasonic wave propagation and density of material for a two-layer system formed by ice and aluminum are presented by Smith et al. [8].

The basic formula of the ultrasonic wave propagation is

$$C^2 = \mu/\rho(1 + \alpha_{1,2}^2) \qquad (7.1)$$

where
C = Ultrasonic wave propagation (m/s)
μ = Poisson ratio
ρ = Density of material of rotor blade, kg/m³
α = Ratio of wave number in z direction perpendicular to plane with respect to wave spread in plane x

Torsion moment of blade rotation was determined in our work [9] as

$$M_x^R = -\rho\omega^2 x \cos\theta r^3/3 \qquad (7.2)$$

where
ρ = Density of material rotor blade
ω = Angular velocity
r = Radius of rotating blade
θ = Angle of the rotating blade

We replace ρ, the density of rotor blade material in Equation (7.2), using Equation (7.1), so the torsion moment of blade rotation would be assigned as:

$$M_x^R = -\frac{\mu}{C^2}\left(1+\alpha_{12}^2\right)\omega^2 x\cos\theta r^3 /3$$

(7.3)

We illustrate this in Figure 7.1.

Following Nicolas et al. [10], interlaminar shear stress correlates with torsion moment and displacement $x - x_0$ fix by piezoelectric or ceramic actuators as:

$$\tau_{xz}^T = \left(M_x^R/P\right)\left(x-x_0\right)\cos 2\theta$$

(7.4)

To solve this problem, shear XZ piezoactuators are used (see Figure 7.2). Ceramic block transfer shear frequencies correlate with interlaminar shear strength.

Our concept of managing interlaminar and plane shear strength is presented in our work [11,12]. A nonthermal-based anti-icing/deicing system for rotary wing aircraft is needed [13].

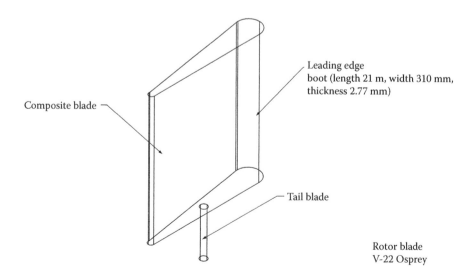

Leading edge boot (length 21 m, width 310 mm, thickness 2.77 mm)

Composite blade

Tail blade

Rotor blade V-22 Osprey

FIGURE 7.1
Skin layer for composite blade fabricated from prepreg carbon fiber and epoxy resin leading edge fabricated from polybutadiene R-45HTLO-polyurethane resin, including milled carbon.

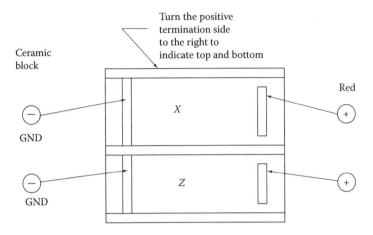

FIGURE 7.2
Shear XZ ceramic piezoactuators.

7.1.3 Design Mechanism for Icing/Deicing Rotor Blades

Neo-Advent Technologies worked on a nonthermal-based anti-icing/de-icing mechanism for rotary wing aircraft. This mechanism is shown in Figure 7.3. It melts ice automatically and drops off water from the blade.

Electromagnetic transducer (pos. 1) driven by input power with ultrasonic frequency 130-kHz spreads waves through the piezoelectric shear actuator (pos. 2). Electromagnetic transducer (pos. 3) received ultrasound waves and spread signals through electrical cable (pos. 4) to encoder (pos. 6). Clutch (pos. 5) was switched on when resonance frequency reached 130 kHz, and electromotor (pos. 7) rotated the blade on an angle of approximately 15° to drop off water.

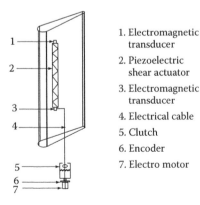

1. Electromagnetic transducer
2. Piezoelectric shear actuator
3. Electromagnetic transducer
4. Electrical cable
5. Clutch
6. Encoder
7. Electro motor

FIGURE 7.3
Rotor blade V-22 Osprey installation nonthermal anti-icing system on rotor blade.

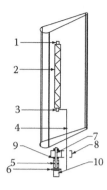

1. Electromagnetic transducer
2. Piezoelectric shear actuator
3. Electromagnetic transducer
4. Electrical cable
5. Clutch
6. Encoder
7. Drive shaft
8. Worm gear
9. Ball bearings
10. Electro motor

FIGURE 7.4
Nonthermal anti-icing mechanism with additional driven shaft on rotor blade.

An additional driven shaft is shown in Figure 7.4. This mechanism has a separate drive shaft with worm gears, which can be installed and removed if necessary. Ultrasonic actuators driven at an input power of 100 watts delaminated accreted ice when it reached a critical thickness of approximately 1.2 mm [38].

7.1.4 Conclusions

1. A nonthermal-based anti-icing/deicing system for rotary wing aircraft was established.
2. The ultrasonic shear wave concept was developed and worked on as correlations between torsion moment of blade rotation and ultrasonic wave propagation for two layers: ice and aluminum.
3. A mechanism for automatic meltoff of ice and dropoff of water from a rotor blade was designed.

7.2 Helicopter Rotor Blade Coatings Development Offers Superior Erosion Resistance and Deicing Capabilities

7.2.1 Introduction

The aim of this section is to find the optimum class of polymers with superior erosion resistance and deicing capability.

The leading edge of the V-22 rotor blade is currently made of titanium and nickel abrasion strips bonded to the composite substrate. Although this bonded metal–composite hybrid is moderately effective for erosion protection and supports the deicing capability of the leading edge, the current configuration limits working strain and fatigue life of the blade [14]. Besides the inability of the existing strips to provide the necessary 250 hours of flying operation, in certain night conditions they also cause a demasking glowing halo around the blades. The 250 hours of flying operation is a Navy requirement parameter correlated with a whirling arm test time to field time.

Summarizing the required performance, the specifications document calls for a wear life of 250 hr in sand, dust, and rain or 125 hr hovering in sand, plus a 125 hr forward flight in rain, but it is highly unlikely that the current rotor blade will achieve this level of durability.

The current strip element also requires frequent inspection for detection of accumulated fatigue cracks in the metal to ensure flight safety. Due to the nonoptimized bonding solution, field removal and maintenance of the strip is unfeasible, increasing operating and service costs of the rotor blade. The Army, for example, is replacing 20 to 25 helicopter rotor blades per month as a result of leading edge and blade tip erosion caused by sand in Iraq and Afghanistan.

The Navy desires to develop an easy-to-apply, high-strain, lightweight, thermally conductive and conformable boot to improve the durability of the leading edges of composite rotor blades. The Bell 210 Medium Utility helicopter is shown in Figure 7.5.

7.2.2 The Problem: Polymer-Based Boot Advantage

The Navy is interested in replacing the metallic leading edge strip with polymer-based, field serviceable, and thermally conductive erosion-resistant boot materials/concepts directly over the existing composite substrate of the rotor blade.

This proposed concept would reduce the overall blade weight and ensure uniform working strain and satisfactory fatigue life in the rotor blade. Simplifying and allowing for repair and reapplication of the leading edge in the field could achieve reduced maintenance requirements and costs. In the case of the V-22, the proposed coating materials/concepts will need to meet the requirement for 250 flight hours of continuous operation in rain, dust, and sand. Specifically, the proposed boot materials/concepts need to demonstrate (via testing in a nationally recognized erosion test facility) a superior resistance to surface abrasion and spoliation caused by high-velocity impact of sand particles and raindrops [14].

Furthermore, for thermal conductivity, the Navy has been estimating the current erosion strip conductivity at 12 W/mk or greater to maintain current deicing performance (the titanium alloy portion is likely to be higher and the

FIGURE 7.5
The Bell 210.

nickel lower), but Bell Helicopter can best say the precise requirement [15]. A piezoelectric shear actuator melts the ice on the blade in wintry environment, and demonstrates the efficiency and field serviceability of the proposed boot materials/concepts in hostile fleet environments.

7.2.3 Reinforced Thermoplastic Materials

The first wearproof materials created, tested, and patented [16,17] consisted of polyvinyl chloride, a butadiene resin mix with epoxy or phenol resin and additives including chipper waste of burned petroleum. These materials cover fiberglass, which replaces aluminum in the fan blades of hovercrafts. Thermoplastic resin has modest abrasion-resistant properties (see Table 7.1). It has recently gained popularity as an easy-to-produce replacement for structural aluminum components to reduce the weight of aircraft elements as well as in automotive industry and construction.

Thermoplastic resins are rapidly replacing reinforced epoxy and polyester thermosetting resins. In the aerospace industry, thermoplastic prepregs have recently been introduced for the Airbus A380 wing's leading edges. The amorphous polymers, such as polyethersulfone (PES) utilized originally were later superseded by crystalline polymers such as polyphenylene sulfide (PPS). Reinforced thermoplastics like polypropylene additive ingredients such as chopped carbon fiber are used. For example, glass-reinforced polyetherimide (PEI) and carbon-reinforced polyetherimide PEI were thermofolded 90° downward at the edges to produce a low-cost, durable closeout for aircraft.

TABLE 7.1

Selected Properties of Thermoplastic Resin for High-Performance Applications

Resin Name	Chemical Resistance	Structure	Moisture Absorption	Density (g/cm³)	Process Temperature (°C)
Polyamide (nylon 6)	Average	Semicrystalline	High	1.15	275
Polyetherimide	Poor	Amorphous	Medium	1.27	315
Polyphenylene sulfide	Excellent	Semicrystalline	Very low	1.35	330
Polymethylmethacrylate	Poor	Amorphous	Very low	1.19	205
Polyetherether ketone	Excellent	Semicrystalline	Very low	1.29	385
Polypropylene	Fair	Semicrystalline	Low	0.91	175
Ultrahigh molecular weight polyethylene	Fair	Semicrystalline	Low		

Thermoplastics hold much promise for aviation applications because of their low density, good mechanical properties, environmental resistance, and cost-effective production.

Our analysis showed that thermoplastics could not reach the performance characteristics required for the erosion resistance in severe rotor blade operational environment.

7.2.4 CombatShield™ (Nonplastic)

Along with 3M tape and L-100 coat, the CombatShield technology was qualified by the Department of Defense (DOD) as a possible solution for the control of rotor blade erosion though only as a mid- to long-term solution [18]. The material is three to four times harder than steel and is made out of advanced alloy compositions. This approach is currently being evaluated by the DOD. Incorporation of this technology for the V-22 will require a fundamental change of the rotor blade manufacturing and service procedures. Apparently, it is a costly process that will not provide a weight-saving advantage.

7.2.5 Polybutadiene Resins

This approach was evaluated by Neo-Advent Technologies (NAT), which formed a team to link requirements between the chemical and aviation industries. The NAT team proposed to create an erosion resistant polymer. Careful analysis identified polybutadiene resins and polyurethane derived from it as promising materials for generating an erosion-resistant polymer framework. Frequent heating of any potential material for deicing purposes presents the challenging task of having a polymer with high stability at elevated temperatures. This task can be addressed by polybutadiene resins [19].

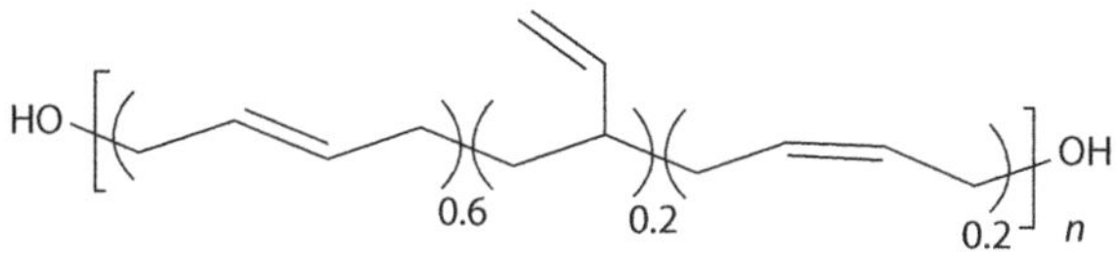

FIGURE 7.6
Chemical structure of the polybutadiene resins ($n = 50$ for polybutadiene R-45HTLO, $n = 25$ for polybutadiene R-20LM).

Polybutadiene resins are a novel class of the hydroxyl-terminated poly-butadiene homopolymers with a structure that is shown in Figure 7.6.

Polybutadiene resins are low molecular weight reactive liquids that offer broad formulating opportunities. Several key attributes are responsible for the formulating advantages provided by polybutadiene resins.

7.2.5.1 Hydroxyl Functionality

Polybutadiene resin end groups are predominantly primary allylic hydroxyl groups (Figure 7.6). These groups have high reactivity with a variety of iso-cyanates to yield polyurethane polymers. The hydroxyl functionalities of the two widely used grades, polybutadiene R-45HTLO and polybutadiene R-20LM, are typically 2.4 to 2.6 per polymer chain.

7.2.5.2 Hydrolytic Stability

Polybutadiene resins have a hydrophobic, nonpolar hydrocarbon backbone that imparts hydrolytic stability to products prepared from it. The stability properties surpass those of polyurethane prepared from other polyols that have ester or ether linkages, which are more hydrophilic and easier to hydro-lyze. Polybutadiene resin-based systems can far exceed the 28-day require-ment of the Naval Avionics test. For example, measuring hardness versus time at 100°C and 95% relative humidity has shown (Figure 7.7) that typical polybutadiene-based urethanes are essentially unaffected, whereas urethanes prepared from other polyols actually liquefy (revert) under test conditions.

The addition of even moderate amounts of polybutadiene resin to polyes-ter polyol-based polyurethane markedly improves the hydrolytic stability of the cured polyurethane. Figure 7.8 shows test results from two comparable polyurethane systems containing a 24.4-wt% poly bd resin/75.6-wt% polyes-ter polyol mixture and based only on the polyester polyols. Figure 7.9 shows change hardness of polybutadiene/polyester/polyurethane composites.

7.2.5.3 High Hydrophobicity

The core polybutadiene aliphatic chain imparts significant hydrophobic properties to polybutadiene-based urethane as compared to the traditional

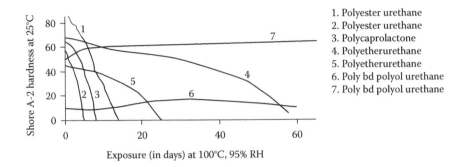

FIGURE 7.7

Comparative hydrolytic stability of polybutadiene urethane versus other materials.

urethanes. Sand bounces off the fractures upon impact, while raindrops change shape and continue to penetrate the substrate. Therefore, improved hydrophobic properties are essential for the protection from rain erosion. Additionally, more hydrophobic materials will be less prone to icing, whereas deicing procedure may be facilitated.

7.2.5.4 Low-Temperature Flexibility

Another attribute of polyurethane systems based on polybutadiene resins is their outstanding low-temperature properties. This characteristic is attributable to the rubbery polybutadiene backbone. Many polyurethane elastomers derived from polybutadiene resins have brittle points as low as –70°C. This characteristic of the polybutadiene resin-containing formulations also contributes to excellent thermal cycling properties.

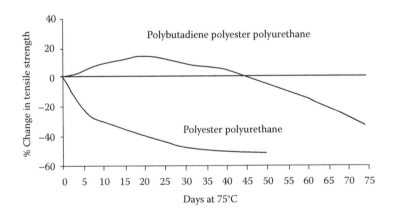

FIGURE 7.8

Change in tensile strength polybutadiene/polyester/polyurethane composites.

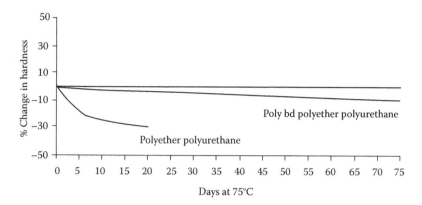

FIGURE 7.9
Change hardness of polybutadiene/polyester/polyurethane composites.

7.2.5.5 Adhesion Properties

Adhesion strength to both galvanized steel and aluminum is high, characterized by a lap shear within $(758–1172) \times 10^4$ Pa (ASTM D102). It changes little with aging, water immersion at room temperature and boiling point, and salt water immersion.

7.2.6 Airthane PET-91A-Based Elastomers

This approach was thoroughly validated by the General Electric Company as a part of the program to construct lightweight aircraft engine blades in the late 1990s [20]. Figure 7.10 represents change in tensile strength, hardness and elongation after water immersion. Through rigorous trials, a material with

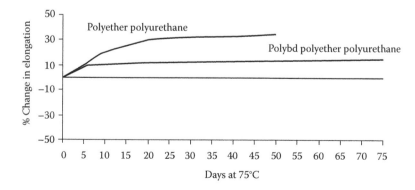

FIGURE 7.10
Change in tensile strength, hardness, and elongation after water immersion.

superior erosion characteristics was obtained from the prepolymer of the TDI and TMEG, a component commercially supplied by Air Products and Chemical, called Airthane PET-91A. When cured by bis-dianiline diamine curative supplied by Lonza under the trademark Lonzacure, the resulting elastomer, combined with N-phenylbenzamine (antioxidant) and Tinuvin 765 (hindered amine light stabilizer), was molded into the blade metal core scaffold. New composite blade led to significant manufacturing cost reduction, weight savings, and erosion protection properties verified by substantial testing panels. Subsequently, the program was terminated because of changed corporate priorities and the technology never commercialized. Some of the properties of the Lonzacure-cured Airthane PET-91A elastomer are given in Table 7.2.

Multiple variants of both prepolymer and cure options are commercially available components. NAT intends to build on the unrealized potential of this group of elastomers based on the commercially available material and enhance their properties for heat conductivity.

7.2.7 Thermally Conductive Materials

Conductive polymer composites are typically formed by the addition of thermally conductive fillers to a polymer matrix. For instance, various carbon fillers are frequently used to increase a composite's thermal and electrical conductivity. Table 7.3 lists the thermal conductivity of some common material relevant to the context of this project [21].

The three fillers preselected for this project were pitch-based carbon fibers, carbon black, and synthetic graphite.

Pitch-based carbon fibers have a unique set of properties that are not easily attainable by polyacrylonitrile (PAN)-based carbon fibers. For example, PAN-based fibers have a maximum modulus of about 650 GPa, whereas pitch-based fibers can reach 1000 GPa. Pitch fibers are also significantly more thermally and electrically conductive. The particular grade of the milled pitch-based carbon fiber proposed for testing is BP/Amoco's ThermalGraph DKD X. This material is graphitized at very high temperatures, which increases thermal and electrical conductivity and modulus of the 28 fibers. Thus, ThermalGraph DKD X in addition to increased heat conductivity could also improve the mechanical stiffness and strength of the resulting composite. The properties of the ThermalGraph DKD X are summarized in Table 7.4 [22].

Carbon black is one of the gold standards for formulating polymer composites with high electrical conductivity. Ketjenblack EC600 JD, manufactured by Akzo Nobel, was chosen in this category in part because it substantially decreases the electrical resistivity of a composite at low level of loading. Another advantage of this material is that it has a highly branched structure, and therefore an extremely high surface interface area that is beneficial for conductive properties. Additionally, according to the literature data, Ketjenblack EC600 JD performed best in a large panel of plastic composites

TABLE 7.2

Processing Conditions

Lonzacure MCDEA level, 95% stoichiometry (%)	17.1
Airthane temperature (°C)	80
Lonzacure temperature (°C)	100
Pot life (min) –80°C	4
Mold temperature (°C)	130
Demold time (min)	30
Post-cure (hours/temperature, °C)	48/130

Selected elastomer properties

Hardness (A/D)	92/42	Compressive Stress (Pa)	
Modulus (Pa) 100% elongation	0.875×10^4	5% deflection	254.5×10^4
200% elongation	1485	10% deflection	419.3×10^4
300% elongation	1870	15% deflection	578×10^4
Tensile strength (Pa)	3.58×10^4	20% deflection	0.76×10^4
Elongation (%)	479	25% deflection	1×10^4
Tear resistance (PLI)	Compressive set (%)	13.8×10^4	
Die C	582	Rebound (%)	39.3×10^4
Split/trouser	57/86	Abrasion index	172.4×10^4

filled by various commercially available carbon black–based conductive polymer additives [23]. Ketjenblack EC600 JD properties are summarized in Table 7.5.

Synthetic graphite is preferred over natural graphite to impart to polymers electro- and heat conductive properties due to higher control of its composition and properties. Synthetic graphite is typically mechanically stronger

TABLE 7.3

Thermal Conductivity of the Selected Common Materials

Materials	Thermal Conductivity (W/mk)
Polymers	0.19–0.30
PAN-based carbon fiber	8–70
Pitch-based carbon fiber	20–1000
Stainless steel	11–24
Aluminum	218–243
Copper	385
Silver	418
Diamond	990

TABLE 7.4

Properties of the ThermalGraph DKD X Carbon Fibers

Tensile strength	>1.39 GPa
Tensile modulus	687–927 GPa
Electrical resistivity	0.0003 Ω cm
Thermal conductivity	400–700 W/mK
Fiber density	2.15–2.25 g/cm^3
Bulk density	0.25–0.55 g/cm^3
Fiber diameter	10 μm
Filament shape	Round
Filament length distribution	<20% <100 μm
	<20% >300 μm
	Average 200 μm
Carbon assay	99+ wt%
Surface area	0.4 m^2/g

and has lower ash content than most natural graphites. Another advantage of synthetic graphite is that its properties are more uniform. Synthetic graphite was used in this project Conoco's Thermocarb Specialty Graphite [22]. This material has high thermal conductivity and is well documented in baseline performance records. Its properties are summarized in Table 7.6.

7.2.8 Technological Process

We developed a technological process for boot manufacturing and selected a pultrusion speed process. Polyester fiber was driven and impregnated by polybutadiene and polyurethane resin and put through a die boot profile. For testing an astronaut boot prototype, we used four layers:

TABLE 7.5

Selected Properties of Carbon Black Ketjenblack EC600 JD

Electrical resistivity	0.01–0.1 Ω cm
Aggregate size	20–100 nm
Specific gravity	1.8 g/cm^3
Apparent bulk density	100–120 kg/m^3
Ash content, max %	0.1
Moisture, max %	0.5
BET surface area	1250 m^2/g
Pore volume	480–510 cm^3/100g
pH	8–10

TABLE 7.6

Selected Properties of Conoco's Thermocarb Specialty Graphite

Ash	0.06 wt%
Sulfur	0.02 wt%
Vibrated bulk density	0.66 g/cm³
Density	2.24 g/cm³
Particle sizing (vol%, by sieve method)	
+48 Tyler mesh	4
–48/+80 Tyler mesh	22
–80/+200 Tyler mesh	48
–200/+325 Tyler mesh	16
–325 Tyler mesh	10
Thermal conductivity at 23°C	600 W/mK on a 0.25-in particle
Electrical resistivity	10^{-5} Ω cm (approximate)
Particle aspect ratio	1.7
Particle shape	Irregular

7.2.8.1 Layer 1: Erosion Resistance

State-of-the-art, commercially available hydrophobic mixed with polyure-thane-butadiene and high-performance polyurethane resin matrixes, which are capable of generating mechanically robust and erosion-resistant elasto-mer frameworks.

7.2.8.2 Layer 2: Thermal Conductivity

Carefully chosen thermoconductive filling components compatible with selected resin compositions. This component renders optimal conductive properties and effectively facilitates deicing of the boot and ice removal as well as aiding maintenance of the material integrity.

7.2.8.3 Layer 3: Electrical Conductivity

Electrical grid/wire as a loop has input and output ends that are inserted in the package and provide current through the length of boot. We can change current and voltage and simultaneously change thermal conductivity.

7.2.8.4 Layer 4: Adhesion

Adhesive enhancement additives secure tight binding of the boot elements to the reinforced epoxy-rotor blade. These elements also warrant cost-effective and prompt field replacement of the boot at the end of a service period.

Optimized blends of layers 1 through 3 will be cured in molded flat sheets of 0.05 to 0.1 in thickness. The boot prototype will be designed as a composite with a steel shim (for better thermal conductivity and structural reinforcement) sandwiched between two prefabricated elastomer sheets.

Test coupons of the boot will be investigated at the certified facilities (Wright-Patterson Air Force Base and Neutzch) for erosion resistance and thermal conductivity. NAT state of the art concept for the boot material will address all the critical technical and economic issues demanded by the Navy, such as:

- Erosion and wear resistance
- Thermal conductivity and stability
- Hydrolytic resistance along with hydrophobicity
- Lightweight
- Cost-effective field replacement
- High adhesion to substrate
- Low cost of manufacturing
- Based on commercially available components

7.2.9 Conductive Resin Formulation

Baseline erosion-resistant compositions using polybutadiene, polyurethane resins, and polyester fibers will be cured by a molding process in 0.1-in sheets. In the initial stages, we are going to establish how carbon filling (10–50%) will affect the overall mechanical properties of elastomer and subsequently optimize electroconductive properties [24,27].

7.2.9.1 Fiber Alignment

As attaining high heat conductivity using randomly oriented conductive fillers is a very challenging task, we proposed a small test panel of elastomers incorporated in oriented carbon rods. Using ordered systems to achieve high conductive characteristics, for example, as a result of orientation induced by extrusion process, has been reported in the literature. Our work [25,26] is an example of such an approach. This approach has been extended to the aligned array of the carbon nanotubes responsible for the preferential electrical and thermal gradients along orientation vectors.

7.2.10 Experimental Conformation Work

Experimental conformation work to verify the proposed model is shown in Figure 7.11.

To heat the top layer of the erosion-resistant boot element, we will use a galvanized steel grid as a built-in heat element residing on the inner surface interface of the heat conductive elastomer sheet. For optimized heat transfer,

FIGURE 7.11

Concept of the boot design: (1) erosion-resistant and heat-conductive elastomer layer, (2) heat-conductive and reinforcing metal grid, (3) insulating elastomer layer, (4) adhesive, (5) rotor blade substrate.

the insulation will be provided by a wrapping base layer of the nonconductive elastomer produced using the same basic formulation, in a sandwich-type three-layer composite. The base layer will also provide an optimal interface with a rotor blade substrate and address adhesive requirements (see Figure 7.1).

Figure 7.12 describes the high-speed boot pultrusion process. Practically, we are going to use a KaZaK composite installation for manufacturing a helicopter leading edge using TEM development technology.

A dry polyester fiber (pos. 2) was installed on the frame (pos. 1) and pulled out to bath (pos. 3–5). Bath 1 (pos. 3) was filled up by polybutadiene resin and bath 2 (pos. 5) was filled up by polyurethane resin. We inserted carbon fiber in a bath 2 (pos. 5). In a heat electrical camera (pos. 6), the mass was melted and penetrated through a die-spinner (pos. 7). Rapid cooling of the boot occurs does in cooling camera (pos. 8). Father boot was pressed by hydraulic cylinders 1 and 2 (pos. 9 and 10). We installed a mandrel (pos. 11) inside the boot. The boot was pultruded by reducer (pos. 12) and electromotor (pos. 13).

7.2.11 Concluding Remarks

1. We selected polyester fiber with an impregnation of polybutadiene and polyurethane resin with 30% carbon fiber as a pultrusion process for boot manufacturing.

2. We modified the thermal conductivity of the selected materials by using pitch-based carbon fiber (thermal conductivity 20–1000 W/mk).

3. We developed a high-speed technological pultrusion process and design capability to create a leading edge of composite blade helicopters (e.g., V-22 Osprey).

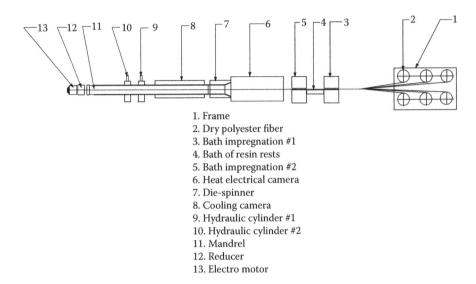

1. Frame
2. Dry polyester fiber
3. Bath impregnation #1
4. Bath of resin rests
5. Bath impregnation #2
6. Heat electrical camera
7. Die-spinner
8. Cooling camera
9. Hydraulic cylinder #1
10. Hydraulic cylinder #2
11. Mandrel
12. Reducer
13. Electro motor

FIGURE 7.12
Pultruded installation for helicopter leading edge.

7.3 Thermoplastic Reinforced Carbon for Large Ground-Based Radomes

7.3.1 Introduction

Sophisticated 3-D air surveillance radar for military and commercial SATCOM and civilian air traffic control applications include weather radar, phased array radar, and secondary surveillance radar.

Standard wind speed design is 150 mph (240 km/h) and optimal designs withstanding up to 250 mph (400 km/h) are available and easy to install hydrophobic coatings for enhanced high-frequency performance in rain, as well as customized shapes for reduced tower loads or radar cross section.

The ground-based radomes heavily rely on the use of thermoset plastics. Thermoset plastics have many inherent problems that make them less than ideal materials for use in the manufacture of radomes. One of the main problems encountered with the use of thermoset plastic (in radomes) is that the resulting structure is extremely susceptible to impact damage and weather-related deterioration (radome delaminating). These compromised radomes will not stop water intrusion, and thus cause transmission loss. This shortcoming makes it imperative that preservation techniques (e.g., the use of primers, topcoats, and abrasive cleaners) be implemented to extend service life. These preservation techniques rely heavily on highly toxic primers that require significant resources and labor for repeated applications and disposal. The purpose of this section is to replace thermoset plastics on the thermoplastics and develop a new, simple, and economically feasible method to research and design thermoplastic materials and processes to manufacture large, ground-based radomes; eliminate delaminating, painting, and hazardous waste; increase life cycles and reduce maintenance costs; while maintaining all structural requirements.

At present, there are no thermoplastic systems available for the manufacture of large, ground-based radome panels. The following issues arise when using thermoplastic polymers reinforced with carbon for such applications:

1. Polyetherether ketone (PEEK) and polyphenylene sulfide (PPS) reinforced with carbon, including carbon nanotubes and glass, give the plastic an additional tough surface and structural stability.
2. The nonrigid nature of thermoplastics implies more movement of the radome as compared to a radome made with thermoset plastic.
3. Weather-related deterioration in the form of UV and wind damage is a concern with thermoplastics as it is with thermosets. This is a reason why we use hydrophobic skin thermoplastics layers, which absorb drops of rain.
4. Thermoplastics also have lower melting points than thermoset plastics and therefore will soften in extremely high temperature environments. The effects of such softening on radar performance have not been fully evaluated.
5. Thermoplastics solidify faster, have no volatility, and do not need autoclave vacuum systems.

7.3.2 Technological Process

Existing thermoset radomes work with advanced composite with materials including prepreg skin materials such as fiberglass, graphite, and Kevlar and a variety of customized core materials including honeycomb, polyurethane foam, and thermoformable foam. Since the 1980s, high-quality thermoplastic materials like PEEK and polyphenylene sulfide (PPS) have become available. These thermoplastic materials with reinforced carbon or glass are very suitable for structural composites [29] and lend themselves well in the manufacture of dielectric space frame (DSF) radome walls.

DSF walls can be constructed as a thin membrane, solid laminate, or as a multilayer sandwich with an internal foam core [30].

We investigated the solid laminate and multilayer sandwich with an internal foam core radome. The solid laminate thermoplastic radomes consist of three layers. The first layer is carbon–Kevlar prepreg, the second layer is glass–Kevlar prepreg, and the third layer is carbon–Kevlar prepreg. All three layers are impregnated with PEEK and PPS resins. The thermoplastic reinforced by carbon and Kevlar aramid fiber manufactured by DuPont Inc. was chosen because it is a semicrystalline polymer that absorbs moisture, decreasing the signal transmission loss in the radome. To quickly evaluate the resin formulations, we used rapid prototyping tools as shown in Figure 7.13.

Kevlar brand fiber is an innovative technology from DuPont that combines high strength with light weight to help dramatically improve the performance of a variety of consumer and industrial products. Groundbreaking research by DuPont scientists in the field of liquid crystalline polymer solutions in 1965 formed the basis for the commercial preparation of the Kevlar aramid fiber. Lightweight and flexible, Kevlar has evolved over four decades of innovation, and its combination with carbon fiber gives thermoplastics a high stiffness and strength.

The preliminary forming of the three-layered panel was done using a hot roller pressure system developed by the principal investigator (Figure 7.14). Figure 7.14 shows a console pressure system for curing composite panels. This novel hot roller system consists of electrical heating elements installed inside the rollers. These rollers heat the panels to a temperature that will facilitate the layup on the rapid prototyping tools.

A schematic of the solution for a dip prepreg operation is shown in Figure 2.1. Dry fiber is dipped into a bath with resin, pulled through the machine for impregnation, and dried in a tower traction unit before it ends up at the roller.

7.3.3 Sandwich Thermoplastic Radome

A sandwich thermoplastic radome consists of skins and core. Surface films have a low surface energy and a high hydrophobic property that absorb raindrops. Skins are made of thermoplastic materials such as PEEK and PPS

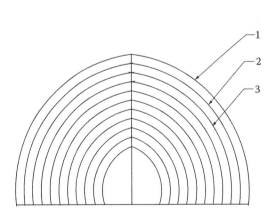

1. Carbon–nylon prepreg
2. Glass–nylon prepreg
3. Carbon–nylon prepreg

FIGURE 7.13
Thermoplastic layers on rapid prototyping tools.

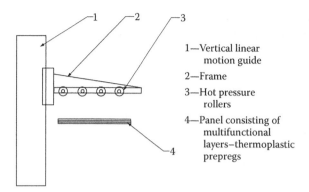

FIGURE 7.14
Hot roller pressure system.

reinforced by carbon or glass, which satisfied current requirements. Cores such as aluminum lattice structures or polyurethane forms are typical examples. A sandwich radome is shown in Figure 7.15.

Hydrophobic film was fabricated using a colander process that mixed carbon nanotubes with thermoplastic pellets. The dry fiber was impregnated with liquid PPS or PEEK resins. Then colander rollers solidified the impregnated fiber.

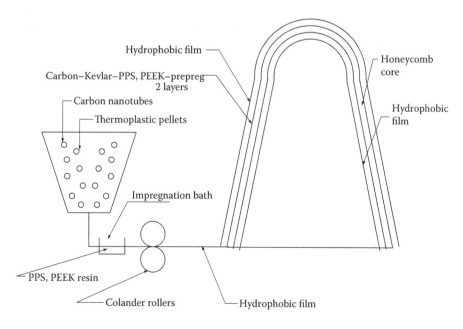

FIGURE 7.15
Sandwich radome.

Carbon–Kevlar–PPS or PEEK preheated layers and honeycomb core lay up on the mandrel (see Figure 7.15). This process is much cheaper than thermoset fiberglass with autoclave polymerization.

7.3.4 Ultrasonic Impregnation Process

Impregnation of the prepreg with the resin system is achieved using the ultrasound process. This operation is done one layer at a time as depicted in the schematic in Figure 7.15. These impregnated prepregs can be laminated to form a multilayer panel that will give strength and stability to the radome.

The impregnation bath consisted of three main parts (see Figure 7.16): impregnation bath (pos. 1) horn (pos. 2), piezoelectric stack (pos. 3), and backing (pos. 4). To optimize the resin distribution through the laminate, we used ultrasound. We achieved resonance by the mechanical attenuation of the horn material. For a 22-kHz resonance frequency a stepped horn of titanium has a length approximately 8 cm.

We proposed a three-layer automation process for manufacturing panels based on flexible impregnation systems (see Figure 7.17). This system will provide the panel with impact resistance.

An impact-resistant system is made up of tough, plastic layers. Every layer has its own frequency and stiffness. The full energy from the impact loads are distributed between tough and plasticity layers. We impregnated carbon–Kevlar prepreg (pos. 8) first and third layers with a thermoplastic resin system PPS or PEEK, and the second layer glass–Kevlar (pos. 9) with PPS and PEEK resins. The tensile/puller rollers (pos. 4) and dancing rollers (pos. 5) connect with transmitting transducers and receiving transducers. This three-layer prepreg is then cut off by a flying cutoff saw (pos. 13). The result is the three-layer prepreg forms as a multifunctional panel (see Figure 7.17).

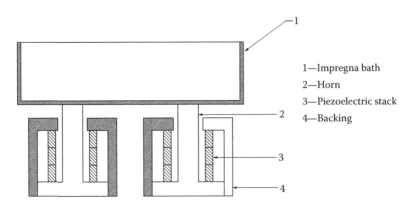

1—Impregna bath
2—Horn
3—Piezoelectric stack
4—Backing

FIGURE 7.16
Two ultrasound transducers installed under the impregnation bath.

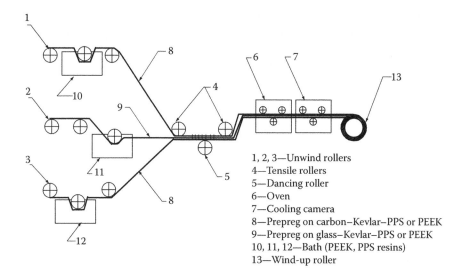

FIGURE 7.17
Automation process manufacturing thermoplastic prepreg based on a multilayered flexible system.

7.3.5 Analysis of the Three-Layer Impregnation Ultrasonic Process

The ultrasonic impregnation process is one of the most widely used NDE techniques for quality control and shear strength evaluation because of its relatively inexpensive cost and the convenience of data acquisition. Generally, in ultrasonic impregnation process, beams of high-frequency sound waves (20 kHz) are introduced into the framework and radome wall. The deflected beam then can be displayed and analyzed to assess the presence of flaws or discontinuities. With UT the size, shape, and discontinuities in the material can be readily identified. Also, the determination of ultrasonic velocities can be used to measure the modulus of elasticity or Young's modulus of materials [31]. Ultrasonic measurements were performed using longitudinal waves. The longitudinal modulus (E) is related to the longitudinal velocity (C) by

$$E_{xy} = C_{xy}^2 \rho (1 - \mu_{xy}\mu_{yx})$$

(7.5)

where C_{xy} is a velocity spread of ultrasonic waves in the x, y directions, ρ is the density of the composite, and μ_{xy}, μ_{yx} is a Poisson's ratio in x, y directions. The modulus of shear elasticity correlates with Young's modulus in $+45°$ as:

$$G_{xy} = E_{xy}^{45} / 2(1 + \mu 45)$$

(7.6)

Dry-coupling longitudinal transducers were used because the samples may be highly porous and the use of a liquid or gel coupling would change the modulus values, thereby contaminating the samples. The dry contact transducers had a thin elastomer face-sheet and coupled the ultrasonic propagation by pressure without the aid of a liquid or gel coupling. The transducers had a crystal size of 19.1 mm and a center frequency of 0.5 MHz [32–35].

The shear stresses are determined as:

$$\tau_{xy} = G_{xy}\gamma_{xy} \tag{7.7}$$

G_{xy} is the modulus of shear elasticity and γ_{xy} is the shear of deformation.

The materials for the three-combination prepreg system is available from Thermo-Lite™ composites (see Tables 7.7 and 7.8) [36].

7.3.5.1 Candidate Materials for Walls and Framework

Water is an excellent electrical conductor; its dielectric constant is 81 and its electroconductivity is $4.2E +8\ \Omega$. For this reason, we selected nylon because of its ability to absorb moisture, maximizing signal transmission.

7.3.5.2 Radome Construction

Ground-based radome structural elements instead of aluminum, which is traditionally used for sandwich radomes, has been proposed to be made from reinforced thermoplastics (see Figure 7.18). The designers can understand the advantage of reinforced thermoplastic structural elements (ribs) assembled with thermal sandwich panels. These structural elements and panels are joined together using resistance or ultrasound welding. Resistance welding is a group of welding processes where coalescence is achieved by the heat obtained from the resistance of the thermoplastic resin when an electric current is passed through it. To assure good adhesion, pressure is applied during the welding process. There are at least seven important resistance-welding processes. These are flash welding, high-frequency resistance

TABLE 7.7

Selected Composite Materials by Thermo-Lite Composites

Carbon–Nylon	Glass–Nylon	Aramid–Nylon
Carbon–polyphenylene sulfide	Glass–polyphenylene sulfide	Aramid–polyphenylene sulfide
Carbon–polyetherimide	Glass–polyetherimide	Aramid–polyetherimide
Carbon–polyetherether ketone	Glass–polyetherether ketone	
Carbon–PFA	Glass–polypropylene	
Carbon–polymethylmethacrylate	Glass–HDPE	
Carbon–HDPE	Glass–HDPE	

Note: PFA and HDPE are the thermoplastic resins. PFA, fluoroplastic polymer; HDPE, high density polyethylene.

TABLE 7.8

Thermoplastic Resin Systems and Basic Properties

Resin Name	Chemical/Solvent Resistance	Structure	Moisture Absorption
Polyamide (nylon 6)	Average	Semicrystalline	High
Polyetherimide	Poor	Amorphous	Medium
Polyphenylene sulfide	Excellent	Semicrystalline	Very low
Polymethylmethacrylate	Poor	Amorphous	Very low
Polyetherether ketone	Excellent	Semicrystalline	Very low
Polypropylene	Fair	Semicrystalline	Low
Polyphenylene sulfide	Excellent	Semicrystalline	Very low

Resin Name	Density (g/cm³)	Process Temperature [°F (°C)]
Polyamide (nylon 6)	1.15	525 (275)
Polyetherimide	1.27	600 (315)
Polyphenylene sulfide	1.35	625 (330)
Polymethylmethacrylate	1.19	400 (205)
Polyetherether ketone	1.29	725 (385)
Polypropylene	0.91	350 (175)
Polyphenylene sulfide	1.35	625 (330)

welding, percussion welding, projection welding, resistance seam welding, resistance spot welding, and upset welding.

These are very similar to laser welding. They are alike in many respects but are sufficiently different. Portable welding ultrasound tools are available in building supply stores. In Figure 7.18 the framework (pos. 1 and 2) scattering loss is several times less than the aluminum radome frame (pos. 4).

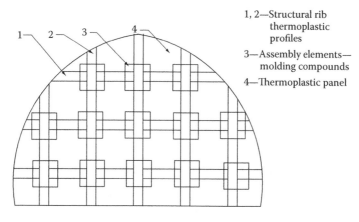

1, 2—Structural rib thermoplastic profiles

3—Assembly elements— molding compounds

4—Thermoplastic panel

FIGURE 7.18
Ground-based radome structural elements.

Assembly elements manufactured the molding compounds (pos. 3) which link the structural rib thermoplastic profiles (pos. 1, 2).

7.3.5.3 Large Shear Deformation Thermoplastic Behavior

The large shear deformation behavior of a thermoplastic three-layer when we impregnate carbon–nylon by polyamid resin is:

$$\left[\gamma_x\right] = \frac{\nabla l_{e1} + \nabla l_{n1}}{L} \; ; \quad \left[\gamma_y\right] = \frac{\nabla w_{e2} + \nabla w_{n2}}{W} \; ; \quad \left[\gamma_z\right] = \frac{\nabla h_{e3} + \nabla h_{n3}}{H} \qquad (7.8)$$

where

L, W, H = Length of panel without elongation in x, y, z directions
$\nabla l_{e1}, \nabla w_{e2}, \nabla h_{e3}$ = Panel elongation from linear elastic response (~10%)
$\nabla l_{n1}, \nabla w_{n2}, \nabla h_{e3}$ = Panel elongation from nonlinear elastic response (~50%)

Virtual deformation approach in multifunctional layers depending on velocity of the impact load, mass, and rigidity panel was shown in our work [37,38]:

$$\gamma_x = \frac{V_{impact}}{2L}\left(\frac{m_x}{E_x L}\right)^{1/2} \; ; \quad \gamma_y = \frac{V_{impact}}{2W}\left(\frac{m_y}{E_y W}\right)^{1/2} \qquad (7.9)$$

$$\gamma_z = \frac{V_{impact}}{2H}\left(\frac{m_z}{E_z H}\right)^{1/2} \qquad (7.10)$$

where

V_{impact} = Velocity of the impact load
E_x, E_y, E_z = Young's modulus of elasticity in x, y and transverse direction z
m_x, m_y, m_z = Mass of fiber in x, y and transverse direction z

We can compare the ability of materials $[\gamma_x, \gamma_y, \gamma_z]$ to the response outside the impact loads, so $[\gamma_x] > \gamma_x$, $[\gamma_y] > \gamma_y$, $[\gamma_z] > \gamma_z$.

7.3.6 Practical Results

We calculate the large deformation nonlinear response of a thermoplastic panel. The large deformation behavior of the thermoplastic panel in diagonal directions +45°, −45° was shown by Mauget et al. [39]. Including the effect of elongation carbon–nylon in +45°, –45° nonlinear deformation will be assigned as:

$$\left[\varepsilon_x^{45}\right] = \frac{\nabla l_{el} + \nabla l_{n1}}{L_{45}}; \quad \left[\varepsilon_y^{45}\right] = \frac{\nabla w_{el} + \nabla w_{n1}}{W_{45}}; \quad \left[\varepsilon_z^{45}\right] = \frac{\nabla h_{el} + \nabla h_{n1}}{H_{45}} \quad (7.11)$$

where

L_{45}, W_{45}, H_{45} = Length of panel without elongation in +45°, –45°, z directions

$\nabla l_{el}, \nabla w_{el}, \nabla h_{el}$ = Panel elongation from linear elastic response (~10%)

$\nabla l_{nl}, \nabla w_{nl}, \nabla h_{el}$ = Panel elongation from nonlinear elastic response (~50%)

Virtual deformation approach in multifunctional layers depending on the velocity of the impact load, mass, and rigidity panels was shown in our work [37]:

$$\varepsilon_x^{45} = \frac{V_{impact}}{2L_{45}}\left(\frac{m_x}{E_x^{45}L_{45}}\right)^{1/2}; \quad \varepsilon_Y^{45} = \frac{V_{impact}}{2W_{45}}\left(\frac{m_y}{E_y^{45}W_{45}}\right)^{1/2} \quad (7.12)$$

$$\varepsilon_z^{45} = \frac{V_{impact}}{2H_{45}}\left(\frac{m_z}{E_zH_{45}}\right)^{1/2} \quad (7.13)$$

where

V_{impact} = Velocity of the impact load

E_x^{45}, E_y^{45}, E_z = Modulus of elasticity in +45°, –45° and transverse directions

m_x, m_y, m_z = Mass of fiber in +45°, –45° and transverse directions

We can compare the ability of materials to response outside impact loads, so $\left[\varepsilon_x^{45}\right] > \varepsilon_x^{45}, \left[\varepsilon_Y^{45}\right] > \varepsilon_Y^{45}, \left[\varepsilon_z^{45}\right] > \varepsilon_z^{45}$.

7.3.7 Conclusions

1. We developed a thermoforming random process using colander and hot roller systems of the three-layer thermoplastic prepreg consolidation.

2. We controlled impregnation of semicrystalline PPS or PEEK resins into the reinforced prepregs carbon–Kevlar and glass–Kevlar

by using the ultrasound resonance frequencies equal to 22 kHz for optimization resin distribution.

3. We laid up panels on the plastic or wood mandrel with structural framework. The structural elements and panels are joined together using plastic structural elements and plastic fasteners.

4. Virtual deformation response panel elongation by linear and non-linear elasticity helps to design panels with impact load resistance.

This section describes the manufacturing process of large ground-based radomes using thermoplastic polymers. The author served as a principal investigator for NAT and developed this work as a DOD proposal. He wishes to thank NAT for their efforts to help solve this problem.

References

1. Coffman, H. J. 1987. Helicopter rotor icing protection methods. *Journal of the American Helicopter Society* 32 (32): 34–39.
2. Flemming, R. J. 2002. The past twenty years of icing research and development at Sikorsky aircraft. In *40th AIAA Aerospace Sciences Meeting*, January 14.
3. Bragg, M. B., T. Bassar, W. R. Perkins, M. S. Selig, P. G. Voulgaris, and J. W. Melody. 2002. Smart icing systems for aircraft icing safety. In *AIAA Aerospace Science Meeting*, January 14.
4. Palacios, J. L., and E. C. Smith. 2005. Dynamic analysis and experimental testing of thin-walled structures driven by shear tube actuators. In *46th AIAA/ASME/ASCE/AHS/ASC Structures, Structural Dynamics & Materials, AIAA-2009–2112*, Austin, TX, April.
5. Kandagal, S., and K. Venkatraman. 2005. Piezo-actuated vibratory deicing of a flat plate. *46th AIAA/ASME/ASCE/AHS/ASC Structures, Structural Dynamics & Materials, AIAA-2009–2115*, Austin, TX, April.
6. Chu, M. C., and R. J. Scavuzzo. 1991. Adhesive shear strength of impact ice. *AIAA Journal* 29 (11): 1921–1926.
7. Gent, R. W., N. P. Dart, and J. T. Candsdale. 2000. *Aircraft Icing*. Defense Evaluation and Research Agency, The Royal Society, Hampshire, UK.
8. Smith, E., J. Rose, J. L. Palacios, and H. Gao. 2006. *Ultrasonic Shear Wave Anti-Icing System for Helicopter Rotor Blades*. American Helicopter Society International, Alexandria, VA.
9. Golfman, Y. 2007. Stress and vibration analysis of composite propeller blades and helicopter rotors. *Journal of Advanced Materials*, Special Edition 3.
10. Pagano, N. J., and S. R. Sony. 1988. Strength analysis of composite turbine blades. *Journal of Reinforced Plastics and Composites* 7 (6): 558–581.
11. Golfman, Y. 1991. Strength criteria for anisotropic materials. *Journal of Reinforced Plastics and Composites* 10 (6): 542–555.

12. Golfman, Y. 2007. The interlaminar shear strength between thermoplastics rapid prototyping or pultruded profiles and skin carbon fibers/epoxy layers. *Journal of Advanced Materials* Special Edition 3.

13. Golfman, Y. 2008. Non-thermal anti-icing/de-icing system for rotor blade leading edges. *JEC Composites* (41): 72–74.

14. Golfman, Y. 1977. Compression molding material for fiberglass protective. Patent USSR 594745, October 28.

15. Golfman, Y. 1980. Method manufacturing large parts. USSR Patent 753859, April 11.

16. Hong, S. C., and T. J. Wiggins. 2001. Advanced rain and sand erosion resistance elastomers. In *Proceedings for the American Helicopter Society, 57th Annual Forum*.

17. Editor of High-Performance Composites. 2004. Bell Helicopter Textron announces new model, increased sales, successful Eagle Eye milestone. *High-Performance Composites*, May.

18. Combat Shield™. 2006. http://www.tapecase.com.

19. Sartomer Company. 2006. Sartomer Application Bulletin. www.sartomer.com, July 26.

20. Evans, C. R. W., D. D. Ward, W. W. Lin, H. S. Chao, J. T. Begovich. 2001. US Patent 6,287,080, November 9.

21. Clingerman, M. L. 1998. Development and modelling of electrically conductive composite materials. Dissertation, Michigan Technology University.

22. Weber, E. H. 1999. Development and modelling of thermally conductive polymer/carbon. Dissertation, Michigan Technology University.

23. The Freedonia Group. 2000. *Conductive Polymers*. The Freedonia Group, Cleveland, OH.

24. Fan, S., P. L. Liu, H. Huang, and Y.-D. Li. 2005. Thermal interface material and method for making same. US Patent 6924335.

25. Golfman, Y. 2004. The interlaminar shear stress analysis of composites sandwich/carbon fiber/epoxy structures. *Journal of Advanced Materials* 36 (2): 16–21.

26. Golfman, Y. 2005. Dynamic local mechanical and thermal strength prediction using nondestructive evaluation of aerospace components ultrasound, x-ray, digital radiography and thermography technologies. *Journal of Advanced Materials* 37 (1): 61–66.

27. MatWeb Material Property Data. 2004. http://www.matweb.com.

28. Golfman, Y. 1978. *Polymer Materials for Hovercraft's Propeller Blades*. Leningrad, Russia: Shipbuilder.

29. Ofringa, A. 2005. Thermoplastics in aerospace, new products through innovative technology. *SAMPE Journal* 41 (7): 19–27.

30. Dielectric Radome Advantages. http://www.radome.net/dsf_adv.html.

31. Golfman, Y. 1993. Ultrasonic nondestructive method to determine modulus of elasticity of turbine blades. *SAMPE Journal* 29 (4): 31–35.

32. Cartz, L. 1995. *Nondestructive Testing*. ASM, Materials Park, OH.

33. Bray, D. E., and D. McBride. 1992. *Nondestructive Testing Techniques*. John Wiley & Sons, New York.

34. Mix, P. E. 1987. *Introduction to Nondestructive Testing*. John Wiley & Sons, New York.

35. ASTM. 1996. *Standard Terminology for Nondestructive Examinations, E1316–96*. ASTM, Philadelphia, PA.

36. Buck, M. E. 2004. Thermo-Lite™ thermoplastic composites. In *Phoenix TPC, Boston SAMPE Meeting*.
37. Golfman, Y. 1996. The interlaminar shear stress analysis of composites in marine front. In *SAMPE Technical Conference*, Arizona.
38. Palacios, J. L., E. C. Smith, and J. L. Rose. 2008. Investigation of an ultrasonic ice protection system for helicopter rotor blades. In *American Helicopter Society 64th Annual Forum*, Montreal, Canada, April 29–May 1, 2008, pp. 1–10.
39. Mauget, B. R., L. Minnetyan, and C. C. Chamis. 2001. Large deformation nonlinear response of soft composite structures via laminate analogy. *Journal of Advanced Materials* 34 (2): 21–26.

Index

Printed and bound by CPI Group (UK) Ltd, Croydon, CR0 4YY

18/10/2024

01776259-0009